Governance & Climate Justice

Julia Puaschunder

Governance & Climate Justice

Wealth of Nature

Second Edition

palgrave
macmillan

Julia Puaschunder
Graduate School of Arts and Sciences
Columbia University
New York, NY, USA

ISBN 978-3-031-86031-7 ISBN 978-3-031-87971-5 (eBook)
https://doi.org/10.1007/978-3-031-87971-5

This Palgrave Macmillan imprint is published by the registered company Springer Nature Switzerland AG
The registered company address is: Gewerbestrasse 11, 6330 Cham, Switzerland

Foreword

In a time when human impact on climate stability becomes more and more apparent but also as the world appears to be on a crucial junction of fast-changing economic conditions due to rising geopolitical tensions, the release of the second edition of *Governance & Climate Justice* is timely and essential.

Julia Puaschunder's work offers an indispensable guide for understanding the ethical, philosophical and economic roots of climate justice—particularly in the context of growing economic disparities in light of global warming. This book grants invaluable insights into a natural drive towards climate justice building from established concepts, such as social responsibility and future orientation, forming intergenerational equity. The book provides a deeply informed perspective on how environmental governance structures can draw from the ethical, economic and political dimensions of climate change in order to alleviate the manifold consequences arising from the variegated climate impacts.

This second edition comes at a pivotal moment. Not only do environmental inequalities become apparent as never before in the history of modern humankind. But also, the world's leading governance bodies have turned to finding a solution for the manifold climate-induced inequalities around the world. The United Nations Sustainable Development Goals and United Nations Climate Change Conferences of the Parties (COP) on Climate Change but also global governance institutions, academic units and the Nobel Peace Prize committee, increasingly draw attention toward sustainability and justice for future generations.

With the Nobel Peace Prize increasingly spotlighting environmental justice and sustainability efforts, and the United Nations Climate Change Conferences of the Parties (COP) on Climate Change reaching for the innovative financialization of economic transformation to mitigate and adapt to climate change, this book provides an innovative way to find the funds for global climate stabilization. Estimating the expected relative economic gains from a warming earth is thereby meant as a gateway to large-scale redistribution opportunities to help those being impacted by global warming the most and the fastest. Quantifying the unequal economic conditions due to climate change over time helps in addition to enact climate justice over time for eternity. Highlighting climate change disparate impacts offers profound insights into how international policies and climate commitments can balance the needs of diverse populations in the common call for responsibility to protect our shared environment.

In illuminating the connections between environmental impacts and socioeconomic disparities, *Governance & Climate Justice* makes a compelling case for addressing climate change through a lens of fairness and accountability. Puaschunder's approach is informed by the latest research and brings innovative ideas to the forefront—from the use of climate taxation and green bonds setting strategic economic market incentives, to the adoption of sustainable finance practices that place value on environmental resources as a key component of economic resilience. The book's emphasis on solutions that integrate environmental, social, and governance (ESG) principles resonates powerfully in today's world, where responsible investment strategies are increasingly seen as crucial economic innovation to achieving a sustainable and just future for all.

From an economic perspective, Julia Puaschunder's work brings to light the urgent need for balancing immediate economic goals with long-term sustainability. The economic analysis provides clear parameters that are attentive to different geopolitical, motivational and financial drivers in order to allow for various global governance redistribution models that work through real-world relevant incentives. The macroeconomic model put forward is also at the forefront of global framework solutions—such as carbon pricing, climate adaptation funding, and international green finance—that help to equitably allocate resources to vulnerable populations and disproportionally affected nations as well as future generations without current active representation. The book's interdisciplinary perspective aligns closely with the latest United Nations Sustainable Development Goals and the COP Paris Agreement as well as subsequent climate stabilization financialization calls, where the equitable distribution of climate change gains can become a crucial contribution.

The second edition of the book also explores the economic valuation of natural resources and the role of environmental preservation in national wealth accounting. Building on the historical foundations of green policies as well as reflecting on the contemporary growing ESG trends in global finance and investing, the book also offers newest insights on quantifying natural capital endeavors at a time when hallmark global governance institutions but also North American and European entities are currently aiming to capture environmental aspects of global wealth. All the shared insights serve as gateways for future research and policy directions on climate justice and economic inequality alleviation.

The book also emphasizes that achieving global climate justice will require a cooperative, multilateral approach. With the growing prominence of international climate accords and the influential platform of the COP conferences, Puaschunder's work provides a roadmap for policymakers, advocates, and citizens who seek to create a more equitable climate stabilization framework. By outlining pathways toward sustainable economic development that do not compromise the livelihoods of the world's emerging economies yet help the most vulnerable populations in a data-driven, targeted approach; *Governance & Climate Justice* serves as both—an urgent call to action and a comprehensive guide for navigating the intricate ethical and political terrain of climate governance.

Julia Puaschunder's *Governance & Climate Justice* stands as a cornerstone of climate justice literature. Its timely, ethical, and practical insights make it an essential reading for global governance leaders, scholars, and activists as they work to address one of humanity's greatest challenges. This book not only informs but also inspires with hope—by empowering readers with the knowledge on anticipated economic gains within the ocean of losses in light of climate change. In its entirety, the book serves to advocate for a world where governance and justice align with our shared environmental responsibilities, fostering a legacy of environmental fairness and natural resource wealth for all and in the future to come.

Lars Oxelheim
Professor of International Business and Finance at the School of Business and Law, University of Agder, Norway
Professor Emeritus, Lund University, Sweden, Honorary Professor, Fudan University, Shanghai
Founding Chairman Swedish Network for European Studies in Economics and Business
Delegate Royal Swedish Academy of Engineering Sciences and the Royal Society of Letters at Lund
Elected Fellow European International Business Academy

Preface

When the initial first edition of '*Governance & Climate Justice: Global South & Developing Nations*' was conceptualized in 2016, the idea to calculate macroeconomic gains of climate change appeared bold but necessary to engage climate change critics or agnostic leaders. The first macroeconomic formalization was brought forward in a simplistic model. From the beginning of this research, it was emphasized that uncovering expected economic gains from a warming globe is only meant to pave a way to quantify a fair redistribution of climate change benefits to negatively affected communities and countries that lose out from climate change the most and fastest. Quantifying some expected windfalls from a warming temperature around the globe was also meant to innovatively raise funds to help conserve the earth's favorable living conditions for distant future generations. The first edition of the book therefore held a climate gains and losses focused model that helped envision redistribution via bonds-and-tax transfer strategies. Much has changed since 2016 and the original publication of the 2020 book.

The second edition of '*Governance & Climate Justice*' now bestows the initial intent to capture the expected gains for a warming globe through real-world endeavors. In the most recent years, there has been a growing acknowledgement of expected gains from a warming globe around the world. Global governance institutions, intergovernmental alliances as well as national governments and several research units now are calculating and estimating what '*Wealth of Nature*' the environment and a stable climate hold. Nowadays also the United Nations Framework Convention on Climate Change (UNFCCC) Conferences of the Parties (COPs) yearly summits urge attention to transfer funds for climate change mitigation and adaptation among nations but also over time to future generations. In a foresighted attempt, the book not only offers novel insights of the philosophical, economic and behavioral foundations of climate justice. The revised monograph also offers a detailed account of financial transfer strategies to help rebalance climate inequalities on the global stage. Based on the geo-impact of climate change, the financial crisis resilience as well as global connectivity and leadership potential of countries around the world, different indices present an implementation plan for enacting climate justice. Variegated input factors as well as multiple indices solutions presented in the book offer a broad range of possibilities to ensure that a fair solution is found in sharing the costs but also expected benefits from the warming of the earth in a fair and sustainable way. The discussion of the results now also further elucidates the economic benefits of a warming climate, which are distributed unequally around the globe.

The second edition therefore offers a new discussion of the *Wealth of Nature*. Starting with a sophistication of the initial idea of climate flexibility as a range of temperatures countries have for production, wealth of natural resources is added as a future country trade and production asset. Current world leader attempts to capture the economic value of natural goods preservation and conservation are outlined by the cases of the White House initiative to measure environmental economics. The United States Green New Deal as well as the European Green Deal are broad-based governance projects to find value in economic aspects of environmentalism. The European Sustainable Finance Taxonomy is discussed as a financial strategy to preserve natural resources and pass on economic gains from a stable environment to future generations. The U.S. endeavors to find environment-economic decision making strategies and statistics to account for nature are additional features of the book's extension. The Environmental, Social and Governance aspects of finance and investments are discussed from a historic perspective and an applied angle. Global endeavors to pay tribute to environmental conditions in economic proceedings are introduced in the United Nations Sustainable Development Goals (UN SDGs) as well as the World Bank and International Monetary Fund work on measuring the economic wealth of nature. All these various contemporary attempts to pay tribute to the wealth in nature in economics will be reflected upon for their contribution to the proposed climate justice implementation scheme. The most recent future-oriented proposals of a Global Green New Deal and its potential for an applied climate justice account will round up the future vision of the ideas brought-forward.

The second edition also features data updates and insights on the most recent geo-political tensions' impact on commodities and therefore also nutrition around the globe. Climate change remains the most crucial influence factor on commodity pricing—but financial instabilities and political unrest impact food prices even in countries that are not directly affected. The Intergovernmental Panel on Climate Change (IPCC) as well as UNFCCC COP negotiations are reflected on for deriving inferences on worldwide climate change mitigation and adaptation policies and practices trends.

The book remains to outline the databases and initial macroeconomic modeling throughout various chapters in order to pay attention to the prevailing trend of online book chapter consumption.

The second edition was conceptualized during COVID-19 while residing in Florida as one of the most environmentally-fragile parts of the United States, which raised awareness for the enormous risk that climate change imbues. The revised version of the book was finalized during a research stay at the International University of Monaco while in Residence at Monaco. It was inspired by the innovative sustainability leadership of the Principality as well as the power of positive incentives to move economic assets.

As always, the author remains highly indebted to all institutions and grant giving authorities who enabled the work on this volume. Most notably, Aix-Marseille University, Albert P. Sloan Foundation, Columbia University in the City of New York, the European Commission, European Master in Law & Economics Programme, European University Institute, Haifa University, Hamburg University, the International University of Monaco, The New School, Omnes Education Group, Principality of Monaco, Rotterdam University, United States Friends of the International Institute for Applied Systems Analysis, United States National Academies of Sciences, Engineering, and Medicine are to be mentioned most gratefully. Without their financial and in-kind support as well as the visionary leadership of Palgrave Macmillan New York, this book and its second edition would not exist. All omissions, errors and misunderstandings remain the author's.

Above all, highest appreciation is expressed to all readers for their most noble gift of care and generous interest to conserve and protect wealth of nature and grow together climate justice.

Monaco
October 2024

Julia Puaschunder

Acknowledgments

The author most gratefully acknowledges the International University of Monaco's support of Residency during the work on the second edition. Intellectual exchange during an academic training enabled by the International Friends of the International Institute for Applied Systems Analysis (IIASA) in cooperation with the United States National Academy of Sciences inspired new parts of the book. The Prize Fellowship in the Inter-University Consortium in cooperation with Columbia University, Princeton University and Yale University granted invaluable access to elite insights for the initial idea of the book. The New School, and in particular the Urban and Environmental Studies Program, provided most generous financial support. Funding is also appreciated of the Alfred P. Sloan Foundation, American Academic Research Conference on Global Business, Economics, Finance and Social Sciences; Austrian Academy of Sciences; European Parliament; Fritz Thyssen Foundation; George Washington University; International Institute for Applied Systems Analysis; Max Kade Foundation; Research Association for Interdisciplinary Studies; The New School (Dean's Office, Department of Economics, Eugene Lang College, Fee Board, The New School for Social Research, The New School for Public Affairs); and the University of Vienna; Vernon Arts and Science and the Vienna University of Economics and Business. The author declares no conflict of interest. All omissions, errors and misunderstandings in this piece are solely the author's.

Contents

List of Figures

List of Tables

CHAPTER 1

Introduction

Based on the idea of climate justice, the following part theoretically outlines the current societal demand for eternal equity. Portraying intergenerational equity as a natural behavioral law strengthens the legal case for codifying intergenerational fairness on a global basis and contributes to interdisciplinary behavioral law and economic models on contemporary climate justice predicaments.

1.1 Climate Justice as Intergenerational Equity Imperative

The most recent decades featured unprecedented economic progress. The post-WWIIeconomic boom heralded golden years of socio-economic advancement and economic capital growth outpacing every measure previous ages had known. Though looking back to an epoch of enormous economic progress in the twentieth century, the improvement of living conditions seemed to be threatened from the turn of the millennium on. The era of globalization, featuring complex interconnections and transactions faster than ever before in history, appeared to hold emergent systemic risks (Centeno & Tham, 2013; Sandbrook, Edelman, Heller, & Teichman, 2007). What happens in one part of the world today has impacts around the globe (Foley, 2009). The global interconnectedness imposing dangers created a need for framework conditions securing from

J. Puaschunder, *Governance & Climate Justice*,
https://doi.org/10.1007/978-3-031-87971-5_1

negative consequences emerging from the new web of social, ecological and fundamental transfers on a grand scale (Klein, 2014).

The following post-2008 era has been dubbed as the age of slowbalisation (The Economist, 2019). Protectionism, trade wars, economic development set back by reshoring and halted Foreign Direct Investments (FDIs) all contributed to a slowing of economic growth (The Economist, 2019). Technology took over human tasks and stagnant economic growth variables in the Western world demanded for measuring economic wealth on more broad-based and inclusive terms (Puaschunder, 2019a). The revolution of Environmental, Social and Governance (ESG) finance contributed to the wish to capture wealth of nations also from an environmental angle. The COVID-19 pandemic of the 2020s exacerbated the call for finding sustainable and resilient market structures. The contemporary geopolitical tensions but also a new era of tariffs imposed around the world have strengthened the call for vigilance and risk reduction in order to maintain a decent standard of living. Technological advancements in the age of AI also draw attention to the environmental impact of economic innovation.

In the eye of climate change and ecologic sustainability, the world appears more vulnerable than ever before. Future generations face declining biodiversity and ecological environmental changes in the wake of climate change. Capitalism'smyopic*short-term*profit focus systematically ignores the broader and longer-ranging implications of its actions and externalities, which results in irreversible ecological destruction (Boscov-Ellen, 2015). Contemporary capitalism raises ecological 'limits' to growth and environment concerns in the eye of unsustainable resource consumption and increasing human-made climate change (Zaretsky, 2015). In 2010, we hit the highest energy resource consumption in the 40-year recording of sustainability. Climate change is going to be the greatest human challenge of the twenty-first century. A warming globe touches on all human rights given the massive and widespread impacts. Substantially-degraded environmental conditions creates irreversible lock-ins for future generations and injustice over time. The destruction of the environment is the most sustainable peril of globalization.

This unprecedented intergenerational kink heralds an overall call for intergenerational equity—the fairness to provide an at least as favorable standard of living as enjoyed today. As we realize that our children may not have the same chances as we do, we must strive for global equity over time. The global challenge is thereby to find sustainable, qualitative economic growth in harmony with human rights of sustainable development over time.

The idea of eternal equity addresses justice over time. As an implicit contract and transfer in-between living and future generations, intergenerational equity not only reduces unfairness for future world inhabitants who are born in less favorable environmental conditions than their parents but also ensures future infrastructure, equal opportunities over time and constant access to social welfare for the youth. Intergenerational equity grants a favorable climate between generations and averts frictions arising from austerity plans, diminishing social welfare standards and declining environmental prosperity. Intergenerational equity avoids discriminating against future generations on the basis of remoteness of the time during which they will live and should have an equal chance to thrive.

While intergenerational equity is as old as humankind, climate change has put a new stance on future generations' right to justice over time (Fraser, 2014; Nafeez, 2013). In light of a warming globe, intergenerational equity is an urgent topic of concern that opens windows of opportunity to implement ecologic sustainability (Magdoff & Foster, 2010). The complex challenges ahead will require heightened attention to future generations' well-being. In the eye of an unprecedented intergenerational equity kink, it has become economically efficient to think about the next generation and future world inhabitants'living conditions and demand foresighted governance.

While the wish for intergenerational equity has sparked, we currently lack a codified legal framework on intergenerational fairness as well as an economic understanding of feasible intergenerational equity models that accurately pay attention to future generations. To measure intergenerational equity, we have to estimate future socio-economic developments around the warming globe. Intergenerational equity requires economic discounting of future events by politicians, policy makers and private individuals who will have to weight in future orientation and social responsibility in current decisions (Puaschunder, 2019b).

Future world inhabitants are put into the focus of today's choices by shifting the current wealth of the elderly to save for future generations by sustainable governance. For this endeavor, the current generation not only has to reflect on contemporary consumption rates and modes to conscientiously transmit the earth's resources to future generations. The current world generation is also called for finding economic alternatives in order to transition the economy and financing climate change abatement. One of the most innovative but just recently discussed avenues for ensuring climate justice lies in highlighting the economic inequality in global warming in order to derive redistribution strategies for ensuring fairness.

The outlined intergenerational equity constraints are complex and their solution interdependent. As the economic and environmental crises have exposed the weaknesses of orthodox economic theory, novel heterodox economic thinking is demanded (Shaikh, 2013). The current world economy opens possibilities to rethink economics in light of threats to future generations.

Eternal equity has always been lived within the family compound and practiced in the wake of humane fairness notions. The human-imbued wish to provide an at least as favorable standard of living to our children stems from evolutionary, social and religious values. Ignorance regarding intergenerational concerns naturally feels wrong and hegemony of now appears like a sin on future generations. Not being intergenerationally conscientious puts offspring's wellbeing at stake and detaches people from their environment. Understanding intergenerational conscientiousness as a natural behavioral humane-imbued law will help integrating future conditions in today's decision making.

Building on Rawls' procedural justice, intergenerational equity ensures fairness between generations based on future orientation and social responsibility for future generations. Pursuing intergenerational equity in the wish to provide a decent standard of living for the upcoming young can be enabled by a mutual transfer between old and young (Goldberger, 1989). Justice can be sought in future outlooks, humane reflexivity and globalized solidarity enabling that one generation does not live at the expense of future generations. When considering climate change, we are already behind the scheduled when it comes to fundamental foresighted preparedness.

The following work thus targets at contributing to eternal equity by introducing the idea that globalization imposes unprecedented intergenerational equity constraints regarding environmental sustainability. Natural resources consumption and climate change capture intergenerational equity as a natural behavioral law. A human-imbued Überethical drive toward intergenerational conscientiousness comprising social responsibility and future orientation is argued as the basis of eternal equity constituting legal foundations, public policies and regulation but also echoing in bottom-up participatory democracy and social representations of intergenerational equity.

The following part focuses on the present and near future regarding an unprecedented intergenerational equity kink in the environmental domain as the basis for a further investigation of climate justice around the world.

1.2 Global Commons and Climate Change

Resources likeair, oceans and water are prerequisites for life. Global common goods are globally linked and accessible for anyone—but owned by no one exclusively. No nation can claim common goods and declare sole access to common goods by itself. Distribution of global common goods has led to legal considerations ever since as fair access to global commons serves as the basis for peace and stability (Moore, 2000).

Free and equal access to common resources appears restrained in the age of globalization. Since the measurement of sustainability, energy consumption reached an all-time-high in the recent decade raising energy procurement costs and resource shortage. Since the recording of glaciers, arctic ice melting was highest during recent summers and the problem is worsening at an exacerbating rate.

While the ecological sensibility toward the environment has improved in recent decades and the Western world appears to have established a basic understanding about the scarcity of resources in line with a social imperative to protect the earth, the threat of climate change implies novel challenges of a lasting environmental decline for humankind.

The—by now—strong evidence of the anthropogenic contribution to climate change demands for concerted global action. The Global Humanitarian Forumsecurity report already claimed 300 million people being affected by climate change, killing estimated 300,000 of them and the total economic expenditures of climate change having reached over 100 billion USD (Parenti, 2011). The people suffering from climate change are expected to rise to 600,000, and the total annual economic costs will hit 300 billion USD within the next decade. In 20 years from now, the deaths caused by natural catastrophe arising with climate change could reach 500,000 civilians per year (Boscov-Ellen, 2015).

As weather patterns become increasingly volatile, the fragile and interdependent ecosystems vital for our survival will begin to fall apart (Boscov-Ellen, 2015). Climate change will determine the way people are living. Climate change precipitation patterns and access to water put agricultural and food preservation at stake. The melting of ice shells will unchangeably raise sea levels. Slugging and storm circuits will lead to natural disasters. The ecosystem and biodiversity are expected to diminish. Health risks, such as malnutrition, water problems and infectious diseases, will spread alongside facing persistent resource and food shortages (Oppenheimer, O'Neill, Webster, & Agrawala, 2007).

While global warming affects on the entire planet, today's climate impacts vary from country to country. Melting of glaciers and ice caps and thermal expansion of the oceans are projected to cause sea levels to rise by more than a meter by the end of the century, which would likely displace about 10% of the earth's human population. Climate refugees will be forced to leave behind their land, ancestry and cultural identity when about 200 million climate refugees will have to move in low-sea-level lands prospected around the next decades. An estimate of 1500 islanders will soon have to be evacuated from small islands in the South Pacific to higher territories. By 2050, it is estimated that there may be between 200 and 700 million climate refugees (Boscov-Ellen, 2015). The warming of ocean water will create hurricanes and storms imposing danger on around 4 billion people in the United States alone, and Inuit in the polar regions will have to move because of a massive ice melting.

Indigenous poor communities, who are already pushed to marginal levels of land, are going to be worst affected. African tribes will have to leave villages as the impact of drought in central African regions, where seasons have been scarce since 6 years. For instance, the Horn of Africa has had the hottest years in succession ever reported since 2011. Severe droughts have also struck Kenya, Ethiopia and Somalia, where attention is drawn to all-time-high food prices resulting in thousands of children starving. Climate justice will become an issue of concern when the major implications of a heating earth will be apparent and scarce resources vanishing potentially lead to violent frictions.

The populations most at risk, who live in the poorest regions of the world, are prone to be hit hardest by climate change, although they are neither emitters, nor central to the political discussions or part of the solution. At the international level, common goods consumption and preservation breed inequality if the lifestyle of some is trading off from the environmental conditions of others (Foster, 2009). The global West appears to cause an unfair impact to human-made climate change. The developed world is estimated to be responsible for 99% of carbon consumption, which raises global justice questions. Less than 1% of the greenhouse gas emissions that caused global warming are offset by fifteen of the poorest nations—including the Middle East, Southeast Asia, Sub-Saharan Africa and small islands developing states.

In the light of irreversible environmental decline, the time is ripe to reflect on intergenerational justice. In March 2015, 400 parts per million of atmospheric CO_2 were measured on earth, which is the highest concentration since the Oligocene period 23 million years ago (Boscov-Ellen, 2015). Storms like supertyphoon Haiyan, which killed or injured 30,000 people and displaced 6 million people, or Cyclone Pam, which decimated Vanuatu and forced thousands of people into relocation, demonstrate this urgency (Boscov-Ellen, 2015). Vanuatu, like many island nations, is already losing coastline and facing significant water and crop shortages as a result of climate change (Boscov-Ellen, 2015). The direct effects of climate change have already displaced around 50 million people (Boscov-Ellen, 2015). Due to the implicit effects of climate change, one-third and one-half of the species of flora and fauna on the planet is projected to go extinct by the end of the century, representing a dying out at about 1000 times the baseline rate (Barnosky et al., 2011).

Already a decade ago, we were facing an estimated 0.8 degrees Celsius of global warming and the speed of temperature change has increased. The Intergovernmental Panel on Climate Change (IPCC) predicts that global average temperatures could rise by up to 6 degrees Celsius by the end of the century. Half of the earth surface is projected to experience persistent droughts by 2100, when suffering from extreme drought is estimated to increase by 900% (Parenti, 2011). Droughts are believed to be accompanied by periodic bursts of heavy rainfall, which will cause flooding and soil erosion making land useless for irrigation (Boscov-Ellen, 2015).

As our current consumption patterns and economic growth endeavors transfer environmental debts to our ancestors or jeopardize the environmental conditions of our future children, there is an urgent need to address climate change. Many of the more difficult issues were concerned in Durban in late December 2011 and the Rio+20 Conference. Debates are also organized by UN agencies, governments or developed and developing nations, corporations, foundations and constituency groups, who increasingly highlight the humane consequences of climate change. The humanitarian dimension of climate change is central to the formal negotiations emphasizing mitigation and security in annual UN Conference of the Parties (COP) framework discussions leading to international agreements (e.g., the Paris Agreement) and the inception of multilateral climate justice programs. During the 2009 Copenhagen meeting, the international focus was placed on emerging economies—such as China and India—as the countries with the highest levels of CO_2emission. Systems of monitoring, reporting and verification of emissions were discussed to be installed to reduce emissions in non-binding agreements based on individual country pledges. As the developing nations claimed a right to economic development as the Western world previously had enjoyed, monetary transfer funds were installed to help with the transition to renewable energy in booming developing economies. The December 2015 Paris UN Framework Convention on Climate Change (UNFCCC) offered an iconic international climate change agreement. Most recent UNFCCC COP agreements have centered around the unequal spread between climate change gains and losses, which exacerbated a redistribution call to unprecedented momentum (Flavelle, 2021a, 2021b). While voluntary climate change aversion has been established during the last few years, it remains unclear how strong the commitment will be in the age of political unrest and eccentric presidential elections in the Western world. Like with any global commons problem, the success of voluntary agreements depends on the consent of all countries around the world. The political tensions exacerbating between the East and West lately put a common climate change accord in question. The most recent UNFCCC Conferences of the Parties (COPs) have further developed a framework of climate change inequalities. Those countries that have caused the problem and are potentially benefiting from a warming globe in the short run are more and more under scrutiny to help those territories on earth that are affected the first and worst by climate change.

The following monograph aims at offering an innovative approach toward intergenerational equity provision. In order to secure the financialization of climate changemitigation and adaptation efforts but also in order to ensure climate equality around the globe and over time, innovative financialization techniques will be grounded on the data-driven finding that climate change gains and losses are distributed unequally around the globe and therefore demand for a redistribution of climate gains in the twenty-first century. The book will also address newest worldwide developments to account for nature in economics and protect the environment via innovative financial means. The discussion closes with an overview of policy recommendations and grants insights into future research avenues.

References

Barnosky, A. D., Matzke, N., Tomiya, S., Wogan, G. O. U., Swartz, B., Quental, T. B., Marshall, C., McGuire,. J. L., Lindsey, E. L., Maguire, K. C., & Mersey, B. (2011). Has the earth's sixth mass extinction already arrived? *Nature, 471*, 51–57. https://doi.org/10.1038/nature09678

Boscov-Ellen, D. (2015, Spring). Lecture notes. *GECO 5250: Rethinking capitalism*. The New School.

Centeno, M. A., & Tham, A. (2013). *The emergence of risk in the global system*. Unpublished Working paper. Princeton University.

Flavelle, Ch. (2021a, December 3). Billions for climate protection fuel new debate: Who deserves it most. *The New York Times*. https://www.nytimes.com/2021/12/03/climate/climate-change-infrastructure-bill.html

Flavelle, Ch. (2021b, December 8). The climate bill includes billions in funding: Will it be spent fairly? *The New York Times*. https://www.nytimes.com/2021/12/03/climate/climate-change-infrastructure-bill.html

Foley, D. K. (2009). The economic fundamentals of global warming. In J. M. Harris & N. R. Goodwin (Eds.), *Twenty-first century macroeconomics: Responding to the climate challenge* (pp. 115–126). Edward Elgar.

Foster, J. B. (2009). Capitalism and ecology: The nature of the contradiction. In *The ecological revolution: Making peace with the planet* (pp. 121–128). Monthly Review Press.

Fraser, N. (2014). Can society be commodities all the way down? Post-Polanyian reflections on capitalist crisis. *Economy and Society, 43*(4), 441–458. https://doi.org/10.1080/03085147.2014.898822

Goldberger, A. (1989). Economic and mechanical models of intergenerational transmission. *American Economic Review, 79*(3), 504–513.

Klein, N. (2014). *This changes everything: Capitalism vs. the climate*. Simon & Schuster.

Magdoff, F., & Foster, J. B. (2010). *What every environmentalist needs to know about capitalism: A citizen's guide to capitalism and the environment*. Monthly Review Press. https://doi.org/10.14452/MR-061-10-2010-03_1

Moore, J. W. (2000). Environmental crises and the metabolic rift in world-historical perspective. *Organization & Environment, 13*(2), 123–157. https://doi.org/10.1177/1086026600132001

Nafeez, A. (2013, June 14). Pentagon bracing for public dissent over climate and energy shocks: NSA prism is motivated in part by fears that environmentally-linked disasters could spur anti-government activism. *The Guardian*.

Oppenheimer, M., O'Neill, B. C., Webster, M., & Agrawala, S. (2007). Climate change: The limits of consensus. *Science, 317*(5844), 1505–1506. https://doi.org/10.1126/science.1144831
Parenti, C. (2011). *Tropic of chaos: Climate change and the new geography of violence*. Nation Books.
Puaschunder, J. M. (2019a, June 10–11). Artificial Intelligence market disruption. *Proceedings of the 13th International RAIS Conference on Social Sciences and Humanities*, organized by Research Association for Interdisciplinary Studies (RAIS) at Johns Hopkins University, Montgomery County Campus (pp. 114–124), Rockville, MD, United States.
Puaschunder, J. M. (2019b). Intergenerational governance and leadership around the world. In J. M. Puaschunder (Ed.), *Intergenerational governance and leadership in the corporate world* (pp. 153–177). IGI Publishing.
Sandbrook, R., Edelman, M., Heller, P., & Teichman, J. (2007). *Social democracy in the global periphery: Origins, challenges, prospects*. Cambridge University Press. https://doi.org/10.1017/CBO9780511491139
Shaikh, A. (2013). Crisis, austerity, and the role of economic theory in policy. *Social Research: An International Quarterly, 80*(3), 653–664. https://doi.org/10.1353/sor.2013.0047
The Economist. (2019, January 26). *The steam has gone out of globalisation: Slowbalisation* (pp. 17–20).
Zaretsky, E. (2015, Spring). Lecture notes. *GECO 5250: Rethinking capitalism*. The New School.

CHAPTER 2

Intergenerational Equity

2.1 Eternal Equity as a Natural Behavioral Law

The idea of intergenerational equity is as old as humankind. Conscientiousness for future generations is imbued in social customs and humane idealism. Intergenerational equity arises from the elderly wanting their offspring to prosper in at least as favorable conditions as experienced. Naturally, parents do not want their child to grow up in worse conditions than they had themselves. The elderly morally feel for future children's well-being. Whether it is the farmer or the forest caretaker, the small entrepreneur or the financial manager, humans are compassionate about future world inhabitants to come. Humans have an inner aversion against making their children worse off and a potentially evolutionary-grown psychological dislike of using up their offspring resources.

All major religions promote intergenerational equity. The more religious people are, the more family members they are likely to have. The stronger the ties between family members, the more intergenerationally responsible they are. Religious roots describe the world being borrowed by current inhabitants, who must preserve the earth for future generations. Religious foundations advocate for the elder generation leaving the earth in a decent state. Judaism advises parents to plant trees for their children. The Old Testament's Commandments call for reciprocal care of children and parents. Protestantism warns about apocalyptic losses if intergenerational care vanishes. Islamic banking favors real values over

J. Puaschunder, *Governance & Climate Justice*,
https://doi.org/10.1007/978-3-031-87971-5_2

speculative credit purchases in order to avoid inequality over generations. Indigenous principles outline one generation may not live at the expense of another.

The natural behavioral law of intergenerational equity was lived for centuries and transpired in the social compound as practiced in ancient, traditional customs ever since. Already the oath of the ancientAthenian city postulates to 'transmit a city not only not less but better and more beautiful than it was transmitted' to ensure justice in-between generations. The Roman Law*prodigus*-clause excluded spendthrifts from economic interaction to save offspring from poverty and promote harmony between young and old (Benke & Meissel, 2008). Intergenerational equity as a natural behavioral law is described as early as 700 BC by Hesiod's parable of a son's concerns over inheritance. In the ancient Greek antique, Zeus symbolizes fatherly justice establishing eternal equity. Cicero's third book of '*De re publica*' introduces justice as a natural law—a humane-imbued virtue that is fortified by the social compound and education.

Intergenerational equity is a natural law in its roots and outcomes. Based on the evolutionary wish to improve the living conditions for descendants, intergenerational equity attributes the ethical obligation to provide an at least as favorable standard of living to future generations as enjoyed today. Intergenerational equity breeds an ethos of justice in-between current world habitants and future unborn following the greater goal to avert imbalances between the young and the elder. Resulting societal benefits comprise common goods preservation, ensure degrees of freedom for posterity as well as promotion of the judicial virtue*suum cuique*—grant everyone what they deserve.

Intergenerational equality is grounded on a human-imbued wish for fairness as there is an ethical preference for fair welfare distribution among different generations. Holding through time and place, evolutionary-grown natural laws of responsibility for the future stem from compassion between overlapping generations. Evolutionary offspring are in an advantaged position if being brought up in socially benevolent environments. Intergenerational conscientiousness as a natural care for offspring is transpired through social influences—for instance, externally nurtured by educational upbringing, role-model learning and critical life events like parenthood. Anthropologists trace back first signs of social conscientiousness in-between generations as soon as human could experience three-generation family members together. Since then elder, younger and youngest could directly transmit knowledge and crystallize wisdom from the old to the young. Stable family ties and salient eternal chains of life made generations feel for another.

Current generations are called upon to make sacrifices today for the well-being of future generations (Sachs, 2014). Advocacy for equity over time is a societal necessity as ever since intergenerational downfalls had disastrous consequences. Whenever inequity and overexploitation of resources emerged, societies broke apart and environmental conditions degraded. Intergenerational clashes fueled revolts of the young versus the old throughout history. A desperate youth in the post-World War I era bred nationalism. European nationalist parties of the young had extremely negative consequences for the community—such as Adolf Hitler's National Party rebelling against the '*Altpartei*' or old party. Communist oppositions promoted themselves as young force juxtaposing old regimes. In the 1968 European protest, the young revolted against the old in the eye of ceasing economic stability given oil and energy crises.

Natural laws determine decisions and drive actions in the social compound (Luf, 2011). As an implicit guide of day-to-day practiced common senseethicality, natural laws are ethical anchors beyond regulatory frameworks and whimsical courtroom decisions (Mamor, 2010). The natural law of intergenerational equity serves as a moral compass in ethical predicaments. As an intrinsic meaning of what is right and what is wrong with a strong emotional message, natural laws unite in times of change through a common sense of fairness. Especially during times of heightened uncertainty or societal upheaval, when people come together to exchange information in their wish to control their world, people orient their decisions on natural laws. Natural laws of inalienable rights are prominent ethical anchors and robust panaceas to avert negative consequences of regulatory downfalls and substitute lacking oversight control (Mayer, 2010). Natural laws serve as inner moral compasses and human-imbued cue on what is right or wrong beyond lagging-behind legal frameworks and policy guidelines during revolutions but also when external shocks come down on society. For instance, less than 24 hours after the 2011 Japan nuclear disaster, newspapers addressed intergenerational conscientiousness. As windows of opportunity, times of crises and external shocks thus prosper human rights formation based on natural laws (Klein, 2014).

Acknowledging intergenerational equity as a natural behavioral law may serve as a legal basis for the codification of human rights to intergenerational justice regarding climate change in the twenty-first century. Globalization, political changes and societal trends have leveraged attention to intergenerational responsibility from the turn of the millennium on. Today's intergenerational challenges of reaching ecological limits regarding climate change put pressure on future generations (Barnosky et al., 2011; World Wide Fund's living planet index). Climate change calls for equity over time. Mass media exposing intergenerational predicaments raises concern for people, planet and profits. Social, environmental and economic harmony is currently demanded for future generations also in the finance sector when considering the growing Environmental, Social and Governance (ESG) investing pledges. In the last years, there is also a rising concern over environmental degradation met with attempts to capture the wealth of nations in environmental terms. The World Bank, the International Monetary Fund, the European Union and the United States have proposed different measurement techniques of natural resources in order to allow for a quantification of nature in economics as the basis for conservation.

2.2 Toward a Human Right of Intergenerational Equity and Intuitionist Sustainability

Intergenerational equity regulation is still hardly codified as the living standards have constantly had an overall improving trend. Given intergenerational equity threats of global climate change, acknowledging intergenerational equity as a natural behavioral law may serve as a first-aid basis for the codification of fairness between generations.

In a history of turning to natural law for solving societal predicaments on a global scale in times of crises, the idea of intergenerational equity as a natural law is targeted at averting contemporary real-world intergenerational imbalances. Describing intergenerational conscientiousness helps alleviate tensions between the old and young in the eye of prospective standard of living declines. New economic thinking may integrate intuitionist and behavioral aspects. Intuitionist sustainability refers to a natural behavioral law of sustainability and common goods preservation, which appear to be prevalent and innate in human beings (Puaschunder, 2017b, 2018). Future research may address external influencing factors on intergenerational conscientiousness and innovatively leverage our understanding how to promote intergenerational harmony in the wake of climate change (Latour, 2014).

Introducing the idea of intergenerational equity as a natural law takes an intuitionist perspective on sustainability in the twenty-first century. Intergenerational equity as a natural behavioral law attributes a dynamic interaction view of intuition and reasoning for sustainability by exploring how subconscious and conscious processes interact to shape socially responsible and future-oriented behavior leading to intergenerationally equitable choices. Intergenerational equity as a natural behavioral law innovatively promotes an intuitionist view of sustainability.

Intergenerational equity as a natural behavioral law comprises pro-active concerns about current actions' future implications and contributions to society. Departing from an ethics field that is primarily focused on avoidance of harm for stakeholders, intergenerational equity as a natural behavioral law describes a pro-active engagement to consider future stakeholders in today's choices. Beyond the duty to not violate the law and generally accepted ethical norms, intergenerational equity as a natural behavioral law outperforms what is expected regarding responsibility in serving the interests of future unborn. Intergenerational equity as a natural behavioral law research thus helps understanding why individuals are more inclined to contribute to the welfare of people or communities with whom they have no contact and pro-actively engage in 'do good' behavior benefactoring temporally distant future stakeholders.

Attributing the pro-active engagement of intergenerational care for posterity, intergenerationally responsible leadership is considered as an 'Überethical' drive (Puaschunder, 2017a) or pro-active overcompliance with contemporary sustainability legislation. Extending ethicality to not simply considering to avoid unethical behavior, intergenerational equity as a natural behavioral law adopts a pro-active ethics lens focused on über-doing compliance and considering interests of a wider range of stakeholders. Understanding intergenerational equity as a natural behavioral law thereby draws a broader social contract between business and society to incorporate novel responsibilities and embrace discretionary activities that contribute to societal welfare and the well-being of future generations.

Contemporary behavioral ethics calls for integrating culture and institutional influences on people's propensity to engage in ethicality as differing legal systems and social norms shape individuals' ethical decision making. Supranational contexts determine the global governance of public and private leaders. A sophisticated conceptualization of 'intergenerational behavior' applicable to culturally diverse stakeholders will allow to unravel the antecedents of responsible and irresponsible intergenerational behavior across national contexts. Public and private concerns over climate instability risks create a need for an international outlook of intergenerational equity in order to solve global common goods predicaments and draw inferences on the harmonization of intergenerational justice on a global scale.

References

Barnosky, A. D., Matzke, N., Tomiya, S., Wogan, G. O. U., Swartz, B., Quental, T. B., Marshall, C., McGuire, J. L., Lindsey, E. L., Maguire, K. C., & Mersey, B. (2011). Has the earth's sixth mass extinction already arrived? *Nature, 471*, 51–57. https://doi.org/10.1038/nature09678

Benke, N., & Meissel, F.-S. (2008). *Übungsbuch Römisches Sachenrecht*. Manz.

Klein, N. (2014). *This changes everything: Capitalism vs. the climate*. Simon & Schuster.

Latour, B. (2014, February 26). *On some of the affects of capitalism*. Lecture at the Royal Academy, Copenhagen.

Luf, G. (2011). *Grundfragen der Rechtsphilosophie und Rechtsethik: Einführung in die Rechtswissenschaften und ihre Methoden*. Manz.

Mamor, A. (2010). *Philosophy of law*. Princeton Foundations of Contemporary Philosophy.

Mayer, H. (2010). *Öffentliches Recht: Einführung in die Rechtswissenschaften und ihre Methoden*. Manz.

Puaschunder, J. M. (2017a). Ethical decision making under social uncertainty: An introduction to Überethicality. *Sustainable Production and Consumption, 12*, 78–89.

Puaschunder, J. M. (2017b). *Global responsible intergenerational leadership: A conceptual framework and implementation guidance for intergenerational fairness*. Vernon Art & Science.

Puaschunder, J. M. (2018). *Intergenerational responsibility in the 21st century*. Vernon Art & Science.

Sachs, J. D. (2014). Climate change and intergenerational well-being. In L. Bernard & W. Semmler (Eds.), *The Oxford handbook of the macroeconomics of global warming* (pp. 248–259). Oxford University Press.

CHAPTER 3

Global Responsible Intergenerational Leadership

Globalization leveraged pressure on contemporary society. In a history of turning to natural law as a human-imbued moral compass for solving societal predicaments on a global scale in times of crises, behavioral socio-economists currently examine the human natural drive toward intergenerational fairness. As an implicit contract and transfer in-between living and future generations, intergenerational equity avoids discriminating against future generations. Intergenerational equity grants a favorable climate between generations and alleviates frictions arising from the negative impacts of intergenerational inequality. Understanding the bounds of human ethicality is key to avoid ethical downfalls on currently emerging environmental decline infringing on intergenerational equity—the fairness to provide an at least as favorable standard of living to future generations as enjoyed today. Pressing social dilemmas beyond the control of singular nation states call for corporate social activities to back governmental regulation in crisis mitigation (Puaschunder, 2016).

The monograph targets at capturing contemporary intergenerational equity constraints in the environmental domain within society, around the globe and over time. Outlining some of the causes of the current international and intergenerational imbalances regarding climate stability prepares for recommendations on how to implement climate justice in the twenty-first century. Theoretically defining intergenerational equity as a natural behavioral law is coupled with providing socio-economic

J. Puaschunder, *Governance & Climate Justice*,
https://doi.org/10.1007/978-3-031-87971-5_3

intergenerational fairnessnudges for environmental ethicality in order to avert climate change over multiple overlapping generations.

For a successful implementation of intergenerational equity transfers, a thorough understanding of contemporary intergenerational imbalances is needed. Investigating intergenerational equity is a formidable task as for touching on unprecedented predicaments comprising manifold stakeholders (Almansa & Martinez-Paz, 2011). An engagement of various stakeholders on the differing intergenerational predicaments results out of a disparity of intergenerational equity approaches. As a first step toward resolving societal losses imbued in the complexity of this novel phenomenon but also to innovatively explore new opportunities to ingrain intergenerational responsibility within globalizing economies, attention to expert opinions and stakeholder facets in the interplay of public and private actors' approaches toward intergenerational equity must be paid. Holistically describing intergenerational equity stakeholder perspectives will help overcoming socio-economic losses implied by various eternal equity societal notions. Averting multi-stakeholder conflicts in the implementation of intergenerational equity and aligning differing stakeholders' view on intertemporal justice will harmonize intergenerational equity endeavors on a grand scale.

Studying foresighted leaders and experts' opinions on current intergenerational frictions captures the contemporary social representation of intergenerational equity. Social representations describe the genesis of collective ideas in order to predict future behavior of social masses during economic upheaval. Through the study of social representations, information about the diffusion of ideas and societal common knowledge on the prevailing Zeitgeist can be obtained (Durkheim, 1898; Moscovici, 1961). As for depicting societal reactions to economic influences over time, social representations grant orientation in times of change (Moscovici, 1984). Social representations emphasize the shared social construction process when individuals try to make sense of external shocks and society collectively debriefs shared events. Social representations thus provide insights into the homogenous social perception of a shared environment. As expressions of social perceptions of socio-economic transition, depicting social representations allows to predict future behavior of social masses and forecast collective societal trends (De Rosa, 1994; Nisbett & Ross, 1980). By comprising a broad range of stakeholders and professions that hold differing notions on the topic of interest, who propose intervention strategies based on their own experience and professional backgrounds, social representations offer a broad societal spectrum to understanding economic phenomena's impact on social masses. The social representations collected thereby offer insights on the subsequent public choices and policy decision making.

The free association method used focuses on qualitatively collecting independent notions that arise when being confronted with the topic of interest. This method grants highest degrees of freedom to choose and express thoughts and prevailing perceptions about a topic of interest. The free association technique thus allows capturing novel social phenomena and the influence of external conditions on the decision making and verbal expression of collectives in an open and multifaceted way. The social representation technique thereby reveals the complex decision making process triggered by external shocks.

Derived from the social discourse on individual needs and shaped by collective moods, social representations portray specific stakeholder facets and emerging societal trends of communities, which allow leadership recommendations on the advancement of ideas and whole-roundedly gain sympathy of diverse constituents for their implementation. While the current prevailing collective soul is mapped based on the social reflection of contemporary ideas, social representations also reveal the different mindsets of various constituents serving a Machiavellian wish to understand the common sense of social masses per stakeholder group.

The communicative character of the free associations collected at a future conference gathering scene sheds light on coping strategies debated in the eye of an unprecedented intergenerational equity kink. The rationale behind using inductive reasoning and the free association method was to capture broad and diverse contents of multiple stakeholders. The exploratory character of free associations provides an authentic snapshot at a given point in time and holds the prevailing Zeitgeist on the topic of interests. The social representation interpretation then allows to outline stakeholder-specific nuances of the topic of concern. The clear merit of the method and material is the authentic nature of the easily collected responses. The access to refined elite knowledge and the reflexive nature of the debated associations provide first-hand, real-world insights on the topic of interest at the pulse of time.

A study of the social representations on intergenerational equity aimed at capturing leaders, experts, practitioners and students' concern about intergenerational balance regarding climate stability as well as intergenerational equity implementation strategies. The social representations of

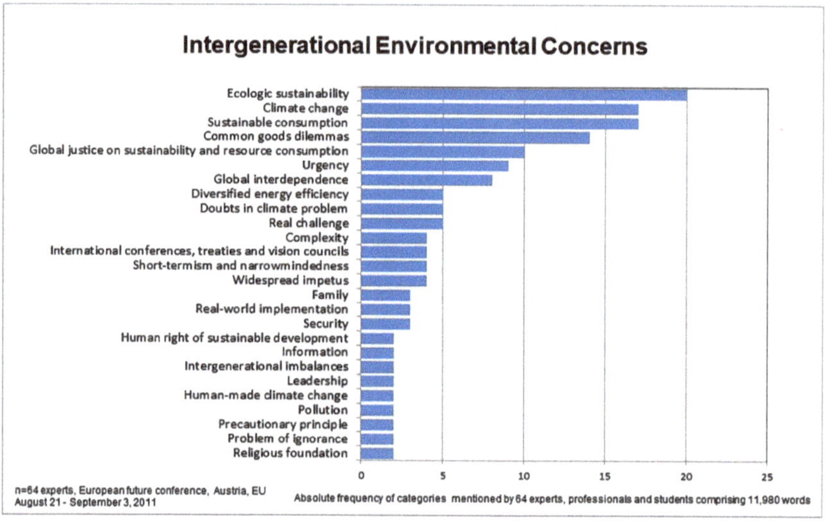

Fig. 3.1 Intergenerational equity environmental concerns

intergenerational equity were retrieved from 110 speeches, interviews and conversations with leaders at a European future conference during the late summer of 2011 (Puaschunder, 2015). The 48,460-word transcript of associations on the topic '*Justice: Responsibility for the Future*' was categorized by the author to abstract the prevailing Zeitgeist of intergenerational equity during the late summer of 2011. All collected associations were screened for contents on environmental sustainability and environmental intergenerational equity constraints (Puaschunder, 2015).

In total, 153 associations comprising 11,980 words of 64 experts, leaders, practitioners or students provided differing information on environmental intergenerational equity constraints. The associations were assigned into a category system that was developed inductively by the author. The contents of the social representations of intergenerational equity environmental concerns were categorized and the frequency of the associations per category calculated for every association that would hold information on intergenerational equity environmental concerns. The categories and frequencies of associations on intergenerational equity constraints in the environmental domain are displayed in Fig. 3.1 featuring the frequencies of the associations per categories.

On intergenerational environmentalism, experts, leaders, practitioners, scholars and students reflected ecological sustainability ($f = 20$; 13.07%), climate change ($f = 17$; 11.11%) and sustainable consumption ($f = 17$; 11.11%). These most frequently mentioned associations are the core of the social representations of environmental intergenerational equity concerns of the respective stakeholders. Urgent environmental concerns require diversified energy efficiency. The prevailing categories also reveal that complex, interdependent common goodsdilemmas demand for global solutions. International conferences, treaties and vision councils legally back intergenerational justice, while family bonds aid real-world intergenerational conscientiousness (see Fig. 3.1).

In total, 398 associations comprising 27,141 words of 89 experts, leaders, practitioners and students provided information on differing intergenerational equity implementation strategies. The contents of the social representations of the implementation of intergenerational equity were categorized. The associations were assigned into a category system that was developed inductively by the author. The frequency of associations in the association process was calculated for every notion that held information on intergenerational equity implementation suggestions. The categories and frequencies of the associations of the experts, leaders, practitioners and students are displayed in Fig. 3.2 holding the frequencies of the association categories.

For a successful implementation of intergenerational equity, experts, leaders, practitioners and students suggested efficiency ($f = 20$; 5.03%) and conscientious consumption ($f = 17$; 4.27%). Long-term solutions ($f = 15$; 3.77%) should comprise institutional regulation and foresighted taxation ($f = 15$; 3.77%) balancing wealth, future investments in education and pension reforms. Open debates will inform global leaders of complex intertemporal frictions and encourage them to find a global solution. International treaties and conventions will spearhead the codification of intergenerational justice as a human right. Innovation and economic growth through investments are market mechanisms to prevent future crises. Essential to eternal equity will be behavioral changes of citizens based—not only on ethicality and religious values, but also on the social glue featuring fairness derived from social responsibility and integrative solidarity. The codification of intergenerational equity must be pursued in international treaties and conventions. Leadership for institutional implementation but also social responsibility, religion and ethicality play an important role in the expression of intergenerational fairness.

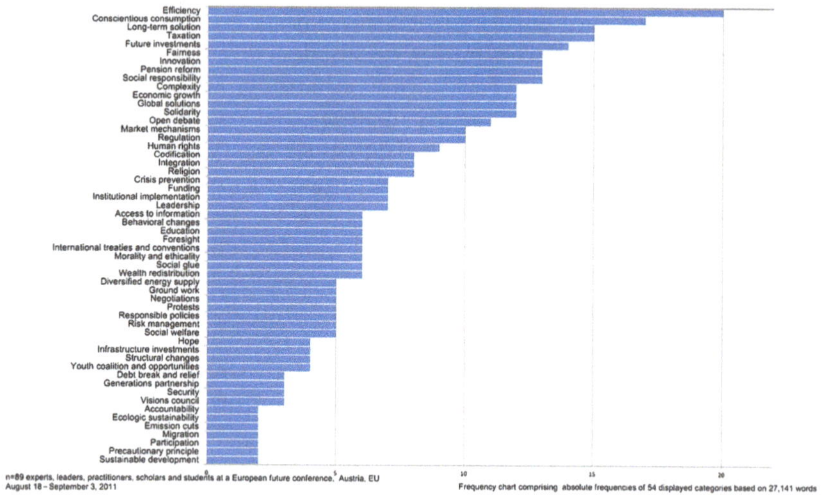

Fig. 3.2 Intergenerational equity implementation

In a stakeholder-nuanced analysis, environmental concerns were found to comprise of climate change and unsustainable resource consumption resulting in potential injustice over time and irreversible lock-ins for future generations. Environmental considerations are foremost on public officials and international organization representatives' agendas calling for leadership on sustainable development. Politics are starting to connect intergenerational justice to human rights. In the environmental domain, ecological sustainability, climate change and sustainable consumption put pressure on future generations. Complex common goods dilemmas demand for global solutions based on legal frameworks. Intergenerational harmony may also be accomplished through behavioral changes in the social compound.

A successful intergenerational equity implementation is envisioned by the experts to feature efficiency, conscientious consumption and long-term solutions such as institutional regulation and foresighted taxation but also future investments in education and pension reforms. Open debates will inform global leaders of complex intertemporal frictions and encourage to find a global solution. International treaties and conventions will spearhead the codification of intergenerational justice as a human right. Innovation and economic growth through investments are market mechanisms preventing future crises. Essential to eternal equity will be behavioral changes of citizens based—not only on ethicality and religious values—but also grounded on the social glue establishing fairness from social responsibility and integrative solidarity.

The presented research offers contributions to the fields of law, economics and governance. Describing social representations spearheads sociology entering economics. Intergenerational equity introduced as a natural behavioral law serves as a basis for future-oriented and socially responsible codification and regulation. Research connecting the individual experience to social responsibility allows to innovatively unravel ethicality nudges (Thaler & Sunstein, 2008). Common goal compliance on intergenerational responsibility is hoped to be modeled by the information retrieved on the social representations of intergenerational equity. Gaining acceptance for future societal contributions will help legislators to automatically nudge decision makers into an intergenerational equitable direction and thereby foster a harmonious government-citizen relation following the greater goal of ensuring a sustainable humankind.

Future research avenues may pay attention to the driving forces of normative ideals of intergenerational equity to understand the meaning and motivations behind the topic of interest, among different stakeholders. For example, more detailed information is certainly needed in order to capture the differences in the social representations of intergenerational equity between economics and finance elites, and public servants and international organizations. Future studies might investigate the change of social representations on intergenerational equity as well as take into account country-level differences in intertemporal policy making during other times of economic cycles. Additional follow-up investigations may also unravel the current dynamics of the local civil society in establishing intergenerational justice (Puaschunder, 2015; Puaschunder & Schwarz, 2012). Attention could be paid to the importance of prevailing county-level mechanisms for the interplay of different institutions, economic and financial market actors, as well as welfare state representatives that respectively affect the representations and implementation of intergenerational equity (Puaschunder, 2015). Cross-national variation in climate prospects is a prime example of different thresholds of intergenerational equity but also potential crowding out effects between public and private sector intergenerational justice implementation strategies that are explored in the following (see Chapters 4, 5, 6, 7 and 8). The following research therefore captures intergenerational conscientiousness with a more quantitative and globally focused view in order to help implementing intergenerational equity and instigate intertemporal transfers on a grand scale.

References

Almansa, C., & Martinez-Paz, J. M. (2011). Intergenerational equity and dual discounting. *Environment and Development Economics, 16*, 685–707. https://doi.org/10.1017/S1355770X11000258

De Rosa, A. S. (1994). From theory to metatheory in social representations: The lines of argumentation of a theoretical-methodological debate. *Social Science Information, 33*, 273–304. https://doi.org/10.1177/053901894033002008

Durkheim, E. (1898). *Représentations individuelles et répresentations collectives: Sociologie et philosophie*. Presses Universitaires de France.

Moscovici, S. (1961). *La Psychanalyse, son image et son public*. Presses Universitaires de France.

Moscovici, S. (1984). The myth of the lonely paradigm: A rejoinder. *Social Research: An International Quarterly, 51*, 939–968.

Nisbett, R., & Ross, L. (1980). *Human inference strategies and shortcomings of social judgment*. Prentice Hall.

Puaschunder, J. M. (2015). On the social representations of intergenerational equity. *Oxford Journal of Finance and Risk Perspectives, 4*(4), 78–99.

Puaschunder, J. M. (2016). The call for global responsible intergenerational leadership in the corporate world: The quest for an integration of intergenerational equity in contemporary Corporate Social Responsibility (CSR) models. In D. Jamali (Ed.), *Comparative perspectives in global corporate social responsibility* (pp. 275–288). IGI Global Advances in Business Strategy and Competitive Advantage Book Series.

Puaschunder, J. M., & Schwarz, G. (2012). *The future is now: How joint decision making curbs hyperbolic discounting but blurs social responsibility in the intergenerational equity public policy making domain*. Working paper. Harvard Law School. http://thesituationist.wordpress.com/2012/12/17/the-situation-of-intergenerational-equity/.

Thaler, R., & Sunstein, C. (2008). *Nudge: Improving decisions about health, wealth, and happiness*. Yale University Press.

CHAPTER 4

Philosophical, Economic, and Behavioral Foundations of Climate Justice

A warming earth under climate change determines current and future generations' living conditions. Environmental economics with particular attention to climate justice as the wish for everyone to enjoy a benevolent environment is based on ethical imperatives and economic calculus laying the foundation to protect climate-disadvantaged current world inhabitants and future generations. Justice and fairness underline inequality alleviation via economic redistribution within society, inbetween the entire world countries community, and over generations.

The philosophical roots for climate justice are derived from Hans Jonas' interpretation of Immanuel Kant's categorical imperative and John Rawls' veil of ignorance. Kant's and Jonas' work is applied to countries and communities around the globe being affected differently by climate change. Given empirical evidence that some countries may economically gain from climate change in the short run, Kant's categorical imperative advises to only actively engage in actions that one passively wants to endure. Hans Jonas' extension of Kant's categorical imperative to the environment suggests a common

J. Puaschunder, *Governance & Climate Justice*,
https://doi.org/10.1007/978-3-031-87971-5_4

but differentiated responsibility to ensure a stable climate for this and future generations. John Rawls' veil of ignorance advocates for analyzing the ethical predicament around climate change without knowing whether one has a short-run gaining or losing economic prospect. The economic foundations of Kaldor's compensation criteria and legal disproportionate impact analyses of unequally-distributed economic gain and loss prospects in light of climate change are the foundation for redistribution to enact climate fairness. Attention is also be paid to the individual decision-maker's choices regarding environmental protection in intertemporal discounting. Behavioral insights are derived from empirical survey studies on how to nudge people into conscientiousness of environmental protection.

4.1 Introduction

Never before in the history of humankind have environmental concerns in the wake of economic growth heralded burden-sharing predicaments as we face today. Climate change affects high- and low-income households, developed and underdeveloped countries, as well as overlapping generations differently (Puaschunder, 2016a, 2016b). Climate justice addresses fairness in everyone's access to a favorable climate in a fair share of benefits and burdens of a warming globe (Sachs, 2021). Global warming presents societal, international, and intergenerational fairness as a challenge for modern economies and contemporary democracies worldwide (Puaschunder, 2017b).

A warming earth under climate change determines current and future generations' living conditions. Environmental economics with attention to climate justice searches for economic means to support the wish for everyone to enjoy a benevolent environment. Ethical imperatives and economic calculus lay the foundation to protect climate-disadvantaged current world inhabitants and future generations. Justice and fairness underline inequality alleviation via economic redistribution within society, inbetween the entire world countries community, and over generations.

This chapter presents the philosophical, economic, and behavioral foundations of climate justice calls. Global warming has an extraordinary impact on the economic, social, and ecosystems. Economic gains and losses in a warming climate differ throughout the world. Philosophical and economic theories provide solid argumentation grounds to consider climate justice based on contemporary empirical findings of multiple inequalities imbued in climate change (Sachs, 2021). Economic calculus can then support international efforts to combat climate change, with particular attention to the role of financing climate change mitigation and adaptation (Sachs, 2021). Individual decision-making and discounting can also be studied in order to strengthen climate justice implementations. Climate change-induced inequalities are proposed to be alleviated with a climate taxation-bonds strategy that incentivizes market actors to transform the energy sector and mitigate as well as adapt to climate change. In the financialization of climate policies, fair climate change benefits and burden sharing within society, between countries worldwide, and over generations will be introduced in a novel taxation-and-bonds strategy.

Overall, this book puts forward novel ideas to capture the dynamic interaction of behavioral nudges for environmental conscientiousness and philosophically inspired economic incentives for the appreciation of environment protection and climate stability. The monograph presents empirical knowledge on how to balance climate change benefits and burdens within society, between differently affected nation-states, and over time among overlapping generations. All the mentioned theories and economic models aim to inform novel policy strategies in the age of environmental, social justice care.

4.2 Climate Justice

Globalization features a globally integrated economy that leveraged pressure on contemporary society. Globalization appears to have brought economic improvements for some, while it also has worsened societal conditions for others. Disparities have been noticed in free trade benefits distributions worldwide (Barry & Reddy, 2008; Miller, 2010; Moellendorf, 2002, 2009a, 2009b). Unfairness toward some of the world's most vulnerable communities in negotiating global agreements that determine economic outlooks and societal conditions has been criticized in recent decades (Barry & Reddy, 2008; Miller, 2010; Moellendorf, 2002, 2009a,

2009b). International distributive justice outlines a gap in the wealth of economic prosperity on a globalized scale (Rawls, 1999).

Fairness is a fundamental interest and innate human feature that structures relationships. Injustice considerations guide individuals' moral compass and interactions with each other (Plotke, 2003). Philosophical imperatives back global and international justice (Rawls, 1999).

In a globalized world, justice pledges range from fairness within a state but also justice for different countries and generations. In the international arena, international law promotes justice and within global governance contexts, the responsibilities of all parties involved are guided by goodwill justice considerations.

Climate change now makes injustices inherent in globalization more apparent than ever before. In the climate change problem, we are all together—but the practices of individuals, institutions, policies, and activities that lead up to global warming influence people, nations, and generations differently. Human-made climate change stems from the actions of international agents and institutions. The practices and activities that cause climate change vary in the international arena. But also, the outcomes of human-made climate change affect the residents in each country varyingly.

Unfairness in the age of climate change may stem from who caused the problem by CO_2 emissions. Inequality also arises in the economic prospects countries face in light of climate change, given different starting grounds on mean temperatures around the world and different productivity factors compositions dependent on the environment. The disparity is also noticed in the societal pressures to live in harmony with the environment. Currently emerging literature points to the disproportionate burden price mark-ups for sustainability place on already marginalized communities (Woo, 2021). Within households, females are more likely traditional guardians of the household and may be affected more by the day-to-day climate-related impetus on a family's nutrition, health, and well-being.

The vast disparities stemming from climate change heighten the call for climate justice (Chang, 2002; Pogge, 2001, 2002; Singer, 2002). Global warming raises concerns about global and international justice. While there is a most critical wealth of research on the negative impacts of global warming, the literature seems scarce but just emerging about the disparate effects, actors, nations, and whole generations face in light of climate change mitigation and adaptation demands (Roston et al., 2022).

Elucidating unnoticed inequalities may shed important light on actors' incentives to take necessary climate stabilization action. Unraveling the potentially underrepresented economic short-term benefits of a warming globe aids in understanding patterns of inequality in international development and climate change-induced migration and investment.

Today's most pressing social dilemmas regarding climate change demand inclusive solutions. Climate justice accounts for the most challenging global governance goal affecting every decision-maker. The burden of climate change abatement and adaptation appears to be weighting disproportionally among different groups within society. In the current climate change mitigation and adaptation efforts, high- and low-income households but also developed and underdeveloped countries as well as generations are affected differently. Understanding the bounds of environmental limits has also become a blatant demand to avoid ethical downfalls beyond the control of singular nation-states infringing on intergenerational equity—the fairness to provide an at least as favorable standard of living to future generations as enjoyed today (Puaschunder, 2019).

Global common goods predicaments demand for intergenerational justice on a worldwide scale. Intergenerational fairness is as old as humankind. Intergenerational equity arises from the elderly wanting their offspring to prosper in at least as favorable conditions as experienced. A pro-active over-compliance with current sustainability legislation may stem from a broader social contract within society (Brunnermeier, 2021).

In a history of turning to natural law as a human-imbued moral compass for solving societal downfalls on a global scale in times of crises; this book covers environmental constraints in light of climate change urging for climate justice around the globe. After introducing the ethical and economic foundations of environmental justice, this chapter concerns intertemporal discounting and environmental influences as prerequisites for environmental justice. These foundations may help implement concrete climate justice solutions within society, between nations, and over time.

The following chapter contributes the philosophical roots of fairness and responsibility as a basis for distributive justice (Rawls, 1999). Fundamental human features, philosophical notions, and economic factors guide our actions toward each other and our collective sustainable lives as overlapping communities. The chapter supports the case for the philosophical and economic foundations of climate justice as the aspirational goal that all world inhabitants are entitled to an equal share of the value of global warming in proportion to their relative losses due to climate change. Empirical work on human-imbued cues and external influences to guide behavior in international and global environmental concerns will be presented. Empirical behavioral microeconomic research strives to outline individual motivational factors for considering intergenerational justice in the environmental domain. An international funding scheme will be introduced to propose fairness strategies to equally share relative economic climate change gains and even out different starting grounds that may have been historically reinforced through globalization. Unequally distributed relative economic climate change gains and losses and the levels of carbon emissions production and consumption patterns of nations as well as historically grown bank lending rates thereby become the basis for redistribution responsibility argumentations to enact climate justice. All these foundations should aid implementing climate justice in the following macroeconomic modeling that maps global climate change-related relative economic gain and loss perspectives in order to derive universally fair climate stabilization strategies. Innovative redistribution schemes will be proposed to share the benefits and burdens of climate change more equally and ensure humankind to feel a fair solution was found to enjoy a stable climate in today's and tomorrow's world.

4.3 Philosophical Foundations

Climate justice is the wish to provide a favorable climate to all world countries. In light of global warming, climate justice connects our common humankind's past to our future in intergenerational fairness (Pope Francis, 2015; Puaschunder, 2016c, 2019). A novel introduction of attention to relative climate change economic benefits drives demand for distributing the gains from warming earth around the globe with respect for offsetting the losses implied by global warming.

Ethical foundations provide the groundwork for countries that reap benefits from a warming earth to be obliged to finance those countries that are impacted negatively by climate change to offset climate change-related losses. Common and international law grounds the basis for argumentations that countries with better means of protection and conservation of the common climate should also face a greater responsibility to protect the earth (Puaschunder, 2016d). In the distribution allocation decision, philosophical foundations serve to back the demand for sharing the positive externalities of climate change and acknowledging the need for climate stabilization as a universal maxim without regarding one's gain or loss prospects.

With reference to Immanuel Kant's Categorical Imperative proposing to "Act only according to that maxim whereby you can at the same time will that it should become a universal law," the climatorial imperative fortifies the idea of common but differentiated responsibility to ensure a stable climate and decent living conditions around the world (Kant, 1785/1993, p. 30; Puaschunder, 2017c). Based on ethical imperatives, the climatorial imperative advocates for the need for fairness in the distribution of the global earth's benefits among nations based on Kant's (1788/2003) imperative to only engage in actions one wants to experience being done to oneself. Passive neglect of action on climate mitigation is considered an active injustice to others (Chichilnisky, 1996; Chichilnisky et al., 1998; Puaschunder, 2017d).

Potential economic climate change benefits should be distributed worldwide to alleviate climate change implied losses in those parts of the world that have already run out of favorable climate time. Since the birth of a nation on earth is involuntary, by birth, one may owe a share of the windfall gains acquired simply by the fate of where one was born to those territories that will be losing in light of climate change in the wish to overcome climate injustice. As outlined by the correlation between Greenhouse Gas emissions and economic climate change gain prospects, countries with the prospect of getting economically more affluent from climate change are also causing the problem. Based on philosophical notions, being born into climate-prosperous regions by fate naturally predestines for an obligation to distribute some climate change-related benefits. A higher contribution to the onset of climate change and passive neglect to change environmentally harmful behavior should be considered when calculating the redistribution transfer payments in relation to other nations.

The German philosopher and New School professor Hans Jonas, a proponent of philosophical biology, addressed the underlying predicament between biological life and economic striving. Jonas (1979) insists on paying tribute to dignity in nature to raise the living human, who only developed within nature. In Jonas' philosophy, humans are bestowed with freedom, but plants and animals are also characterized by their freedom and striving. Humans thus have to pay tribute to the ethics of human responsibility to nature, which implies an underlying affinity in the relation of humans with nature (Jonas, 1979). Jonas thereby builds on Kant but rethinks ethicality as an absolute, unconditional responsibility for children and nature. Jonas advocates for acting in such a way that makes survival possible.

Jonas breaks with a long dualistic tradition in philosophical thought that stresses the elevating difference between humans and nature. Jonas is thereby critical of anthropocentric philosophical positions that portray humans as independent of nature and self-sufficient in their thoughts. In outlining the alienation from nature prevalent in the philosophy of his time, he established forethinking philosophical leadership on the embeddedness of humans in natural, social, environmental, and cultural conditions that become the spring feather of environmentalism, ecological economics, and behavioral insights on environmentalism. The self-realization of humans and their relation to—but also dependence on—nature create an obligation or responsibility to protect and conserve nature. Responsibility includes taking care of the continuous existence and well-being of the interconnected nature.

Hans Jonas builds on Immanuel Kant's categorical imperative but extends the scope of responsibility and ethics to the entire biosphere (Kant, 1785, 1993). Individuals should only act in harmony with continuous life and sustainable living conditions. Jonas (1979) advocates for self-realization based on *Bildung* (education) of humans. Intelligence is pegged to morality and power to responsibility. In harmony with the overall ecosystem, *Bildung* is nurturing a natural inclination toward preserving and enriching the life of this generation and future living beings. The connection of life to others and the broader social and environmental contract create a natural dependency to care for each other and a moral obligation to preserve the life and livelihood of others. Self-realization is naturally transformed into a claim for preservation. The power dominance of humans over nature coupled with self-realization forms a natural responsibility to protect the earth in which humans are embedded. The strongest in the biosphere also have the starkest ethical obligation to protect the weakest, including other humans, the environment, and future living and being on earth (Jonas, 1979).

Hans Jonas also addresses the problem of future generations lacking the retaliation principle for non-cooperation. In standard economic games, retaliation can improve social norms of cooperation. As future generations are non-existent, future world inhabitants have no means of enacting retaliation and therefore need future orientation and social responsibility notions of today's world inhabitants as a prerequisite of intergenerational fairness (Puaschunder, 2019). In Jonas' philosophy, extending parental care for children includes the future orientation of care for future existence. Ethical considerations also involve the well-being of future generations within environmentally-favorable conditions and the prosperity of humankind. Moral obligations arise from the co-existence and the mutual care and inequality between humans and nature in power and self-awareness.

Jonas (1979) also forethinks the conflict of technological capacities' effect on the biosphere. Technological innovation is assessed and valued for the well-being of humanity and the sustainability of the biosphere. Human knowledge and technological power to affect the condition of the entire biosphere have created a natural responsibility and obligation to protect and conserve. The biosphere is elevated from being just an instrument for human purposes to an end in itself that needs to be cherished, conserved, and protected. Human existence and well-being are portrayed as dependent on the responsibility to preserve whole-roundedly and future-orientedly.

Building on Kant (1785/1993) and Jonas (1979), Professor Nikulin (2021) pays attention to modern radical changes that lead to an inability to control the future and predict its consequences. Nikulin (2021) attributes the principle of responsibility and draws inferences about why and how to act to be responsible. Nikulin (2021) stresses that one should act morally and politically in such a way as to make the life of future generations of living beings possible. Nikulin (2021) derives the duty to existence from the right to life.

Climate change heralds a call for ethics of responsibility. Nikulin's (2021) mandate to act now on climate change is motivated by the heuristics of fear, as climate change is considered a threat to the continuation of a good life and nature's existence. In the eye of contingency factors of human survival, natural interconnections, and historical political landscapes, Nikulin draws attention to the complexity of the problem. At the same time, for climate stabilization, the world will need concerted action. It will be hard to figure out a standard way to act and respond. Therefore, Nikulin advises for the precautionary principle regarding the outcomes of climate change. A political framework will be needed that proposes reasonable solutions backed by political support. Universal maxims and philosophical imperatives that every human may follow easily may guide the way.

The realization that the climate change problem connects the globalized world in a common stratosphere and the global character of the ecosystem underline the international impetus of the global warming problem. The existence of tipping points and irreversible lock-ins implies an urgency in the climate agenda to act on climate change mitigation and adaptation for the entire world community. Based on the political insight of the Conferences of the Parties (COP) agreements in the United Nations Framework Conventions that all world leaders have to come together to ratify climate agreements, one realizes the importance of widespread support of climate stabilization around the globe. Since the COP26 meeting of 2021, different hierarchies of climate change causing countries have been debated and the call to estimate the economic benefit from global warming in the short run has been launched as a basis for fair redistribution strategies (Sachs, 2021).

Based on social contract ideas in the tradition of Thomas Hobbes, Jean-Jacques Rousseau, and John Lock, John Rawls' (1971) idea of the veil of ignorance is a thought experiment to structure a free, equal, and moral society. Parties should thereby choose principles of justice free from consideration of their advantageous or disadvantageous positions in relation to each other. Similar concepts have been argued in Adam Smith (1761/2014) with an impartial spectator who is not incentivized by market mechanisms.

Climate change simultaneously produces relative economic climate change windfall gain reaper and victim countries in the short run. John Rawls' veil of ignorance (1971) can aid society to agree on supporting climate stabilization efforts without considering the position one may find her- or himself in a warming temperature. John Rawls' veil of ignorance suggests that one should not weigh whether to be a relative climate change economic gains reaper or a climate loser when analyzing the overall climate problem. In light of the widespread damage caused by global warming, one should abandon considering the personal gain and loss perspective. The idea of a veil of ignorance over the economic gains of climate change pays homage to behavioral economics attempts to bring ethics and social care into the standard utility function.

The veil of ignorance and risk aversion can improve equality and reduce uncertainty within society and inbetween overlapping generations. Evaluating the overall climate justice problem behind a veil of ignorance leads to the conclusion to take action concertedly against a warming globe as soon as possible. At the same time, shedding light on the economic gain and loss prospects of global warming can help find a well-balanced redistribution system that bestows fairness perception to all parties involved. Arthur Schopenhauer and Buddhist notions of all parts of the world having an equal value back the urge to only consider the short-term prospected economic gains of a warming earth in order to show how to redistribute economic benefits from global warming to losing areas in light of climate change.

As for redistributing the economic gains of a warming globe to offset losses incurred by global warming, a climate change bonds-and-tax finance strategy is proposed in the following to bear the burden of climate change in a right, just, and fair way within society, around the globe and over time. Bonds are ethically justified for solving future generations' missing advocacy and retaliation option. Bonds are also powerful economic means offering targeted incentives to redistribute economic gains and losses between market actors and serve as an innovative burden-sharing strategy over time.

4.4 Economic Foundations

Economic foundations of intergenerational transfers touch on the concept of a social contract (Brunnermeier, 2021). Such voluntary contracts emerge from recognizing the impact on others (Brunnermeier, 2021). Social contracts fulfill evolutionary-favorable qualities, foremost to prevent society from the harmful effects of adverse shocks. Social contracts are thus a prerequisite for resilient communities (Brunnermeier, 2021).

If adverse impacts on others get unnoticed, economists detect problems of negative externalities (Brunnermeier, 2021). Negative externalities can lower social welfare without a social contract (Brunnermeier, 2021). Fairness notions play a crucial role for people to appreciate and follow social norms that build a favorable social contract that fosters societal welfare (Brunnermeier, 2021; Puaschunder, 2015). Social contracts thus limit negative externalities and enhance societal resilience based on a quasi-insurance to compensate for losses in the wake of external shocks (Brunnermeier, 2021). Global governance institutions are predestined to implement social contract-driven ideas by breeding favorable social norms (Brunnermeier, 2021).

Social contracts can foster social cohesion based on trust and reciprocity within society (Puaschunder, 2015). Social contracts are primarily facilitated by positive social norms and reinforced by governments. Following the rules of social contracts is often based on experiences within a given society. For instance, receiving governmental aid during crises may breed future reciprocity to help ensure unlucky fellow citizens later. Or having inherited favorable living conditions may overall create a positive propensity to form a wish to leave the earth in an at least as favorable state as found, if not even breeding the hope for a better world for future generations.

To alleviate inequalities in climate change impacts between countries, Kaldor-Hicks compensation criteria can guide a prospective redistribution scheme (Law & Smullen, 2008). The Kaldor-Hicks test for improvement potential within a society aims to move an economy closer toward Pareto efficiency (Law & Smullen, 2008). Kaldor-Hicks criteria assume that any change usually makes some people better off and others worse off at the same time. The Kaldor-Hicks criteria then test if this imbalance can be alleviated by relative winners compensating relative losers for the change in conditions. In the Kaldor-Hicks criteria, both winners and losers must also agree that the benefits exceed redistribution costs.

The Kaldor-Hicks compensation can be applied to environmental constraints regarding climate change. As relative economic gains and losses from a warming earth are distributed unequally around the globe, ethical imperatives lead to the pledge to redistribute profits to losing territories in the quest for climate justice. Following the rationale of the Kaldor-Hicks compensation and alleviating climate injustices, this book aims to introduce individual motivating drivers for sustainability and a novel climate taxation-and-bonds strategy. Climate justice can enact fairness between countries and over generations in a unique and unprecedented tax-and-bonds climate change gains and losses distribution strategy. Climate change winning countries are thereby advised to use taxation to raise revenues to offset the losses incurred by climate change. Relative economic climate change windfall gain reapers could share their economic growth via taxation transfer to global warming losing territories that could be redistribution payment recipients.

Governments in global warming loser countries should receive tax transfers in the present from the winning countries. The climate change loser countries could become beneficiaries of transfer payments that fund loans, or the issuance of bonds could be enacted to be paid back by future generations. Future generations thereby can reduce long-term climate change damage risks and irreversible environmental lock-ins. This tax-and-transfer policy appears as a Pareto-improving solution worldwide and over time.

Tax-and-bonds transfers could also incentivize industry actors to choose clean energy. The revenues raised from taxation-and-bonds would subsidize corporations choosing clean energy. This market incentive could shift the general race-to-the-bottom regarding price-cutting behavior and choosing dirty, cheap energy to a race-to-the-top hunt for subsidies for going into clean energy and production.

Climate justice within a country should also pay tribute to the fact that low- and high-income households share the same burden proportional to their disposable income, for instance, enabled through progressive carbon taxation. Those who caused climate change could be regulated to bear a higher cost through carbon tax combined with retroactive billing through a corporate inheritance tax to reap benefits from past wealth accumulation that contributed to global warming (Roston et al., 2022). Finding the optimum balance between consumption tax adjusted for disposable income through a progressive tax scheme promises to foster tax compliance in the sustainability domain. For the Kaldor compensation to work effectively, relative economic climate change windfall gain reaper and victim countries must also agree that the benefits of a commonly agreed-upon compensation scheme exceed the costs of such action. This could be accomplished if all parties feel that a fair redistribution was pursued based on economic calculus and rational choice that benefits all parties involved in the long run.

4.5 Behavioral Foundations

From a microeconomic perspective, the application of ecological economic principles to environmental problem-solving is presented in human decision-making and everyday choices. The relationship between the economy and the environment breaks down to us all choosing on a constant basis given the natural conditions around us. Behavioral insights stress the role of economic analysis in understanding how individuals decide given their environment and how to create positive incentives to the layperson valuing the environment and its conservation over time.

One of the hallmarks of innovative sciences in the field of economics was the behavioral economics revolution (Puaschunder, forthcoming).

Starting in the 1950s as a theoretical critique of rational choice models, which assumed all human beings strive for efficiency and rationality all the time, behavioral economics first drew attention to deviations from rationality and the non-applicability of rational choice models for depicting the actual behavior of humans (Puaschunder, 2024; Simon, 1956; Simon & Bartel, 1986). During the 1970s, Amartya Sen formalized the rational choice critique and published powerful examples of how economics needs a reality check (Puaschunder, 2024). Sen (1971, 1977) advised for back-testing of economics' core axioms of rationality, efficiency, and time consistency for actual real-world relevancy and external validity of the standard rational choice claims (Puaschunder, 2024). In subsequent work over decades, Amartya Sen debunked some of the hallmarks of economic science assumptions (Puaschunder, 2024; Sen, 1993, 2002a, 2002b).

By 1979, two psychologists, Daniel Kahneman and Amos Tversky presented a line of laboratory experiments at universities that proved the rational choice theory to be inaccurate to explain the decision-making patterns of individuals on a constant basis (Kahneman & Tversky, 1979; Puaschunder, 2024). The following 1980s and 1990s saw a revolution of behavioral aspects for economic sciences (Kahneman & Tversky, 1983, 1992; Kahneman et al., 1991; Puaschunder, 2024). First laboratory experiments at universities but later also field experiments became the norm to cartograph a new order by which humans are 'economic'—often striving for efficiency and economic ideals, but in reality, oftentimes muddling through a complex world and thereby only taking the quick best alternative of choices given the complex world human face and in light of their natural cognitive limitations (Kahneman & Tversky, 2000; Puaschunder, 2024). The change in the field stemmed from academic institutions that rewrote economics for accuracy and predictability for actual behavior (Kahneman & Tversky, 2000; Puaschunder, 2024). Several Nobel Prizes crowned the accomplishment to change a field as dominant as economics from within but also from different disciplines adding to bundle efficiency with accuracy and predictability (Puaschunder, 2024). First sociologist George Akerlof (2001), Kahneman and Vernon L. Smith (2002), political scientist Elinor Ostrom (2009), and Richard Thaler (2017) all gained Nobel Prizes to rewrite economics and pursuing a goal to find ways how to 'nudge' individuals, communities, and—most recently—leaders to help others to make better choices by the guidance of behavioral insights (Akerlof, 2001;

Kahneman, 2008; Nobel Prize, 2002; Ostrom, 2009; Puaschunder, 2020, 2024; Thaler, 2017; Thaler & Sunstein, 2008).

In the most recent decades, human decision-making heuristics were studied to show how nudging and winking can help citizens to make rational choices (Puaschunder, 2017d, 2023). Behavioral economics started with outlining human decision-making deviations from rationality, so-called heuristics or mental shortcuts (Puaschunder, 2023). Heuristics were perceived as failures in the North American Behavioral Economics School, while the European tradition saw human decision-making as successful strategy to cope with an overly complex world (Puaschunder, 2020, 2023). Mental heuristics and biases were studied in field and laboratory experiments but also by the help of big data and online observations to retrieve powerful nudges to benefit from life and improve environmental conditions (Puaschunder, 2023). Over time, a broad range of nudges and winks were developed to curb the harmful consequences of human decision-making or improve human fitness to live in harmony with the environment (Puaschunder, 2023). Some techniques were communicated openly, while other behavioral insights informed more subliminal change strategies, so-called nudges and winks (Puaschunder, 2023). Nudges are strategic environmental conditions that help people naturally fall for the more optimal choices, while winks are visual cues that guide toward better decisions.

Since a bit more than 10 years, the behavioral economics approach was applied to political context in the behavioral insights revolution and therefore also fostered environmental policy making (Puaschunder, 2020, 2023). Behavioral Economics and Finance Leadership demonstrated how economics can be employed for the greater societal good (Puaschunder, 2020, 2023). Most recently, leadership and followership directives on nudging and winking for the promotion of Corporate Social Responsibility and Socially Responsible Investment favorable choices emerged (Puaschunder, 2023). The Environmental, Social, and Governance (ESG) mandates added to the trend of promoting socio-psychological aspects in business and finance endeavors.

In the extension of behavioral economics on environmental justice different methods have come to play. For over a decade universities laboratory and field experiments helped find powerful 'nudges' for instigating sustainable choices (Puaschunder & Schwarz, 2012; Puaschunder, 2015, 2017a). For instance, a laboratory experiment at a North American Behavioral Decision Science Laboratory, in which students played a standard economic trust game followed by a common goods game revealed that trust and reciprocity lead to individuals contributing more to common goals (Puaschunder, 2015; Puaschunder, 2024). Educating future leaders at universities in a climate of trust and reciprocity may thus help trickle down in the economy and nurture a broad-based consensus on sustainable development in the twenty-first century (Puaschunder, 2024).

In addition, a field experiment outlined the role of social status prospects for pro-social choices and provided evidence that in particular social status drops are associated with sustainability compliance (Puaschunder, 2017a; Puaschunder, 2024). The field experiment was recently independently replicated in a large-scale field experiment in a different population and setting, validating the core findings beyond the university context (Gangl et al., 2022; Puaschunder, 2024). This serves as a powerful example of how universities can serve as breeding grounds for large-scale change in the general population and around the world (Puaschunder, 2024).

An international survey study was conducted at three different universities in paper-and-pen format as well as online (Puaschunder, 2024). Behavioral international online and paper-pen campus survey studies that were prepared in Europe and held in Asia and international online showed that when individuals visually bundle different choices and review them together next to each other, they become more intergenerationally balanced in the resource allocations (Puaschunder, 2024; Puaschunder & Schwarz, 2012). The joint decision-making advantage can thus improve sustainability (Puaschunder, 2024). The study conducted in different settings on three continents underlines the impetus of universities for science diplomacy in mutual understanding of scientific facts, leaving aside national interests or cultural boundaries (Puaschunder, 2024).

All these examples offer invaluable insights how behavioral 'nudges' can change individuals' behavior toward sustainability. Future avenues in behavioral sciences comprise of a shift from 'nudges' to self-determined leadership in behavioral economics in order to foster an even stronger sustainable society.

References

Akerlof, G. A. (2001). Nobel prize lecture. Behavioral macroeconomics and macroeconomic behavior. December 8, 2001, Aula Magna, Stockholm University. https://www.nobelprize.org/prizes/economic-sciences/2001/akerlof/lecture/

Barry, C., & Reddy, S. (2008). *International trade and labor standards*. Columbia University Press.

Brunnermeier, M. K. (2021). *The resilie2nt society*. Colorado Springs: Endeavor Literary.

Chang, H.-J. (2002). *Kicking away the ladder: Development strategy in historical perspective*. Anthem.

Chichilnisky, G. (1996). *Development and global finance: The case for an international bank for environmental settlements*. United Nations Development Programme, Office of Development Studies.

Chichilnisky, G., Heal, G., & Vercelli, A. (1998). *Sustainability: Dynamics and uncertainty*. Kluwer.

Gangl, K., Walter, A., & Van Lange, P. A. M. (2022). Implicit reminders of reputation and nature reduce littering more than explicit information on injunctive norms and monetary costs. *Journal of Environmental Psychology, 84*(101914), 1–9.

Jonas, H. (1979). *Das Prinzip Verantwortung: Versuch einer Ethik für die technische Zivilisation*. Suhrkamp.

Kahneman, D. (2008). *Thinking, fast and slow*. Farrar, Straus & Giroux.

Kahneman, D., Knetsch, J. L., & Thaler, R. H. (1991). The endowment effect, loss aversion, and status quo bias. *Journal of Economic Perspectives, 5*(12), 1325–1347.

Kahneman, D., & Tversky, A. (1979). Prospect theory: An analysis of decision under risk. *Econometrica, 47*(2), 263–291.

Kahneman, D., & Tversky, A. (1983). Choices, values, and frames. *American Psychologist, 39*(4), 341–350.

Kahneman, D., & Tversky, A. (1992). Advances in prospect theory: Cumulative representation of uncertainty. *Journal of Risk and Uncertainty, 5*, 297–324.

Kahneman, D., & Tversky, A. (2000). *Choices, values, and frames*. Cambridge University Press.

Kant, I. (1785/1993). *Grounding for the metaphysics of morals*. Hackett.

Kant, I. (1788/2003). *Critique of practical reasons*. Cambridge University Press.

Law, J., & Smullen, J. (2008). *A dictionary of finance and banking*. Oxford University Press.

Miller, R. (2010). *Globalizing justice: The ethics of poverty and power*. Oxford University Press.

Moellendorf, D. (2002). *Cosmopolitan justice*. Westview.

Moellendorf, D. (2009a). *Global inequality matters*. Palgrave.

Moellendorf, D. (2009b). Treaty norms and climate change mitigation. *Ethics and International Affairs, 23*(3), 247–265.

Nikulin, D. (2021). Responsibility and hope. *Bollettino della Società Filosofica Italiana*, Carocci editore Roma, Nuova Serie n. 232, Gennaio/Aprile 2021.

Nobel Prize. (2002). *The Sveriges Riksbank Prize in Economic Sciences in Memory of Alfred Nobel 2002*. https://www.nobelprize.org/prizes/economic-sciences/2002/summary/

Ostrom, E. (2009). Nobel prize lecture: Beyond markets and states: Polycentric governance of complex economic systems. 8 December 2009 at Aula Magna, Stockholm University. https://www.nobelprize.org/prizes/economic-sciences/2009/ostrom/lecture/

Plotke, D. (2003). Democracy and groups. *Social Research, 70*(2), 463–498.

Pogge, T. (2001). Priorities of global justice. *Metaphilosophy, 32*(1/2), 6–24.

Pogge, T. (2002). *World poverty and human rights*. Polity.

Pope Francis. (2015). *Encyclical Letter Laudato Si of the Holy Father Francis on care for our common home*. http://w2.vatican.va/content/dam/francesco/pdf/encyclicals/documents/papafrancesco_20150524_enciclica-laudato-si_en.pdf

Puaschunder, J. M. (2015). Trust and reciprocity drive social common goods allocation norms. *Proceedings of the 2015 Cambridge Business & Economics Conference*. University of Cambridge.

Puaschunder, J. M. (2016a). Intergenerational climate change burden sharing: An economics of climate stability research agenda proposal. *Global Journal of Management and Business Research: Economics and Commerce, 16*(3), 31–38.

Puaschunder, J. M. (2016b). Mapping Climate Justice. *Proceedings of the 2016 Young Scientists Summer Program Conference*, International Institute for Applied Systems Analysis (IIASA), Laxenburg, Austria.

Puaschunder, J. M. (2016c). On eternal equity in the fin-de-millénaire: Rethinking capitalism for intergenerational justice. *Journal of Leadership, Accountability and Ethics, 13*(2), 11–24.

Puaschunder, J. M. (2016d). The call for global responsible intergenerational leadership in the corporate world: The quest for an integration of intergenerational equity in contemporary Corporate Social Responsibility (CSR) models. In D. Jamali (Ed.), *Comparative perspectives in global corporate social responsibility* (pp. 275–288). IGI Global Advances in Business Strategy and Competitive Advantage Book Series.

Puaschunder, J. M. (2017a). Climate in the 21st century: A macroeconomic model of fair global warming benefits distribution to grant climate justice around the world and over time. *Proceedings of the 8th International RAIS Conference on Social Sciences and Humanities organized by Research Association for Interdisciplinary Studies (RAIS) at Georgetown University*, Washington, D.C., United States, March 26–27, pp. 205–243.

Puaschunder, J. M. (2017b). *Global Responsible Intergenerational Leadership: A conceptual framework and implementation guidance for intergenerational fairness*. Vernon Art & Science Series in Economics.

Puaschunder, J. M. (2017c). Mapping climate in the 21st century. *Development, 59*(3), 211–216.

Puaschunder, J. M. (2017d). The climatorial imperative. *Agriculture Research and Technology, 7*(4), 1–2.

Puaschunder, J. M. (2019). *Intergenerational equity: Corporate and financial leadership*. Edward Elgar.

Puaschunder, J. M. (2020). *Behavioral economics and finance leadership: Nudging and winking to make better choices*. Springer Nature.

Puaschunder, J. M. (2023). *The future of resilient finance: Finance politics in the age of Sustainable Development*. Palgrave Macmillan.

Puaschunder, J. M. (2024). Behavioral foundations of sustainable development at universities. In W. Leal Filho, T. Dibbern, S. Ruiz de Maya, M. d. C Alarcón-del-Amo & L. M. Rives, (Eds.), *The Contribution of Universities Towards Education for Sustainable Development: World Sustainability Series* (pp. 489–505). Springer Nature.

Puaschunder, J. M., & Schwarz, G. (2012). The future is now: How joint decision-making curbs hyperbolic discounting but blurs social responsibility in the intergenerational equity public policy domain. The Situationist Law and Mind Sciences. https://thesituationist.wordpress.com/2012/12/17/the-situation-of-intergenerational-equity/

Rawls, J. (1971). *A theory of justice*. Harvard University Press.

Rawls, J. (1999). *The law of peoples*. Harvard University Press.

Roston, E., Kaufman, L., & Warren, H. (2022, March 24). How the world's richest people are driving global warming. *Bloomberg*. https://www.bloomberg.com/graphics/2022-wealth-carbon-emissions-inequality-powersworld-climate/

Sachs, J. D. (2021, November 17). Fixing climate finance. *Social Europe: Politics, economy and employment & labor*.

Sen, A. K. (1971). Choice functions and revealed preference. *The Review of Economic Studies, 38*(3), 307–317.

Sen, A. K. (1977). Rational fools: A critique of the behavioral foundations of economic theory. *Philosophy & Public Affairs, 6*(4), 317–344.

Sen, A. K. (1993). Internal consistency of choice. *Econometrica, 61*(3), 495–521.

Sen, A. K. (2002a). Consistency of choice. In A. K. Sen (Ed.), *Rationality and freedom* (pp. 207–224). Harvard University Press.

Sen, A. K. (2002b). Goals, commitment, and identity. In A. K. Sen (Ed.), *Rationality and freedom* (pp. 35–56). Harvard University Press.

Simon, H. A. (1956). Rational choice and the structure of the environment. *Psychological Review, 63*(2), 129–138.

Simon, H. A., & Bartel, R. D. (1986). The failure of armchair economics. *Challenge, 29*(5), 18–25.

Singer, P. (2002). *One world: The ethics of globalization*. Text.

Smith, A. (1761/2014). *Theory of moral sentiments*. https://matej.ceplovi.cz/cizi/moral.pdf

Thaler, R. H. (2017). Nobel prize lecture: From cashews to nudges: The evolution of behavioral economics. 8 December 2017 at the Aula Magna, Stockholm University. https://www.nobelprize.org/prizes/economic-sciences/2017/thaler/lecture/

Thaler, R. H., & Sunstein, C. R. (2008). *Nudge: Improving decisions about health, wealth, and happiness*. Yale University Press.

Woo, A. (2021, April 16). Environmental fascism and the zero waste movement. *Panel on Monitoring and Evaluation in Environmental Economics*. The New School New York.

CHAPTER 5

Mapping Climate Justice

Climate stability accounts for one of the most pressing problems of our time. Climate change has leveraged intergenerational equity as contemporary challenge of modern democracy and temporal justice an ethical obligation for posterity. As never before in history since the birth of the earth, there is an environmental sensitivity to economic growth calling for concurrent climate change mitigation and adaptation approaches (Centeno et al., 2013; Grasso & Markowitz, 2015; Rovenskaya, 2005; The World Economic Forum Report, 2015). As a complex, multilayered problem, ensuring climate justice is a formidable task. The perils of the common good of climate stability appear to challenge contemporary law and economics frameworks—demanding for a whole-rounded systems analysis solution.

Climate justice accounts for a difficile global governance goal as the burden of climate change abatement and adaptation appears to be placed disproportionally in society. In the current climate change mitigation and adaptation efforts, high- and low-income households but also developed and underdeveloped countries as well as this generation and the following are affected differently. The following chapter therefore proposes to map international climate change mitigation and adaptation regimes in order to derive universally fair climate stabilization strategies based on innovative compensation schemes to share the burden of climate change more equally *within* society, *between* countries as well as *over time*. Legal and

J. Puaschunder, *Governance & Climate Justice*,
https://doi.org/10.1007/978-3-031-87971-5_5

economic analyses will unravel how to allocate the burden of climate protection in a fair manner and ensure humankind to feel a fair solution was found to enjoy a stable climate in today's and tomorrow's world.

Climate justice within *society*: In order to finance climate change mitigation and adaptation efforts, a diversified taxation scheme is proposed. To find a fair and just distribution of the burden of climate change, a taxation mix of (1) consumption tax, (2) progressive tax and (3) inheritance tax is recommended. Carbon emission consumption tax appears fair in light of placing a heavier burden on those who engage in activities that offset climate change. Consumption tax can also curb harmful emissions and directly nudge behavior toward sustainability. Yet to place a fair share of the burden of climate change mitigation upon society, these taxes have to be adjusted to the disposable income level of individuals in order to not charge low-income households heavier. Retroactive taxation of past wealth accumulation at the expense of environmental damage through inheritance tax will be proposed. In addition, the redistribution of benefits of climate change will be discussed.

Climate justice between *countries*: When comparing the burden of climate change distributed throughout the world, legal arguments from the past may be subsumed to ensure a fair and just solution for tomorrow's climate stability. In the past, private property rights were in some cases weakened for establishing common goods for the greater good. If the possessor of goods was not alive anymore and previously private goods would be turned into public goods for the greater good, private property rights could be overruled. Using argumentation points of the unknown possessors of tomorrow's stable climate but also drawing attention to the benefits of the common good of a favorable climate will be main argumentation points in order to draw a case how to distribute the burden of climate stability among different countries in a fair manner throughout the world. Countries with better starting conditions to benefit from a warming globe in the short run and territories that are also more prone to cause climate change by CO_2 emissions will be argued to necessarily having to bear a higher share of climate change abatement. Based on legal subsumptions, argumentations of some parts of the world having better means of protection and conservation of a stable climate will lead to different weights in responsibility to act on climate change.

Climate justice over time: In order to finance climate change abatement, a climate bonds financing mix will be proposed to subsidize the current world industry for transitioning to green solutions. Sachs (2014) introduced to fund today's climate mitigation through an intertemporal fiscal policy mix backed by climate bonds and carbon tax. Bonds are primarily used by companies, municipalities, states and sovereign governments to raise money and finance future-oriented long-term projects. Through debt investments, investors loan money to an entity in bonds, which borrows the funds for a defined period of time at a variable or fixed interest rate. This solution appears as real-world relevant means to tap into the worldwide USD 80 trillion bond market in order to fund the transition to a sustainable path (World Bank, 2015). Sharing the costs of climate change aversion between and across generations appears as important strategy to immediately instigate climate change mitigation and adaptation. Overall, this turns climate change burden sharing into a Pareto improving option over time.

Mapping climate justice within society, between countries and over time is targeted at finding a fair, just and feasible climate solution.

5.1 Climate Justice Within Society

Global warming has become reality in temperature anomalies, extreme weather events, unprecedented hurricane seasons and up to 50 inches sea level rise predicted until the end of the century. History has also reached an iconic agreement on global warming mitigation at the UN Paris Conference of the Parties on Climate Change (COP21) climate change conference. The Paris Climate Agreement signed in 2015 by 195 United Nations Framework Convention on Climate Change (UNFCCC) member countries recognizes the importance of combining mitigation, adaptation, and finance to effectively handle global warming and its negative consequences worldwide. International governance underlines the importance of funding climate change mitigation and adaptation strategies multifacetedly and steering markets towards a zero-carbon emissions direction via renewable energy solutions (Graph 5.1).

Regarding creative financing strategies, a focus group was staged during August 2016 at the International Institute for Applied Systems Analysis (IIASA) in Laxenburg, Austria, with thirty-nine young scientists representing diverse disciplines, well-balanced gender composition and differing nationalities. In addition, several sustainable development financing approaches were discussed during the 2016 Alpbach Retreat comprising open source investment platforms, innovative public-private partnership plans as well as self-financing tools to create constant revenue

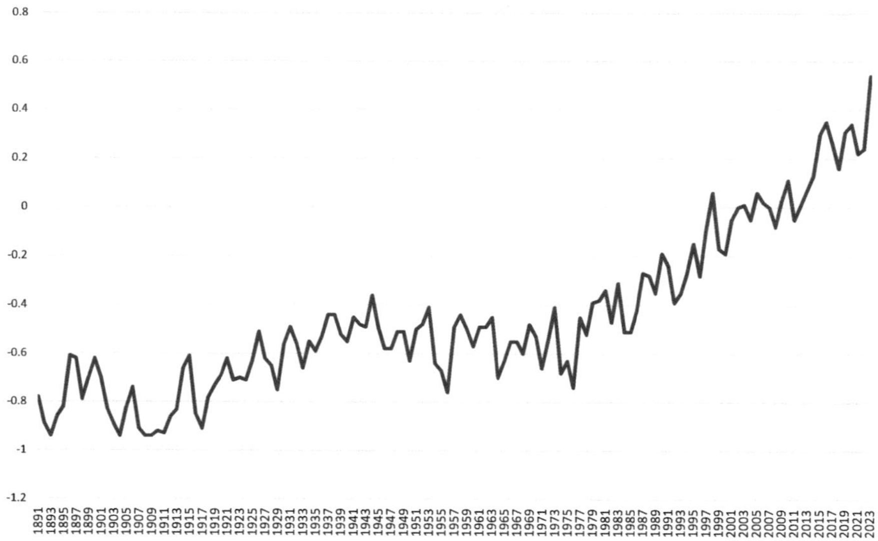

Graph 5.1 World surface temperature anomalies 1891–2023 (*Data* Tokyo Climate Center https://ds.data.jma.go.jp/tcc/tcc/products/gwp/temp/list/year_wld.html)

streams to settle expenses for a transition to renewable energy long-term. One of the most prominent forms to create revenues for public causes are taxes. Since the UN Paris Conference of the Parties on Climate Change (COP26) underlined widespread inequalities inherent in the climate change problem, the world now faces the most urgent need for funding climate change mitigation and adaptation policies with attention to climate justice.

Taxation is codified in all major societies and a hallmark of democracy. Aimed at redistributing assets to provide public goods and ensure societal harmony, taxation improves societal welfare and fairness. Tax compliance is a universal phenomenon based on cooperation for enhancing the social compound. Taxpayers voluntarily decide to what extent to pay or avoid tax that limits the personal freedom. Taxpayers thereby contribute to higher social goals over time in a trade-off with their immediate personal gain. Fairness notions within society and a universal insurance to make society resilient together are underlying drivers of tax compliance. In a social dilemma, individual interests are in conflict with collective goals. From a mathematical and economic perspective, the optimal strategy of rational individuals would be to not cooperate. Short-term single civilian tax contributions do not make a significant difference in the overall maintenance of public goods—if only a few taxpayers evade taxes, public goods will not disappear or be reduced. But if a considerable number of taxpayers do not contribute to tax over time, common goods are not guaranteed and ultimately everyone will suffer from suboptimal societal conditions in the long run (Dawes, 1980; Stroebe & Frey, 1982).

Historic economic research has focused on costs and risks of tax evasion (Tyler & De Cremer, 2006). Coercive means—such as audits and fines—were found to crowd out tax morale and ultimately result in greater non-compliance as people feel controlled and not being trusted (Cialdini, 1996; Feld & Frey, 2002; Frey, 1992; Hasseldine, 1998). Andreoni, Erard, and Feinstein (1998) recognize the importance of incorporating morals and social dynamics in economic theory on tax behavior as positive drivers of tax compliance to overcome the 'burden of taxes' and associations of losses (Kirchler, 2007). Convincing the public that taxes are fair and lead to a public good appears to require a multidimensional conception of taxes as a dynamic phenomenon (Braithwaite, 2003).

When analyzing tax behavior, behavioral economics insights have drawn attention to social influences. Behavioral economists widen the lens of incorporating sociological and socio-psychological understandings of tax compliance (Kirchler, 2007; Puaschunder, 2020). Social dilemmas have been replicated in laboratory settings that capture participants' decisions to cooperate or defect common pools in order to derive theoretical models on tax compliance (Davis & Holt, 1993; Green et al., 1994; Puaschunder, 2015).

Social nudge theories apply behavioral economics insights to change people's tax behavior by studying the impact of the situational setting on tax compliance (Braithwaite & Ahmed, 2005; Eichenberger & Frey, 2002; Frey, 1997; Lewis, 1982; Torgler, 2002). From the psychological perspective, individuals are altruistic but human comparisons with each other influence tax compliance (Puaschunder, 2015). Apart from governmental control and sanction mechanisms, the social situation was found to determine cooperation on tax payments (Hanson, 2012; Poppe, 2005). In social dilemma experiments, participants' cooperation increases if their behavior is publicly known, if they are allowed to communicate with each other and if mutual sympathy, trust and reciprocity are established (Dawes, 1980; Van Lange et al., 1992; Puaschunder, 2015).

Social comparisons determine social norms that define internalized standards how to behave. Social norms are therefore potential tax ethics nudges. Taxpayers internalize social norms and act accordingly. Yet internalized social norms are based on comparisons with others that then may determine tax morale (Frey, 1997; Mumford, 2001; Schmölders, 1960). Social norms elicit concurring behavior when taxpayers identify with the goals of a group but also if they feel they are treated in a fair manner by that group. If taxpayers believe that non-compliance is a widespread and socially accepted, then it is more likely that they will not comply as well. Non-compliance may stem from the notion of unfairness in how the tax burden is weighted heavier on some parts of society.

People tend to adhere to a social contract in the relationship with their government. According to this psychological bonding with the community and relation to the government, taxpayers feel an obligation and willingness to comply with tax requirements (Puaschunder, 2015). Governmental actions are believed to ensure tax compliance. Individual taxpayers tend to adhere to a social contract in their relationship with their government. According to this psychological bonding with the community and relation to the government, taxpayers feel an obligation and willingness to comply with tax requirements (Puaschunder, 2015). In a tit-for-tat strategy, taxpayers' compliance may depend on public goods provided by the government as well as procedural and distributive justice.

Based on the interaction between tax authorities and taxpayers, taxpayers perceive themselves as members of a social group, whose social norms are to be followed. Fairness in tax codes may also bestow the feeling of being treated in a respectful way by the government.

Research on tax compliance has focused on personal ethics and subjective perceptions of behavioral habits in taxpayers' reference group (Puaschunder, 2015). Innovative tax psychology studies refer to social situations breeding tax ethics (Kirchler, 2007; Puaschunder, 2015). Citizens' approval of tax politics is hypothesized to depend on fairness notions and cooperation between citizens and the community. Social fairness considerations in a tax reference group may breed taxpayer compliance. Laboratory experiments outline that individuals' approval of common goods contributions depends on trust, fairness, and cooperation between citizens and the community (Puaschunder, 2015). In a tax reference group, social fairness considerations breed taxpayer compliance (Puaschunder, 2015).

When bearing the burden of climate taxes in a fair and just way, finding the optimum balance between consumption taxes adjusted for disposable income through a progressive tax scheme will aid to elicit tax compliance in the sustainability domain (Puaschunder, 2015). If climate taxation is perceived as fair and just allocation of the climate burden, this could convince tax payers to pay one's share (Puaschunder, 2015). From a governmental perspective, a 'service-and-client' atmosphere could promote taxpayers as cooperative citizens who are willing to comply if they feel their share is a fair contribution to the environment. Taxpayers, as cooperative citizens, who perceive redistribution mechanisms as fair and overall accepted, are more willing to comply voluntarily, following the greater goal of promoting taxpayer collaboration and enhancing tax morale in the environmental domain (Puaschunder, 2015).

Regarding concrete climate taxation strategies, a carbon tax on top of the existing tax system was proposed to reduce the burden of climate change and encourage economic growth through subsidies (Chancel & Piketty, 2015). Besides progressive taxation schemes to imbue a sense of fairness in climate change burden sharing, corporate inheritance taxation, when there is a change in the corporate board structure or a merger and acquisition wealth peak of a corporation, is also a flexible means to reap past wealth accumulation, which potentially caused environmental damage. Future research could outline how to allocate the burden of climate change mitigation and adaptation fairly within society. International comparisons of tax behavior also reveal tax norms being related to different stages of institutional development of the government, which is an essential consideration in sharing the climate change burden in a fair manner *between* countries.

In the most recent COP26 debate about redistributing funds for climate control to alleviate climate injustices, a focus on innovative financing strategies following disparate impact analysis schemes was demanded (Sachs, 2021). The climate change predicament is manifold: The climate change scientific community is convinced that urgent actions are needed to prevent further global warming and that keeping the global average temperature below 1.5°C above the pre-industrial level is necessary to avoid catastrophic damages (Schleussne et al., 2016). Economists are concerned with finding feasible incentives and financing mechanisms of emission reduction; while public policy specialists are already now dealing with the consequences of future damages (High-Level Commission on Carbon Prices, 2017; Stern, 2008). Therefore, innovative strategies and unconventional compensation schemes to raise funds for climate stabilization are currently required by public and private sector entities informed by evidence of individual motivating mechanisms and data-driven climate injustices accounts. When it comes to climate, the effects of climate change are distributed unequally around the world (Puaschunder, 2017a, 2017b). Pareto optimality for the world society differs from the individual nations' optimal climate conditions. To alleviate climate injustice imbued by the disparate impact of climate change within organizations, between nation-states, and over time, taxation and bonds solutions provide a unique transfer scheme. Paying attention to the stratified effects of climate change will allow to find solutions that are perceived to implement fairness.

Behavioral insights on navigating climate change within the nuclear family become fundamental for developing an idea of averting and adapting to climate change efficiently. Heterodox Law and Economics approaches offer the benefits of a quantified disparate impact analysis with attention to all facets of society (Howell, 2021). For instance, current trends in sustainable development argue for attention to worldwide differences and advocate for self-imposed country goals. Law and Economics can address particularly vulnerable groups on whom sustainability pledges place a disproportionate burden on the societal level. Think about low-income households and zero-carbon emission pledges that drive up the prices of goods. Especially Black, Indigenous, and People of Color (BIPOC) communities have been found particularly vulnerable to sustainability pressures that may place a disproportionately-higher burden on constraint households (2021 Global Environmental Justice Conference at the Yale School of the Environment, 2021; Monitoring & Evaluation Collaborative Association MECA at The New School, 2021; Woo, 2021). Within these households, the traditional caretakers, who are women may face the highest burden to implement climate goal-driven consumption patterns (2021 Global Environmental Justice Conference at the Yale School of the Environment, 2021; Monitoring & Evaluation Collaborative Association MECA at The New School, 2021; Woo, 2021). Behavioral insights about individual decision makers' choices being influenced by external conditions and life-changing events can help derive recommendations how to make people pay attention to environmental causes. At a very granular but widespread level, these insights into the individual choice propensities regarding environmental care help to improve this generation and the following's living conditions. Coupled with innovative strategies for allocating the burden of climate change mitigation and adaptation in a fair way within society to breed tax compliance, these measures can help to alleviate climate injustices. Addressing the unequally-distributed economic gains and losses of global warming within society and around the world aids in finding fair climate change gains redistribution schemes based on taxation and climate bonds. Climate bonds particularly aid in finding the balance of climate change burden-sharing between generations, which will be proposed in more detail in the following chapters in macroeconomic modeling.

5.2 Climate Justice Between Countries

The legal argumentation about climate stability touches on different fields of law. While climate is historically understood as a common good, which is non-excludable and non-rivalrously shared and beneficial for all world members, this perception is currently challenged by climate change. If climate becomes less stable, climate may not be considered as a common good anymore. For instance, if climate instability impacts certain world areas by weather extremes - such as sea level rise, flooding, severe droughts, desert formation, storms and hurricanes - more severe than other parts of the world that may even benefit from a warmer climate (e.g., consider Russia's and Canada's access to natural resources when ice is melting in their Northern territories), a rising quest for living in a beneficial area of the world may leverage climate as a quasi-luxury. Living in a favorable climate may become an exclusive privilege that is rivalrously contested. Under these circumstances, climate would not be considered as classical common good. But also take the example of rising sea levels. If climate gets warmer, sea levels rise and put landscape under water. Predictions of Venice disappearing and almost half of Manhattan being under water in 200 years[1] if we continue on a business-as-usual path underline the direct impact of the common good climate on private property rights. If temperature rises, private property will be destroyed or more expensive to maintain.

In these regards, climate justice concerns directly touch on human rights. Climate justice links human rights and sustainable development to safeguard the rights of those affected by climate change. The claim for a human right of access to a stable climate ensures intergenerational fairness. The goal of climate justice is that not one generation creates irreversible lock-ins for future generations. But considerations of human rights in the light of a changing sea level will also determine the conditions of climate refugees and their access to decent living conditions. From the Arctic to the Indian Ocean to the South Pacific, small island states and coastal lines have nowadays become home to the most vulnerable communities. The threat of rising sea levels in the wake of climate change pushes populations to relocate to safer areas. Contemporary legal frameworks do not recognize and protect the rights of climate migrants, as climate refugees do not fall under the Geneva Convention[2]protectorate—which rules immediate asylum for only those fleeing from armed conflict. Climate justice may thus be leveraged as a quasi-human right in the years to follow. Capturing the underlying causes for widespread population flows based on global warming reduces human tragedy, social conflict, and economic inefficiencies during the relocation processes (Burke et al., 2015).

As the destabilization of climate will directly destroy, damage or intensify maintenance costs of private property, the climate stabilization efforts therefore also have entered private property considerations.

[1] http://www.businessinsider.com/map-shows-how-much-of-nyc-could-be-underwater-in-200-years-2015-7

[2] https://www.icrc.org/en/war-and-law/treaties-customary-law/geneva-conventions

In the financing of climate stabilization, private market rationale has been proposed following World Bank and United Nations approaches to pricing natural resources. These financialization and commodification of nature efforts, however, have just started. In the commodification of climate, economic rationale should be applied but with the caution of legal oversight. The basic economic rules of supply and demand suggest that an over-demand of climate elevates the price of a stable climate. Legal subsumptions attribute a higher responsibility to protect the climate to those countries that have a rising economic prospect due to a warming temperature. Fair climate change burden sharing between countries could therefore comprise of the following argumentations.

Climate justice implementation in ensuring a stable climate throughout the world presents itself as a dilemma to grant fair climate change burden-sharing strategies between countries. First, those countries that benefit more from a stable climate should also bear a higher degree of the burden of climate change mitigation and adaptation efforts. Second, those countries that reap short-term windfall economic benefits from a warmer earth temperature should redistribute some of the wealth accumulation due to climate change to offset the costs arising from global warming in other parts of the world that suffer from a decline in economic prospects and thus decreasing living conditions due to rising temperatures. Third, those countries that cause climate change in CO_2 emissions and consumption levels, as well as those countries that show no intent over time to curb their harmful CO_2 emitting action, should be obliged to fund climate justice by offsetting the losses in other parts of the world due to climate change. Fourth, countries with a historical advantage in raising funds in access to favorable bank lending conditions may face a higher responsibility to protect the common good of a stable climate and redistribute some of their advantageous finance sources to those countries with unfavorable financialization conditions.

The legal foundation for this rational can be found at the heart of the climate problem having been declared as one in which countries have 'common but differentiated responsibilities,' which was first discussed in Principle 7 of the Rio Declaration at the first Rio Earth Summit in 1992. New to this rational, however, is the argumentation that the benefits of global warming—which are real and exist, for instance, in melting ice allowing unprecedented access to below surface resources and larger arable landscape—should be shared globally to offset costs and harms produced from a higher temperature at other parts of the world.

In addition, building on case and international law, those countries that have better means of protection or conservation of a common climate should also face a greater responsibility to protect the earth. The legal basis for this argument stems from an inverted subsumption of the argumentation whether climate stability is a common good or impacts on private property. Legally, one can draw from historical cases of legally justified expropriation. Private property rights are some of the starkest legal claims existent around the world. Private property rights hold through time, distance and space. If neighbors go on vacation, one cannot simply move in their home and claim oneself as sole possessor. But there is one interesting case in history, where private property rights could be neglected for the sake of creating common goods. In history, the private property claim of a country was legally justified neglected if (1) those who possessed a good were not alive anymore and direct attribution of the possessor was non-existent; (2) the former private property was turned into a common good and (3) the new possessors had better means of protection and conservation than the good had experienced before. Take historical examples such as the Stone of Rosetta and the 'Elgin' Parthenon Marbles as part of the Athenian Acropolis (Downs, 2008). Historically, the stone of Rosetta was discovered by Napoleon Bonaparte in 1799 in the small town of Rosetta in Egypt, but is now exhibited in the British Museum for more than 200 years.

There are several arguments for not restituting the former private property to the countries of their origin as in the case of the Stone of Rosetta: (1) The *actual former possessors* are not alive anymore and unknown, which turns any direct claim to the international law domain, hence Egypt contests Great Britain over possession. Great Britain claims (2) to turn the former private property to a common good through granting a *large and diverse group access*. London—as a vibrant metropolitan hub with a vast array of visitors—is a better stage to explain the meaning of the stone to a broader public than Rosetta, a small town near Alexandria Egypt. The display of the stone in a museum adds additional value: (3) The British Museum has better *means of protection, preservation and conservation* of the stone than if it were displayed by itself in a desert town. (4) The surrounding in a museum grants the stone meaning and a *Gestalt* bringing to life the mysteries of ancient Egypt and the stone's importance as a historical artifact in deciphering Greek, Hieroglyphs and demotic Egyptian (Downs, 2008). Economically, displaying the stone in an outpost may also attract visitors to travel to Egypt for more historic artefacts.

Despite raising critical questions whether former colonial regimes have the right to retain misplaced cultural heritage, the previous argumentations for legally-backed displacement of cultural heritage could be subsumed to the case of climate justice. Considering that some developing nations and transition economies have argued a right to economic growth by the same—unfortunately climate-change causing—means of economic transition as the first world had in previous centuries, the burden for climate stabilization could be placed on those parts of the world that have caused the problem with their actions in the past. In the arguments whether the first and the second world should bear the same burden of climate change mitigation and adaptation efforts as developing nations and transitioning economies, one could subsume from the above argumentation: (1) That climate change will potentially infringe on private property rights of *future possessors* who are currently unknown. (2) Those countries should bear a higher burden of climate change mitigation and adaptation that have better means of *protection, preservation and conservation*, hence the first world may have to take on a greater responsibility to avert climate change. (3) There is a natural *Gestalt* over time regarding climate. Over time, the sum of a stable climate is more precious than the individual generation's costs incurred to maintain a favorable climate. Shedding light at these deficiencies underlines the need for considering climate justice as an eternal natural law that connects the past to our future.

In finding a legally-sound justification for global warming benefit transfers, one can argue that climate change will potentially infringe on the private property rights of future owners who are currently unknown. Those countries with more access to a favorable and economically-beneficial climate should bear a higher burden of climate change mitigation and adaptation efforts. So-called relative economic climate change windfall gain reapers that have better economic and climatic starting grounds on a warming earth but also those countries that have better means of protection, preservation, and conservation of a stable climate in light of global warming and more time ahead until the temperature makes them unproductive, should take on greater responsibility in averting climate change. In addition, there is a natural Gestalt over time regarding climate as the sum of a stable climate is more precious than the individual country's, industry's, or generation's costs to maintain a favorable climate. In a legal subsumption, the international law of disowning a country to create a common good by those countries that have better means of protection, preservation, and conservation can back climate justice redistribution argumentations of today.

5.3 Climate Justice Over Time

Society as a whole outlasts individual generations. Pareto optimality for society over time differs from the aggregated individual generations' preferences. As the sum of individual generations' preferences does not necessarily lead to societally favorable outcomes over time (Bürgenmeier, 1994; Klaassen & Opschoor, 1991), discounting based on individual generations' preferences can lead to an unjust advantage of living generations determining future living conditions (Rawls, 1971). In the climate domain, intertemporal questions arise whether to invest in abatement today—in order to prevent negative effects of global warming—or to delay investment until more information on climate change is gained (Rovenskaya, 2008). In general, resources are balanced across generations by social discounting to weight the well-being of future generations relative to those alive today. Regarding climate justice, current generations are called upon to make sacrifices today for future generations to cut carbon emissions to avert global warming (Sachs, 2014). Intergenerational balance is therefore accomplished through individual environmental conservation decisions of the present generation (Bauer, 1957). Policies curbing preferences and taxes distributing welfare between the present and future generation may, however, decrease economic growth. Climate change mitigation at the expense of lowered economic growth seems to pit the current generation against future ones. Costly climate change abatement prospects are thus hindering currently necessary action on climate change given a shrinking time window prior to reaching tipping points that make global warming irreversible (Oppenheimer et al., 2011).

As an innovative angle in this debate of economic growth versus sustainability that seems to pit the current generation against the future, a novel climate change mitigation approach with bonds funded through debt and taxation imposed on future generations will be proposed in the following. In order to avoid governmental expenditure on climate change hindering economic growth but also to instigate immediate action on climate change abatement (Barro, 1990), Sachs (2014) proposes to fund today's climate mitigation through an intertemporal fiscal policy mix backed by climate bonds and carbon tax (Marron & Morris, 2016). Bonds are debt investments in which investors loan money to an entity, which borrows the funds for a defined period of time at a variable or fixed interest rate. Bonds are primarily used by companies, municipalities, states and sovereign governments to raise money and finance a variety of future-oriented long-term projects and activities. A climate bonds financing could subsidize the current world industry for transitioning to green solutions as a real-world relevant means to tap into the worldwide USD 80 trillion bond market in order to fund the incentives to a transition to a sustainable path (Puaschunder, 2016; World Bank Report, 2015).

Bonds are a strategy to balance climate justice between generations. A tax-and-transfer mitigation policy is proposed, in which all generations are better off with mitigation through climate bonds as compared to the business-as-usual (BAU) non-mitigation scenario (Sachs, 2014). While future generations enjoy a favorable climate and averted environmental lock-ins; the current populace does not face drawbacks to economic growth. Sharing the costs of climate change aversion between and across generations is a vital strategy to instigate immediate climate change mitigation action by incentivizing emission reduction and providing capital for a fast transition to renewable energy.

Taxes as a redistribution mechanism could be levied in countries with a natural advantage on climate change, such as a cooler starting temperature or different temperature zones. Taxes could also become essential market incentives curbing harmful emissions if countries that are emitting more CO_2 compared to other countries or consume goods that have led to CO_2 emissions in their production and are not willing to change harmful emission patterns are taxed – especially if taxes in particular target emitting industries. Taxation is also a way to even out historical hierarchies in access to capital when funds are raised in countries that have low bank lending rates as an indicator of better conditions to acquire seed funding for grand causes to grant bonds in countries that have historically more challenging times to accept starting grants for economic transitions.

In countries with a natural disadvantage on climate change as for a hotter starting temperature or limited temperature ranges, bonds could be used to receive transfer funds for climate change mitigation and adaptation. Bonds could also fund those countries that emit less CO_2 than other countries and relatively do not consume many goods and therefore CO_2 and/or are willing to change harmful emission patterns. Bonds funded via taxation are also a way to block historical hierarchies in capital access when supporting countries with high bank lending rates.

The bonds are proposed to be financed through individual and institutional investors but also through taxation, such as carbon or inheritance tax. Carbon taxes can raise substantial revenue until the economy is largely decarbonized (Marron & Morris, 2016). In Sachs (2014) two-period model, climate change mitigation is financed by debt to be repaid by tax revenues on labor income in the future. Leaving the current generation with unchanged disposable income allocates the burdens of climate change mitigation across generations without the need to trade off one generation's well-being for another's. While today's young generation is left unharmed, the second period young generation is made better off ecologically. Taxes on later generations are justified as for the assumed willingness of future generations to avoid higher costs of climate change prevention and environmental irreversible lock-ins. Shifting the ultimate costs of climate change aversion to later generations leverages climate stability into a Pareto-improving strategy for humankind.

Overall, in this tax-and-transfer mitigation policy, all generations are better off with mitigation through climate bonds as compared to the business-as-usual (BAU) non-mitigation scenario (Sachs, 2014). While future generations enjoy a favorable climate and averted environmental lock-ins, the current populace does not face drawbacks on economic growth. Sharing the costs of climate change aversion between and across generations appears as important strategy to instigate immediate climate change mitigation through incentivizing emission reduction and providing funds for adaptation. The following chapters provide an empirical introduction of climate justice within society, between countries and over time. Empirical analyses reveal novel innovative climate justice approaches and climate stabilization financialization in order to aid global governance policy makers to efficiently herald climate justice in the twenty-first century.

5.4 Mapping Climate Justice

In the following, the contemporary evidence on the burden of climate change being unevenly distributed within society, between countries and over time will be introduced empirically. Presentations of climate change winning and losing areas around the globe will inspire novel insights on how to implement climate justice and finance climate stabilization. Based on GDPoptimal temperature measurements, outlining climate change winners and losers around the globe is targeted at deriving universally fair climate stabilization strategies. The results will help working out innovative compensation schemes to share the burden of climate change more equally *within* society, *between* countries as well as *over time*.

The following empirical analyses reveal complex interaction and feedback systems of climate, climate change mitigation, and adaptation choices. Macroeconomic modeling will help recommend optimal climate change policies to financialize climate change mitigation and adaptation strategies. Concerning country-specificities and regional risk estimates, a well-balanced climate change policy mix to fund climate gains redistribution based on climate taxation and green bonds will be brought forward with attention to regional peculiarities and real-world relevance. The complexity and number of interactions will also require a qualitative analysis on the challenges of climate policy funding. Mapping solutions will help in monitoring, inspection and surveillance of climate change impacts around the globe. These findings target at innovatively helping to derive adaption and mitigation policies. Graphical solutions will allow for employing further macroeconomic modeling approaches that help recommend optimal climate change policies with respect for country-specificities in order to design and calibrate a well-balanced climate change adaptation and mitigation policy mix with real-world relevance. In addition, mapping climate change winners and losers will help calibrate and estimate the development level on the mitigation versus adaptation policy mix scale in order to retrieve well-balanced, efficient climate modeling strategies around the globe. Outlining costs and benefits of mitigation and adaptation is key in determining security strategies for vulnerable cities, communities and countries, which helps protect from variegated climate change risks (Nordhaus, 1994). The link of climate change mitigation and adaptation initiatives at the regional level aids develop real world-relevant climate change policy prescriptions for governments, private sector stakeholders as well as Intergovernmental Panel on Climate Change IPCC executives. A framework of costs and benefits of climate change mitigation and adaptation infrastructure serves to derive implications how to balance global warming responses with economic growth postulates. Redistribution solutions will feature indices and graphical presentations of the relation of countries to each other in enacting climate change mitigation.

The presented tax-and-bonds strategy enacts climate justice over time by incentivizing to shift from being a net grantor of common climate bonds in taxation revenue payments to fund countries that are net beneficiaries of bonds premium payments based on a country's economic prospect to be benefitting from global warming and the country's CO_2 emission production and consumption levels as well as the willingness of a government to change CO_2 emission levels and historically-grown bank lending rates. This tax-and-bonds strategy breaks novel ground on how to enact climate justice between countries and over time.

The innovative tax-and-bonds approach is meant to inform policymakers and global governance experts about the possibility of implementing climate fairness based on the overall starting ground of countries' temperature-related economic growth prospects but also the country's contribution to the climate change problem and willingness to change CO_2 emitting behavior and even out historical bank lending inequalities. Deriving respective policy recommendations for the broader climate change community in the discussion of the results aims to share the burden and the benefits of climate change within society in an economically efficient, legally equitable, and practically feasible way.

All the mentioned efforts target at helping multivariate stakeholders for shaping economic growth and sustainable development. The described models will become the basis for climate change burden sharing through bonds. Another important aspect of this type of work is that it allows for compensation if the cost of mitigation has very uneven distributional effects. Legal and economic analyses will be steered on how to allocate the burden of climate protection in a fair manner and ensure humankind to enjoy a stable climate in today's and tomorrow's world.

In its entirety and interdependence, the mapping climate justice approach will allow to generate recommendations on how to share the benefits and burdens of climate change in a fair and just manner within society, between countries and over time. All these efforts will help leading to important political contributions for international climate negotiations. The over-arching mapping endeavors are targeted at redesigning of governance structures and institutional arrangements that harmoniously distribute the costs of climate change mitigation and adaptation within society, between countries and over time. In the following, cutting-edge behavioral economics, 'wink' communication strategies but also the most novel educational case-study-based training tools will be employed to innovatively communicate the findings to a broad array of stakeholders and leave a lasting impact for this generation and the following (Kahneman & Tversky, 1979; Thaler & Sunstein, 2008).

Overall, the followingly presented results will innovatively develop new interpretations, understandings and concepts of climate risks within society, around the globe and over time. The mapping of the derived results will help guideline balanced approaches to implement climate justice through mitigating climate change and adapting to global warming. In compiling scholarship and theories on risk mitigation strategies in the climate action domain coupled with insights on how to finance climate justice, the innovative research will help create a central reference point and resources on aggregate information about the implementation and sophistication of climate justice.

The findings are also meant to elevate the importance of climate justice scholarship while deriving implications for climate stability. Emphasizing areas where to apply climate mitigation and where to promote climate adaptation strategies will lead to practical implications for the private industry and public policy sector. Understanding the different climate risk attitudes but also shedding light on previously unknown climate mitigation and adaptation interdependencies will aid environmental sustainability to ensure a future humankind.

For practitioners, the results will help lowering institutional downfalls of increasingly interconnected and fragile global networks. For academia, the book will spearhead interdisciplinary research on climate justice in academic journal articles, conferences and university teaching. For the general public, literature compilations as well as empirically-backed resources on the coping strategies in light of a warming earth will hold short-term innovative practical advantages as well as long-term historic invaluable assets to capture attempts to cope with a warming earth in the twenty-first century.

The novel, innovative climate gains redistribution approach aims to aid global governance policymakers inefficiently herald climate justice in the twenty-first century. The results are targeted at helping the individual decision-maker make a pro-environmental choice and embrace multivariate stakeholders for shaping economic growth in harmony with sustainable development. A heterodox approach will innovatively develop new interpretations, understandings, and concepts of climate-induced economic gains and losses. The findings will help work out innovative compensation schemes to share the burden of climate change more equally within society, between countries as well as over time. The findings are meant to elevate the importance of climate justice scholarship while deriving implications for climate stability. Emphasizing areas where to apply climate taxation and where to promote climate bonds strategies will lead to practical implications for global governance on environmental justice. The overall book pursues a multi-disciplinary approach to motivate people on the individual decision-making level for future discounting of environmental values. In addition, top-down strategies will be outlined to steer market dynamics towards a transition to renewable energy and enact climate justice. Highlighting the regional differences in the impact of climate change allows for compensation for the uneven effects. Understanding the different starting grounds on climate-related economic growth prospects and shedding light on previously-unknown climate benefits will aid in the redistribution to share the burden of a warming globe and thereby ensure environmental sustainability for the current and future humankind.

References

Andreoni, J., Erard, B., & Feinstein, J. S. (1998). Tax compliance. *Journal of Economic Literature, 36*(2), 818–860.

Barro, R. (1990). Government spending in a simple model of endogenous growth. *Journal of Political Economy, 98*, 103–125. https://doi.org/10.1086/261726

Bauer, P. T. (1957). *Economic analysis and policy in underdeveloped countries.* Duke University Press.

Braithwaite, V. (2003). Dancing with tax authorities: Motivational postures and non-compliant actions. In V. Braithwaite (Ed.), *Taxing democracy. Understanding tax avoidance and tax evasion* (pp. 15–39). Ashgate.

Braithwaite, V., & Ahmed, E. (2005). A threat to tax morale: The case of Australian higher education policy. *Journal of Economic Psychology, 26*(4), 523–540. https://doi.org/10.1016/j.joep.2004.08.003

Bürgenmeier, B. (1994). Environmental policy: Beyond the economic dimension. In B. Bürgenmeier (Ed.), *Economy, environment, and technology: A socio-economic approach* (pp. 166–175). Armonk Sharpe.

Burke, M., Hsiang, S. M., & Miguel, E. (2015). Global non-linear effect of temperature on economic production. *Nature, 527*, 235–239.

Centeno, M. A., Creager, A. N., Elga, A., Felton, E., Katz, St. N., Massey, W. A., et al. (2013). *Global systemic risk: Proposal for a research community.* Working paper. Princeton Institute for International and Regional Studies, Princeton University.

Chancel, L., & Piketty, T. (2015). *Carbon and inequality: From Kyoto to Paris*. Paris School of Economics.

Cialdini, R. B. (1996). Social influence and the triple tumor structure of organizational dishonesty. In D. M. Messick & A. E. Tenbrunsel (Eds.), *Codes of conduct* (pp. 44–58). Sage.

Davis, D. D., & Holt, C. A. (1993). *Experimental economics*. Princeton University Press. https://doi.org/10.1515/9780691233376

Dawes, R. M. (1980). Social dilemmas. *Annual Review of Psychology, 31*(1), 169–193. https://doi.org/10.1146/annurev.ps.31.020180.001125

Downs, J. (2008). *Discovery of Rosetta: The stone that unlocked the mysteries of ancient Egypt*. Constable & Robinson.

Eichenberger, R., & Frey, B. S. (2002). Democratic governance for a globalized world. *Kyklos, 55*(2), 265–288. https://doi.org/10.1111/1467-6435.00186

Feld, L. P., & Frey, B. S. (2002). Trust breeds trust: How taxpayers are treated. *Economics of Governance, 3*(2), 87–99. https://doi.org/10.1007/s101010100032

Frey, B. S. (1992). Tertium datur: Pricing, regulation and intrinsic motivation. *Kyklos, 45*(2), 161–184. https://doi.org/10.1111/j.1467-6435.1992.tb02112.x

Frey, B. S. (1997). *Not just for the money: An economic theory of personal motivation*. Edward Elgar.

Grasso, M., & Markowitz, E. M. (2015). Climate change: Multidisciplinary perspectives on climate ethics. *Climatic Change, 130*(3), 327–334.

Green, D., Kahneman, D., & Kunreuther, H. (1994). How the method and scope of public funding affects willingness to pay for public goods. *Public Opinion Quarterly, 58*, 48–67. https://doi.org/10.1086/269407

Hanson, J. D. (2012). *Ideology, psychology, and law*. Oxford University Press. https://doi.org/10.1093/acprof:oso/9780199737512.001.0001

Hasseldine, J. D. (1998). Prospect theory and tax reporting decisions: Implications for tax administrators. *International Bureau of Fiscal Documentation, 52*(11), 501–505.

High-Level Commission on Carbon Prices. (2017). *Report of the high-level commission on carbon prices*. World Bank.

Howell, D. R. (2021, September). Low pay in rich countries: Institutions, bargaining power and earnings inequality in the U.S., U.K., Canada, Australia and France. Washington Center for Equitable Growth: Evidence for a stronger economy, Working paper. https://equitablegrowth.org/working-papers/low-pay-in-rich-countriesinstitutions-bargaining-power-and-earnings-inequality-in-the-u-s-u-k-canada-australiaand-france/

Kahneman, D., & Tversky, A. (1979). Prospect theory: An analysis of decision under risk. *Econometrica, 47*(2), 263–291. https://doi.org/10.2307/191418510.2307/1914185

Kirchler, E. M. (2007). *The economic psychology of tax behaviour.* Cambridge University Press. https://doi.org/10.1017/CBO9780511628238

Klaassen, G. A. J., & Opschoor, J. B. (1991). Economics of sustainability or the sustainability of economics: Different paradigms. *Ecological Economics, 4*, 93–115. https://doi.org/10.1016/0921-8009(91)90024-9

Lewis, A. (1982). *The psychology of taxation.* Martin Robertson.

Marron, D. B., & Morris, A. C. (2016). *How to use carbon tax revenues.* Tax Policy Center Urban Institute & Brookings Institution. https://doi.org/10.2139/ssrn.2737990

Mumford, A. (2001). *Taxing culture.* Ashgate.

Nordhaus, W. D. (1994). *Mapping the global commons: The economics of climate change.* MIT Press.

Oppenheimer, M., O'Neill, B. C., Webster, M., & Agrawal, S. (2011). Climate change: The limits of consensus. *Science, 317*(5844), 1505–1506. https://doi.org/10.1126/science.1144831

Poppe, M. (2005). The specificity of social dilemma situations. *Journal of Economic Psychology, 26*(3), 431–441. https://doi.org/10.1016/j.joep.2004.12.006

Puaschunder, J. M. (2015). Trust and reciprocity drive social common goods allocation norms. In *Proceedings of the 2015 Cambridge Business & Economics Conference.* University of Cambridge.

Puaschunder, J. M. (2017a, March 26–27). Climate in the 21st century: A macroeconomic model of fair global warming benefits distribution to grant climate justice around the world and over time. *Proceedings of the 8th International RAIS Conference on Social Sciences and Humanities*, organized by Research Association for Interdisciplinary Studies (RAIS) at Georgetown University, Washington, D.C., United States, pp. 205–243.

Puaschunder, J. M. (2017b). Mapping climate in the 21st century. *Development, 59*, 3, 211–216.

Puaschunder, J. M. (2020). *Behavioral economics and finance leadership: Nudging and winking to make better choices.* Springer Nature.

Puaschunder, J. M. (2016). Intergenerational climate change burden sharing: An economics of climate stability research agenda proposal. *Global Journal of Management and Business Research: Economics and Commerce, 16*(3), 31–38.

Rawls, J. (1971). *A theory of justice.* Harvard University Press. https://doi.org/10.4159/9780674042605

Rovenskaya, E. (2005, September). Sensitivity and cost-benefit analyses of emission-constrained technological growth under uncertainty in natural emission. *International Institute for Applied Systems Analysis Interim Report.* Laxenburg, Austria.

Rovenskaya, E. (2008, January). Optimal economic growth under stochastic environmental impact: Sensitivity analysis. *International Institute for Applied Systems Analysis Interim Report.* Laxenburg, Austria.

Sachs, J. D. (2014). Climate change and intergenerational well-being. In L. Bernard & W. Semmler (Eds.), *The Oxford handbook of the macroeconomics of global warming* (pp. 248–259). Oxford University Press.

Sachs, J. D. (2021, November 17). Fixing climate finance. *Social Europe: Politics, Economy and Employment & Labor.*

Schleussner, C.-F., Lissner, T. K., Fischer, E. M., Wohland, J., Perrette, M., Golly, A., Rogelj, J., Childers, K., Schewe, J., Frieler, K., Mengel, M., Hare, W., & Schaeffer, M. (2016). Differential climate impacts for policy-relevant limits to global warming: The case of 1.5°C and 2°C. *Earth Systems Dynamics, 7*, 327–351.

Schmölders, G. (1960). *Das Irrationale in der öffentlichen Finanzwirtschaft.* Suhrkamp.

Stern, N. (2008). The economics of climate change. *The American Economic Review, 98*, 1–37.

Stroebe, W., & Frey, B. S. (1982). Self-interest and collective action: The economics and psychology of public goods. *British Journal of Social Psychology, 21*(2), 121–137. https://doi.org/10.1111/j.2044-8309.1982.tb00521.x

Thaler, R., & Sunstein, C. (2008). *Nudge: Improving decisions about health, wealth, and happiness.* Yale University Press.

The World Bank. (2015). Green bonds attract private sector climate finance, World Bank Brief.

The World Economic Forum Report. (2015). Davos, Switzerland: World Economic Forum.

Torgler, B. (2002). Speaking to theorists and searching for facts: Tax morale and tax compliance in experiments. *Journal of Economic Surveys, 16*(5), 657–683. https://doi.org/10.1111/1467-6419.00185

Tyler, T. R., & De Cremer, D. (2006). How do we promote cooperation in groups, organizations, and societies? In P. A. M. van Lange (Ed.), *Bridging social psychology. Benefits of transdisciplinary approaches* (pp. 427–433). Lawrence Earlbaum.

Van Lange, P. A. M., Liebrand, W. B. G., Messick, D. M., & Wilke, H. A. M. (1992). Social dilemmas: The state of the art. In W. B. G. Librand, D. Messick, & H. A. M. Wilke (Eds.), *Social dilemmas: Theoretical issues and research findings* (pp. 3–28). Pergamon.

Woo, A. (2021, April 16). Environmental fascism and the Zero Waste Movement. *Panel on monitoring and evaluation in environmental economics*. The New School New York.
World Bank Report. (2015). Washington, DC: World Bank.

CHAPTER 6

Global Climate Justice

6.1 From Climate Change Burden Sharing to Fair Global Warming Benefits Distribution

Climate stability accounts for the most challenging global governance goal. In the current climate change mitigation and adaptation efforts, high- and low-income households but also developed and underdeveloped countries as well as various overlapping generations are affected differently (Puaschunder, 2016b). This part of the book investigates climate change winners and losers around the world until the end of this century. Innovatively differing from the mainstream contemporary climate change literature and discussions centered around the problem of a warming earth and losses arising from climate change risks demanding for a viable burden sharing strategy, this chapter will shed novel light on the expected benefits of climate change in order to derive fair distribution mechanisms. In the current frustration over the neglect of interest in ratifying intergovernmental climate agreements, this innovative way to frame the prisoner's dilemma of a warming common climate proposes a completely new incentive structure based on favorable benefits share prospects. As such, this chapter opens up a black box of unidimensional climate negotiations that are one-sidedly focused on problem solving through burden sharing.

Methodologically, state-of-the-art welfare function measurements and economic productivity parameters become the basis for conclusions about

J. Puaschunder, *Governance & Climate Justice*,
https://doi.org/10.1007/978-3-031-87971-5_6

a fair spread of the gains and benefits from a warming earth. Contemporary Gross Domestic Product (GDP) measurements are taken into account to estimate the productivity of the agriculture, industry and service sectors around the globe in a business-as-usual path of rising temperatures. Based on the optimal temperatures for the agriculture, industry and service sector productivity, the average temperature per country around the world as well as climate projections of the year 2100 under the business-as-usual path, this chapter reveals relative climate change winners and losers around the world. Overall and simply seen from a narrow-minded GDP perspective, the world will macroeconomically benefit more from climate change until 2100 than lose. A world CO_2 emissions dataset outlines winning and losing from a warming earth is significantly positively correlated with the Paris COP 21 self-reported Greenhouse Gas (GHG) emissions per country, leading to the conclusion that the national motivation to do something about the global climate change predicament differs based on the short-term benefits and time until a warming climate becomes an obstacle.

The chapter concludes with a prospect how to phase in climate-related risks and introducing the climatorial imperative—advocating for the need for fairness in the distribution of the global earth benefits among nations based on Immanuel Kant's (1785/1993) imperative to only engage in actions one wants to experience being done to oneself. Countries passive or agnostic about global warming mitigation that reap benefits from a warming earth should be obliged to finance international aid for those that are impacted negatively by climate change, for example, climate refugees. In addition, building on international law, those countries that have better means of protection or conservation of the common climate should also face a greater responsibility to protect the earth (Puaschunder, 2016b).

While the method to measure the gains from climate change can certainly be refined in future investigations, this chapter is meant as very first preliminary step to open a gate to future research, legal codifications and policy work to settle for a right, just and fair distribution of benefits and gains from our warming common mother earth. All these endeavors are targeted at helping to build a more whole-rounded participatory incentive structure as the basis for contemporary climate change negotiations. The introduction of the gains from climate change is a novel approach that should solely be seen in connection to the imperative to distribute the gains in a fair manner among all contemporary and future world inhabitants. The gain prospect should only be seen as means to hopefully draw attention to climate change of agnostic market actors or those who shy away from action given the negative connotation of burden sharing and loss aversion. The provocative results therefore have to be viewed with the caution of attention to fairness in between society and the generations to come.

All these efforts should alleviate the contemporary global governance predicament that seems to pit today's generation against future world inhabitants in a trade-off of economic growth versus sustainability. Deriving respective policy recommendations for the wider climate change community around the globe and over time is aimed at ensuring that the shared benefits of climate change reach all contemporary and future world inhabitants around the globe in an economically efficient, legally sound and equitable but also practically feasible way.

6.1.1 The Problem of Climate Change

Global commons are globally linked and accessible for anyone—but owned by no one solely. No nation can claim common goods and declare sole access to common goods by itself. Distribution of global common goods has led to legal considerations ever since as fair access to global commons serves as the basis for peace and stability. But fair, equal and free access to a stable climate is getting harder in the age of globalization leveraging common goods problems into international dilemmas given unprecedented levels of interconnectedness and ecological pressures. Since the measurement of sustainability, energy consumption reached an all-time-high in recent decades, while resources deplete and the global temperature is rising (Puaschunder, 2016c). The rising trend of world surface temperatures around the world is exhibited in Graph 6.1.

Rules of global access and use of global common goods are captured in international laws. Global common conscientiousness and solidary sharing of resource preservation is an ethical imperative of sustainable development. In the international compound, nation states have common but differentiated responsibilities in their differing contributions to global environmental protection. Common responsibilities comprise of an equal share of environmental protection costs, yet our knowledge of benefits distribution of a stable climate and favorable time prospect to 'climate extinction' to this day is limited. Global warming benefits, however, could

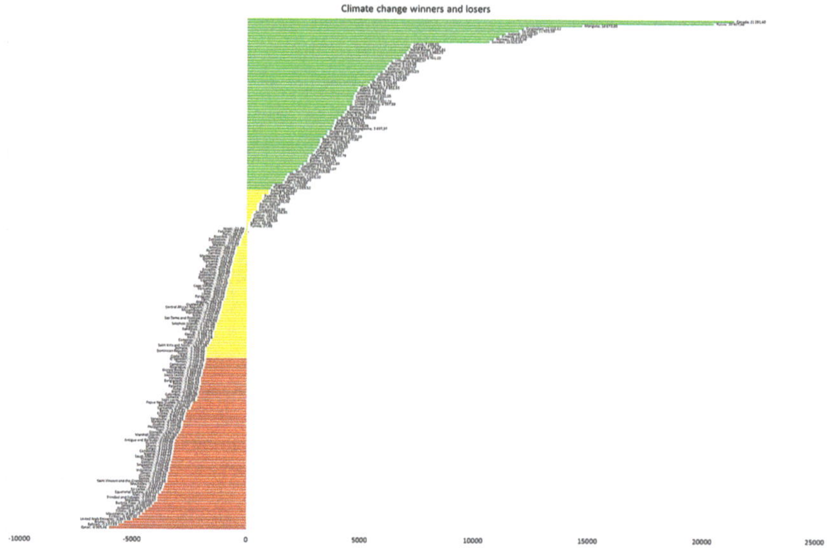

Graph 6.1 Climate change winners and losers (*Source* Author's own)

be used to offset some of the costs of climate change mitigation, damage alleviation and adaptation.

In the eye of current production levels leading to ecologic decline, all nations of the world will have to come together and form a global thinking on fair access to a global common good of a stable climate. Fair access to global benefits of climate stability will alleviate global emergent risks as the core of international security politics and sustainable development. Global common good climate stability prevention will be the basis of societal advancement of a sustainable humankind.

While the ecological sensibility toward the environment has improved in the last years and the world appears to have established a basic understanding about the scarcity of resources and the social responsibility to protect the earth, the threat of climate change implies novel and lasting environmental challenges for humankind. By now there is strong evidence of the anthropogenic contribution to climate change. Today's global climate change impacts vary from country to country. The Global

Humanitarian Forum Security Report 2009[1] claimed already 300 million people being affected by climate change and the total economic expenditures of climate change having reached over 100 billion USD. The people suffering from climate change were expected to rise to 600,000, and the total annual economic costs will hit 300 billion USD. In 10 years from now, the deaths caused by natural catastrophes arising with climate change could reach 500,000 civilians per year (Puaschunder, 2016c).

Climate change will determine the way people are living. Climate change precipitation patterns and access to water put agricultural and food preservation at stake. The melting of ice shells will unchangeably raise sea levels. Slugging and storm circuits will lead to natural disasters. The ecosystem and biodiversity are expected to diminish. Health risks, such as malnutrition, water problems and diseases, will spread. Climate refugees will be forced to leave behind their land, ancestry and cultural identity when about 200 million climate refugees will have to move in low-sea-level lands prospected around the next decades (Ferreira, 2017). An estimate of 1500 islanders will soon have to be evacuated from small islands in the South Pacific to larger islands and higher territories. The warming of ocean water will create hurricanes imposing danger on around 4 billion people in the United States alone. But also Inuit in the Arctic will have to move because of a massive ice melting.

Indigenous poor communities, who are already pushed to marginal levels of land, are going to be affected the worst. African tribes will have to leave villages as the impact of drought in central African regions, where seasons have become scarce and the hottest years in succession ever reported are happening now. Severe droughts have also struck Kenya, Ethiopia and Somalia, where attention is drawn to all-time-high food prices resulting in thousands of children starving. Climate justice will become an issue of concern when the major implications of a heating earth will be apparent and scarce resources vanishing potentially lead to violent frictions (Puaschunder, 2016c).

The populations most at risk, who live in the poorest regions of the world, are prone to be hit hardest by climate change, although they are not major contributors to causing the problem. At the international level, common goods consumption and preservation breed inequality if the lifestyle of some is trading off from the environmental conditions of

[1] http://www.ghf-ge.org/human-impact-report.pdf

others. The global Western World appears to cause an unfair impact to human-made climate change. The developed world is estimated to be responsible for 99% of carbon consumption, which raises strong global justice questions.[2] Less than 1% of the greenhouse gas emissions that caused global warming are offset by 15 of the poorest nations—including the Middle East, Southeast Asia, Sub-Saharan Africa and small islands developing states. People in the poorer regions will go on the street and fight for their rights in the eye of missing future orientation or social equity regarding current ecologic problems (Puaschunder, 2016c).

In the light of irreversible environmental decline, the time is ripe to reflect on intergenerational justice to avert environmental decline in a cost-effective way. Our current consumption patterns should not transfer environmental debts to our ancestors or jeopardize the environmental conditions of our future children.

Many of the more difficult issues were concerned in the Rio+20 Conference and the yearly United Nations Framework Convention on Climate Change (UNFCCC) Conferences of the Parties on Climate Change (COPs). Debates are also organized by UN agencies, governments or developed and developing nations, corporations, foundations and constituency groups, who increasingly highlight the humane consequences of climate change. The urgent need to address climate change is attributed in annual UN framework discussions leading to international agreements (e.g., the Montreal Protocol) and the inception of multilateral climate change aversion programs (e.g., the annual UN negotiations at the end of each year). The humanitarian dimension of climate change is central to the formal negotiations emphasizing mitigation and security. The international focus is placed on emerging economies—such as China and India—as the countries with the highest levels of CO_2 emission but viewed in light of the relatively low per capita emissions of these countries.

The United Nations Conference of the Parties (COP) Framework Convention on Climate Change (UNFCCC) serves as an international environmental treaty negotiated at the Earth Summit in Rio de Janeiro in 1992 to stabilize greenhouse gas (GHG) concentrations in the atmosphere at a level that prevents dangerous anthropogenic interference with the climate system. The 197 countries strong framework provides a stage to negotiate to set binding limits on GHG. The annual meetings since

[2] http://unfccc.int/resource/docs/2015/cop21/eng/10.pdf#page=30

1995 have produced nameable results ranging from the 1997 Kyoto Protocol establishing legally binding obligations for developed countries to reduce their GHG emissions and the 2010 Cancun agreement stating that future global warming should be limited to below 2 degrees Celsius. The 1997 Kyoto Protocol initially regulated GHG reduction measures until 2012 and was extended until 2020 in the Doha Amendment. The ad hoc working group on the Durban Platform for Enhanced Action governs climate change mitigation measures from 2020.

The COP21 Paris Agreement under the United Nations Framework Convention on Climate Change, which was drafted in December 2015 and signed in April 2016 by 194 nations' representatives, focuses on GHG mitigation, adaptation and finance staring in 2020. The COP 21 Paris Agreement was adopted in December 2015 by consensus by all of the 195 UNFCCC participating member states and the European Union to reduce emissions and carbon output as soon as possible and to do the best to keep global warming well below 2 degrees Celsius, which equals 3.6 degrees Fahrenheit. The aim of the convention was to hold the global average temperature well below 2 degrees Celsius above preindustrial levels, pursue efforts to limit the temperature increase to 1.5 degrees Celsius above preindustrial levels, increase the ability to adapt to the adverse impacts of climate change and foster climate resilience and low GHG emissions development, making finance flows consistent with a pathway toward low GHG emissions and climate-resilient development.

The agreement entered into force with 55 countries that produce at least 55% of the world's GHG emissions ratifying, accepting, approving or acceding to the agreement. In total, 174 nations and the European Union signed the treaty on the first date it was open for signature, and more than 20 countries issued a statement of their intent to join as soon as possible with a view to joining in 2016. In October 2016, enough countries had ratified the agreement that produce enough of the world's GHG for the agreement to enter into force. The Agreement obtained enough parties to enter into effect as of November 2016. As of December 2016, 194 UNFCCC members have signed the treaty, 134 had ratified it. Since the UN COP21 Paris was ratified in 2016, systems of monitoring, reporting and verification of emissions are discussed to be installed to reduce emissions in non-binding agreements based on individual country pledges. The Paris Agreement enabled countries to self-govern emission reductions from 2020 on through commitments of countries in nationally-determined contributions. The implementation of the agreement by all member countries together is evaluated every 5 years.

The innovative elements of the Paris Agreement are the nationally-determined contributions and country limits. Countries were given freedom and flexibility to ensure country-sensitive climate change mitigation and adaptation endeavors in nationally-determined contributions. This bottom-up structure in contrast to previous top-down international environmental law treaties was meant to enhance compliance and motivation of countries to adopt a GHG conscientious economic growth strategy. Building on consensus, the Paris Agreement is unique in its voluntary and nationally-set targets, which should encourage politically rather than using legally-binding implementation structures. Because there are no legal mitigation or financial targets, the agreement is an executive agreement, which requires all countries to submit emission reduction plans. Cooperation is needed of parties to achieve the nationally-endeavored carbon emissions reduction, accounting and trading goals. In the further goal of climate stability financing, the demand to determine the individual country contributions to the climate solution arises. Ethical questions emerge on what determinants' contribution obligations should be pegged. Although the Paris Agreement attempts to call for a balance of climate finance between mitigation and adaptation, in-between country differences in common but differentiated responsibilities were not clarified. A division of costs and climate justice in the burden sharing is essential given the ambitious goals to raise funds for climate preparedness and climate risk insurance and the launching of a Climate Risk and Early Warning System as demanded by the agreement.

The COP26 meeting of 2021 has emphasized hierarchies of climate change causing countries, which are also more likely to economically benefit from global warming in the short-run (Sachs, 2021). As the COP26 underlined widespread inequalities inherent in the climate change problem, the world now faces the most urgent need for funding climate change mitigation and adaptation policies with attention to climate justice. Worldwide redistributing funds for climate control are established to alleviate climate injustices with a focus on innovative financing strategies following disparate impact analysis schemes (Sachs, 2021). In the aftermath of the COP26 meeting, it has been argued that the advanced countries have an obligation and responsibility to finance the adaptation to global warming of the low-income countries through direct transfers and credit guarantees (Sachs, 2021). Now the most pressing question arises, what measures should the world community take to determine the relative economic windfall gain reapers and most blatant victims of climate change.

In the aftermath of the COP26 annual climate meeting of the United Nations, Jeffrey Sachs put forward an idea of funds for climate change mitigation and adaptation that should be raised by climate tax-funded grants provided by some countries as transfer payments, while other countries should be recipients of green bonds granted to low-income countries (Sachs, 2021). While Sachs (2021) argues that half of the funds raised should be grants (transfers) and half green bonds that help transition to renewable energy in low-income countries, this book will offer an empirical strategy to refine prioritizing which countries should be grantors and which recipients based on data-driven criteria. For instance, country's starting ground on the relative economic climate gains and losses spectrum, a country's climate flexibility, and a country's CO_2 emissions contributions to production and consumption levels, as well as a country's CO_2 emissions levels changes and the bank lending rate, will determine whether a country will be on the taxation regime for funding mutual climate stabilization or whether a country will be on the receiving end of climate bonds solutions. Pegging the country's situation to the initial starting levels on rational grounds and relations to each other allows for turning the general price-cutting competitive race-to-the-bottom into a race-to-the-top for the mutual climate fund allocations. Integrating the CO_2 emissions levels changes over time imbues incentives to change harmful behavior to this common climate fund solution.

Overall, it can be concluded that while voluntary climate change aversion has been established during the last few years, it remains unclear how strong the commitment will be. The success of voluntary agreements depends on the consent of all countries around the world. Problematic appears that the voluntary agreements offer no binding limits on GHG emissions for individual countries and contains no enforcement mechanisms or concerted implementation strategies. The currently perceived lack of progress may be due to the burden sharing framing and cost imposing predicament of climate change mitigation. The following part therefore introduces a novel inventive approach through benefits distribution prospects of a warming earth. While the climate justice fund raising communication was previously mainly focused on losses and damages from climate change, this chapter will draw attention to the expected economic gains of global warming as innovative asset sources in order to find fair distribution strategies for all.

6.1.2 *From Burden Sharing to Benefits Distribution*

Climate change accounts for one of the most pressing contemporary problems as for exacerbating more complex risks than ever before. As never before in history since the birth of the earth, there is an environmental sensitivity to economic growth (Centeno & Tham, 2012; The World Economic Forum Report, 2015). While classic economics portrayed balancing the interests of different generations as ethical problem of competitive markets requiring governance for intergenerational transfers and some economists even opposed discounting of future utilities (Allais, 1947; Harrod, 1948; Ramsey, 1928), climate change has leveraged intergenerational equity as contemporary challenge of modern democracy and temporal justice an ethical obligation for posterity.

Uncertainty arising in assessing economic growth in relation to climate change creates an unprecedented predicament for scientists and global governance technocrats. Intertemporal questions are posed as to whether invest in abatement today—in order to prevent negative effects of global warming—or to delay investment until more information on climate change is gained (Rovenskaya, 2008). In general, resources are balanced across generations by social discounting to weight the well-being of future generations relative to those alive today. Regarding climate justice, current generations are called upon to make sacrifices today for future generations to cut carbon emissions to avert global warming (Sachs, 2014). The implementation of climate change avoidance and the adaptation against the coming climate risk were previously described as to pit today's generation against future world inhabitants in the trade-off of economic growth versus sustainability (Puaschunder, 2016a; Sachs, 2014).

In this framing, the problem of climate change is therefore mainly associated with risks and burden sharing costs, which may have caused a lethargy on action (Puaschunder, 2016a). Climate change burden sharing strategies have been thematized alongside mitigation and adaptation policies against climate risks (Puaschunder, 2016b). Recent IPCC research, international conferences on climate change and fund raising activities to combat global warming stress that it is advisable to pursue both mitigation as well as adaptation now. While climate justice will require both climate change mitigation and adaptation, concurrently, no macroeconomic model exists to date on the short-term benefits of a warming earth.

Shedding unprecedented light on the advantages of global warming will help to derive real-world relevant climate justice implementation recommendations. Contrary to negative connotations of burden sharing on climate change, outlining the short-term economic benefits of global warming will help gain attention from agnostic market actors. As a positive incentive, expected economic gains will help raise awareness for the issue at stake and ensure that the positive advantages of a warming earth can be distributed based on right, just and fair ethical principles. Without knowledge and quantification of the gains of climate change, climate inequality may become unnoticed (Chancel & Piketty, 2015). Only by the sound understanding of who will gain what on a warming earth, justice can be established—very likely insofar as to let the entire world benefit from the gains of a warming globe in a right, just and fair way. Knowledge of the concrete benefits based on contemporary finance and growth models will help maximizing utility over the world and implement climate justice between countries but also balance gains inbetween generations over time.

The measurement and description of the short-term climate change winners and losers of a warming earth is an innovative and novel angle toward accomplishing climate justice that is introduced in order to find a behavioral economics solution to steer action on climate change through positive incentives. The following empirical part therefore elucidates short-term climate change gains around the world in order to find right, just and fair benefit sharing strategies and mechanisms, which will be proposed in the following chapters (Puaschunder, 2017).

As the very first preliminary attempt in the benefits distributions direction, the chapter provides real-world relevant means how to implement climate justice around the globe. The climate model will help analyze how global governance experts can distribute the gains of a warming earth in the international compound. Outlining the distribution of benefits of climate change is key in determining redistribution strategies for vulnerable cities, communities and countries and protecting them from variegated climate change risks (Nordhaus, 1994). The prospective results of a climate change gains analysis will therefore help multivariate stakeholders achieving compensation for a transition to renewable energies and thereby foster sustainable development. Winking with insights on the benefits of climate change appears as novel, feasible and easily implementable solution to gain interest from the very many contemporary stakeholders we need to address in order to avert a warming of the earth. Informing about gains expected from global warming will also help nudging overlapping generations toward future-oriented sustainability. All these endeavors follow the greater goal to make the world a fairer place for this generation and the following.

6.2 Method

6.2.1 Overall Model Assumptions

Economic output was measured under projected conditions of a warming earth. For 189 world countries, the Gross Domestic Product (GDP) of agriculture, industry and service sector composition was retrieved from the Central Intelligence Agency (CIA)World Factbook.[3] *Agricultural output* is the component of the GDP of a nation that describes the process of producing food, feed, fiber and other goods by the systematic raising of plants and animals. *Industry* is the GDP segment of the economy concerned with production of goods including fuels and fertilizers, mining and extraction sectors. The *service* sector is the non-material equivalent of a good. Service provision is defined as an economic activity that does not result in ownership of physical goods. It is a process that creates benefits by facilitating either a change in customers, a change in their physical possessions or a change in their intangible assets. Service output is a component of the GDP of a nation that also includes—but is not limited to—farm and factory-related activities.

Contemporary climate change projections estimate a mean temperature rise of approximately 4.24 degrees Celsius. The cardinal or optimal temperature per sector was related to the prospected temperature in 2100 per country. The cardinal age or optimal temperature for agriculture is 28.5 degree Celsius for growing wheat and rice.[4] The cardinal temperature for the industry sector comprising of construction, automotive, airframe, locomotive, petroleum, coal and chemicals was found to be 21 degree Celsius (Russell, 1957). The cardinal temperature for

[3] https://www.cia.gov/library/publications/the-world-factbook/

[4] http://agritech.tnau.ac.in/agriculture/agri_agrometeorology_temp.html

the service sector lies at 25 degree Celsius (Somanathan, Somanathan, Sudarshan, & Tewari, 2014).

From the current world temperature mean per country and the estimate of a 4.25 degree Celsius climate change until 2100, the closeness to cardinal temperature per sector was calculated for each country by using the following formula (6.1):

$$T_A = C_A - T_e, \tag{6.1}$$

whereby T_A represents the closeness to the optimum cardinal temperature for agriculture calculated by the cardinal temperature for agriculture C_A, which equals to 28.5 degrees Celsius, subtracted by T_e as the temperature estimated for the year 2100 per country.

For calculating each country's time to optimum economic output in the industry sector, Eq. (6.2) was used:

$$T_I = C_I - T_e, \tag{6.2}$$

whereby T_I represents the closeness to the optimum cardinal temperature for the industry sector calculated by the cardinal temperature for the industry sector C_I, which equals to 21 degrees Celsius, subtracted by T_e as the temperature estimated for the year 2100 per country.

For calculating each country's time to optimum economic output in the service sector, Eq. (6.3) was used:

$$T_S = C_S - T_e, \tag{6.3}$$

whereby T_S represents the closeness to the optimum cardinal temperature for the service sector calculated by the cardinal temperature for service sector C_S, which equals to 25 degrees Celsius, subtracted by T_e as the temperature estimated for the year 2100 per country.

The result for the distance to the optimum temperature for each sector for each country was then multiplied by the GDP contribution percentage of the sector using the following formulas (6.4), (6.5) and (6.6):

For the agricultural sector, formula (6.4) reads

$$WL_A = T_A \times GDP_A, \tag{6.4}$$

whereby WL_A stands for the climate change winner and loser index for agriculture per country composed of the distance to the optimum cardinal temperature per GDP sector T_A multiplied by the percentage of agriculture contributing to the overall GDP indicated by GDP_A.

For the industry sector, formula (6.5) reads

$$WL_I = T_I \times GDP_I, \tag{6.5}$$

whereby WL_A stands for the climate change winner and loser index for industry per country composed of the distance to the optimum cardinal temperature per GDP sector T_I multiplied by the percentage of industry contributing to the overall GDP indicated by GDP_I.

For the service sector, formula (6.6) reads

$$WL_S = T_S \times GDP_S, \tag{6.6}$$

whereby WL_A describes the climate change winner and loser index for service per country composed of the distance to the optimum cardinal temperature per GDP sector T_S multiplied by the percentage of the service sector contributing to the overall GDP indicated by GDP_S.

The overall Winner-Loser WL_T index was calculated per GDP sector leading the sector-specific Winner–Loser indices WL_A for the agriculture sector, WL_I for the industry sector and WL_S for the service sector, that were then added up into the Winner–Loser total index WL_T based on the following formula (6.7):

$$WL_T = WL_A + WL_I + WL_S, \tag{6.7}$$

whereby WL_A denotes the Winner–Loser index for the agriculture sector, WL_I the index for the industry sector and WL_S the index for the service sector. All indices were calculated per country for the year 2100 under a business-as-usual projection for climate change.

Overall, a positive *WL* index result would indicate a long distance to the optimum from now until the year 2100, whereas a negative index would be considered as negative prospect. Basically, the more positive the index, the longer time the country could expect to be having a favorable climate for agriculture, industries or service production until peak condition as measured by mean temperature under climate change. And the more negative the index, the sooner the country would (have) run out of efficiency time under business-as-usual assumption given global warming. In sum, the Winner–Loser index is an indicator how much cool time a country still has ahead from now in prospect of an optimum GDP productivity temperature and the assumption that the earth is warming.

6.2.2 Model Variants

In order to derive the overall GDP gains and losses until the year 2100 and paying attention to literature controversies over the mathematical representation of wins and losses, three model variants of linear model, prospect model and hyperbolic model are calculated (Burke, Hsiang, & Miguel, 2015).

6.2.2.1 Linear Model

Model 1 assumes linear gain and loss functions around the optimum temperature for production to capture the earth's warming-related GDP prospects based on the following formulas (6.8), (6.9) and (6.10). The winners' index W is calculated by multiplying the distance to the optimum cardinal temperature per GDP sector T_A for agriculture and GDP_A for GDP contribution to agriculture, T_I for industry sector and GDP_I for GDP contribution to industry and T_S for service sector and GDP_S for GDP contribution to service using the formulas (6.1), (6.2) and (6.3) multiplied by the respective GDP per sector contributions as exhibited in formulas (6.8), (6.9) and (6.10):

$$W_A = {T_A}^1 \times GDP_A \tag{6.8}$$

$$W_I = {T_I}^1 \times GDP_I \tag{6.9}$$

$$W_S = {T_S}^1 \times GDP_S \tag{6.10}$$

For the losers' index L, a linear growth is calculated by formulas (6.11), (6.12) and (6.13) for each respective sector:

$$L_A = -{T_A}^1 \times GDP_A \tag{6.11}$$

$$L_I = -{T_I}^1 \times GDP_I \tag{6.12}$$

$$L_S = -{T_S}^1 \times GDP_S, \tag{6.13}$$

per country and per GDP sector. The overall Winner–Loser WL index was calculated per GDP sector leading to the sector-specific Winner–Loser indices WL_A for the agriculture sector, WL_I for the industry sector and

WL_S for the service sector, that were then added up into the Winner–Loser total index WL_T based on formula (6.7).

6.2.2.2 *Concave Gains and Convex Losses Prospect Model*

Model 2 assumes concave gain and convex loss functions around the optimum temperature for production to capture the earth's warming-related GDP prospects based on the following formulas (6.14), (6.15) and (6.16). The winners' index W is calculated by multiplying the distance to the optimum cardinal temperature per GDP sector T_A for agriculture, T_I for industry sector and T_S for service sector using the formulas (6.1), (6.2) and (6.3) multiplied by the respective GDP per sector contributions as exhibited in formulas (6.14), (6.15) and (6.16):

$$W_A = {T_A}^{0.5} \times GDP_A \tag{6.14}$$

$$W_I = {T_I}^{0.5} \times GDP_I \tag{6.15}$$

$$W_S = {T_S}^{0.5} \times GDP_S \tag{6.16}$$

For the losers' index L, a convex growth is calculated by formulas (6.17), (6.18) and (6.19) for each respective sector:

$$L_A = -{T_A}^2 \times GDP_A \tag{6.17}$$

$$L_I = -{T_I}^2 \times GDP_I \tag{6.18}$$

$$L_S = -{T_S}^2 \times GDP_S, \tag{6.19}$$

per country and per GDP sector. The overall Winner–Loser WL index was calculated per GDP sector leading to the sector-specific Winner–Loser indices WL_A for the agriculture sector, WL_I for the industry sector and WL_S for the service sector, that were then added up into the Winner–Loser total index WL_T based on formula (6.7).

6.2.2.3 *Hyperbolic Model*

Model 3 assumes hyperbolic gain and loss functions with the cardinal temperature equaling the maximum point to capture the earth's warming-related GDP prospects based on the following formulas (6.20), (6.21) and (6.22). The winners' index W is calculated by multiplying the distance to the optimum cardinal temperature per GDP sector T_A for agriculture, T_I for industry sector and T_S for service sector using the formulas (6.1), (6.2) and (6.3) multiplied by the respective GDP per sector contributions as exhibited in formulas (6.20), (6.21) and (6.22):

$$W_A = {T_A}^2 \times GDP_A \tag{6.20}$$

$$W_I = {T_I}^2 \times GDP_I \tag{6.21}$$

$$W_S = {T_S}^2 \times GDP_S \tag{6.22}$$

For the losers' index L, a hyperbolic growth is calculated by formulas (6.23), (6.24) and (6.25) for each respective sector:

$$L_A = -{T_A}^2 \times GDP_A \tag{6.23}$$

$$L_I = -{T_I}^2 \times GDP_I \tag{6.24}$$

$$L_S = -{T_S}^2 \times GDP_S, \tag{6.25}$$

per country and per GDP sector. The overall Winner–Loser WL index was calculated per GDP sector leading to the sector-specific Winner–Loser indices WL_A for the agriculture sector, WL_I for the industry sector and WL_S for the service sector, that were then added up into the Winner–Loser total index WL_T based on formula (6.7).

6.3 Results

6.3.1 *Overall GDP Gains and Losses Until 2100*

6.3.1.1 *Linear Model*

When unidimensionally focusing on estimated GDP growth given a warmer temperature and estimating a linear growth of expected economic losses and wins, overall the world will be gaining more than losing until 2100. Based on the WL index of 198 countries of the world and under the assumption of linear gains and losses in light of climate change, less countries ($n = 79$) will win more from global warming until 2100 than more countries ($n = 119$) will lose from a warming earth. In particular, 79 countries of the world will gain $WL_L = 78{,}139.08$ in GDP output, whereas 119 countries of the world will lose estimated $WL_L = -52{,}061$.

6.3.1.2 *Prospect Convex Losses and Concave Gains Model*

When unidimensionally focusing on estimated GDP growth given a warmer temperature and estimating an exponential growth of losses and concave wins, overall the world will be losing more than gaining until 2100. Based on the *WL* index of 188 countries of the world, more countries ($n = 113$) will lose more from global warming until 2100 than less countries ($n = 75$) will win from a warming earth. In particular, 75 countries of the world will gain $WL_P = 22{,}717.161$ in GDP output, whereas 113 countries of the world will lose estimated $WL_P = -353{,}175.32$.

6.3.1.3 *Hyperbolic Model*

When unidimensionally focusing on estimated GDP growth given a warmer temperature and estimating a hyperbolic growth, overall the world will be gaining more than losing until 2100. Based on the *WL* index of 188 countries of the world, less countries ($n = 79$) will gain more from global warming until 2100 than more countries ($n = 109$) will lose from a warming earth. In particular, 79 countries of the world will gain $WL_H = 1{,}037{,}192$ in GDP output, whereas 109 countries of the world will lose estimated $WL_H = -352{,}088$.

6.3.1.4 *Total Estimate*

The total Winner–Loser *WL* index was calculated per country based on the mean of the overall Winner–Loser index WL_L for the linear model, the overall Winner–Loser index WL_P for the prospect model and the overall Winner–Loser index WL_H for the hyperbolic model that were then added up into the Winner–Loser total index WL_{TT} and divided by the number of models calculated (n) to gain the mean based on the following formula (6.26):

$$WL_{TT} = (WL_L + WL_P + WL_H)/n \qquad (6.26)$$

whereby WL_L denotes the Winner–Loser index for the linear model, WL_P the index for the prospect model and WL_H the index for the hyperbolic model and $n = 3$. All indices were calculated per country for the year 2100 business-as-usual projection.

When unidimensionally focusing on estimated GDP growth given a warmer temperature, over all calculated models, the world will be gaining more than losing until 2100. Based on the *WL* index of 188 countries of the world, less countries ($n = 77$) are prospected to gain more from global warming until 2100 than more countries ($n = 111$) will lose from a warming earth. In particular, 77 countries of the world are estimated to gain $WL_H = 354{,}039.6345$ in GDP output, whereas 111 countries of the world are estimated to lose $WL_H = -235{,}726.6280$.

6.3.2 Country Differences

6.3.2.1 Climate Change Winners

Based on the WL_{TT} index, which measures the distance to cardinal temperature per GDP sector in 188 countries of the world, all climate change winners ($n = 77$) and losers ($n = 111$) are outlined in Graph 6.1 (author's own).

Based on a country ranking, Graph 6.1 highlights the top one-third countries with longest time prospect in green color and the one-third countries that have run out of time in red color.

Based on the WL_{TT} estimate, Graph 6.2 (author's own) highlights climate change winners in green and yellow and losers in orange and red geographically around the world.

Based on the WL_{TT} estimate, Graph 6.3 (author's own) reveals that Africa will be holding climate change losses until the year 2100.

Based on the WL_{TT} estimate, Graph 6.4 (author's own) features climate change wins and losses until the year 2100 in Asia.

Graph 6.5 (author's own) shows only wins in Europe from climate change until the year 2100 based on the WL_{TT} estimate.

Graph 6.6 (author's own) shows only wins in North America from climate change until the year 2100 based on the WL_{TT} estimate.

Graph 6.7 (author's own) holds climate change wins and losses for South America until the year 2100 based on the WL_{TT} estimate.

Oceania will experience climate change wins and losses until the year 2100 based on the WL_{TT} estimate as Graph 6.8 (author's own) exhibits.

Graph 6.9 (author's own) holds the top one-third countries with quantitatively highest gain perspective in green color and the one-third countries that lose the most in red color. One-third of all total losses are indicated in red, one-third of all total wins in green.

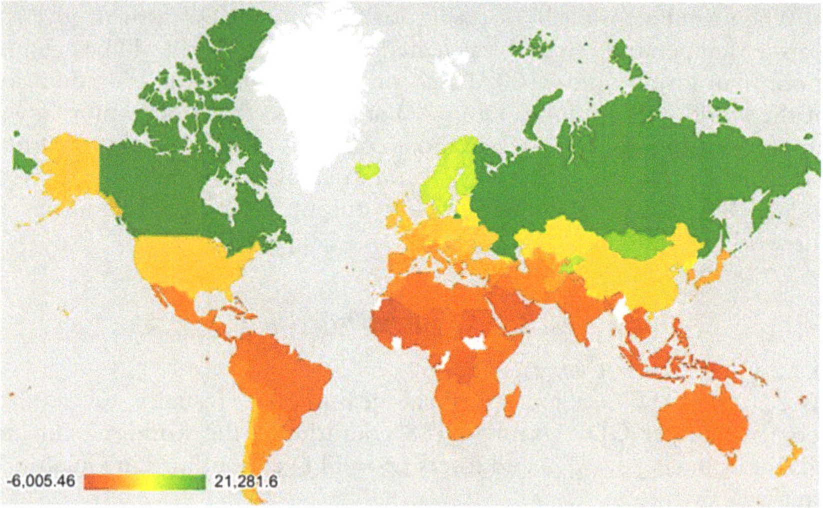

Graph 6.2 Climate change winners and losers around the world (*Source* Author's own)

Graph 6.3 Climate change losers in Africa (*Source* Author's own)

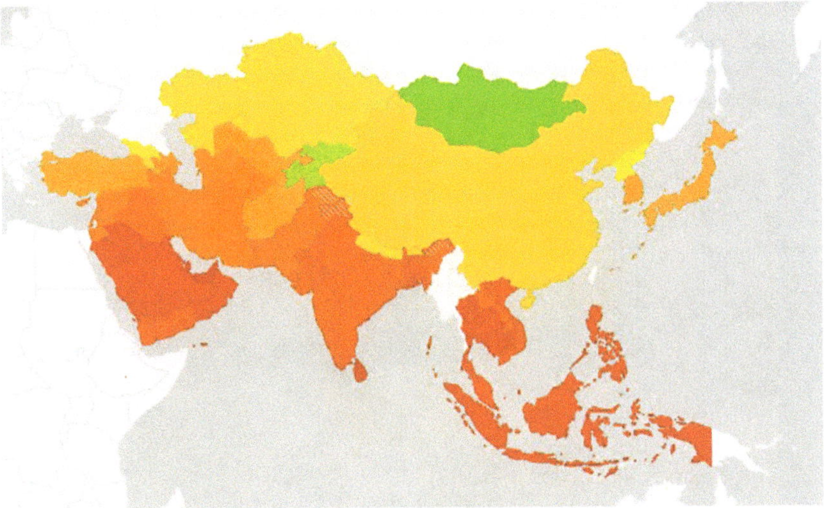

Graph 6.4 Climate change winners and losers in Asia (*Source* Author's own)

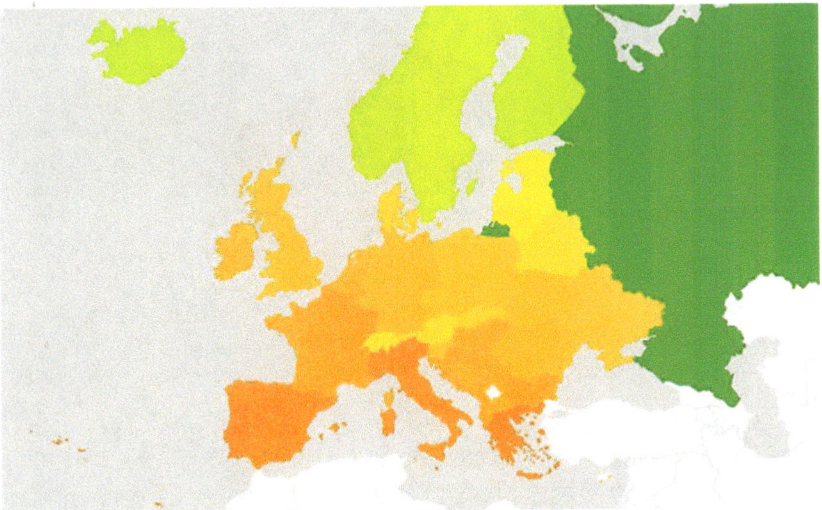

Graph 6.5 Climate change winners in Europe (*Source* Author's own)

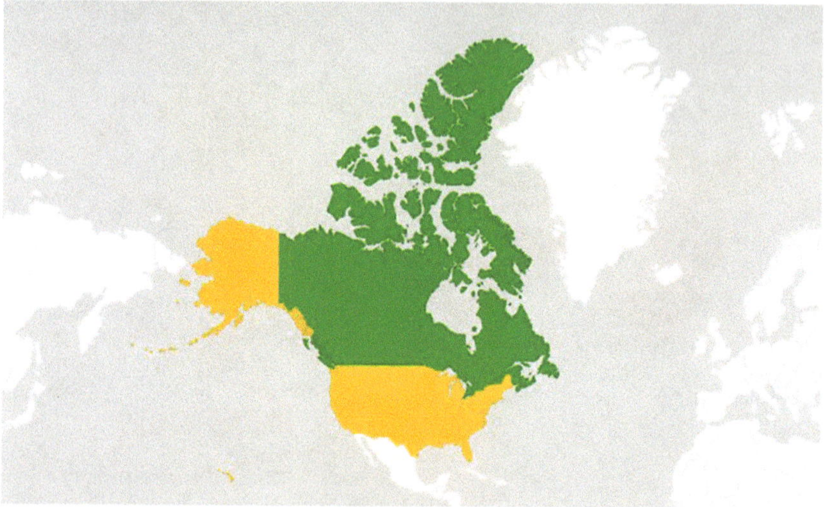

Graph 6.6 Climate change winners in North America (*Source* Author's own)

Graph 6.7 Climate change winners and losers in South America (*Source* Author's own)

Graph 6.8 Climate change winner and losers in Oceania (*Source* Author's own)

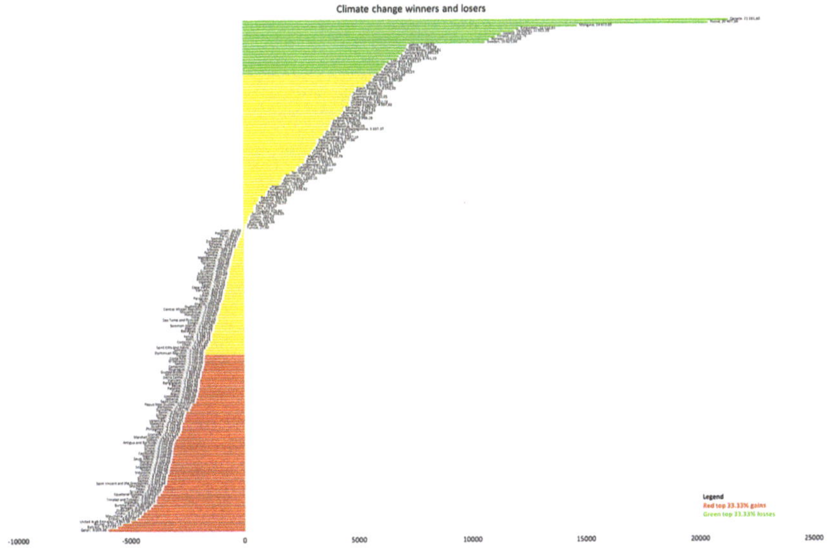

Graph 6.9 Climate change winners and losers (*Source* Author's own)

Graph 6.10 (author's own) features 77 countries with most time to a favorable climate. The top climate change winners are Canada (WL_{TT} = 21,281.6), Russia (WL_{TT} = 20,407.66), Mongolia (WL_{TT} = 14,673.85), Kyrgyzstan (WL_{TT} = 12,112.61), Tajikistan (WL_{TT} = 11,923.39), Iceland (WL_{TT} = 11,335.87), Finland (WL_{TT} = 11,074.76), Norway (WL_{TT} = 10,812.56), Sweden (WL_{TT} = 10,625.84), Latvia (WL_{TT} = 7188.83), Georgia (WL_{TT} = 7149.79), North Korea (WL_{TT} = 7080.81), Switzerland (WL_{TT} = 6980.25), Estonia (WL_{TT} = 6830.75), Liechtenstein (WL_{TT} = 6761.1), Lithuania (WL_{TT} = 6382.37), Nepal (WL_{TT} = 6214.06), Austria (WL_{TT} = 6131.72), Belarus (WL_{TT} = 6056.21) and Kazakhstan (WL_{TT} = 5945.25).

The top global warming winners per inhabitants are outlined in Graph 6.11 (author's own), which ranks 77 countries around the world on the per inhabitant GDP productivity time in light of climate change. The best off are inhabitants of Liechtenstein (WL_{TTi} = 0.179), San Marino (WL_{TTi} = 0.063), Monaco (WL_{TTi} = 0.046), Iceland (WL_{TTi} = 0.034),

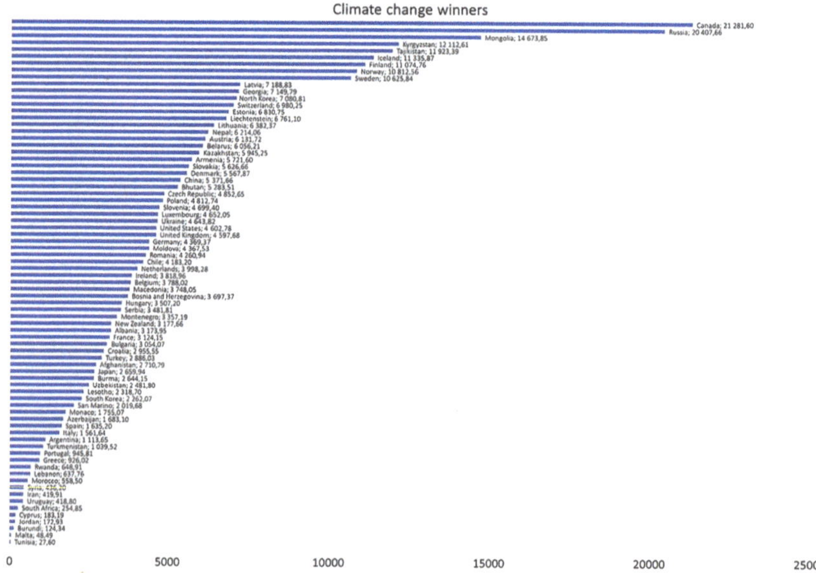

Graph 6.10 Climate change winners (*Source* Author's own)

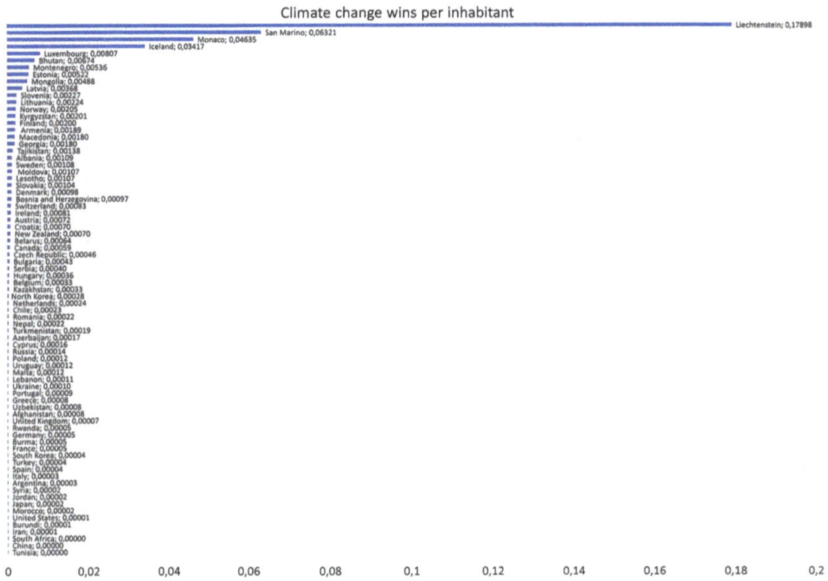

Graph 6.11 Climate change wins per capita (*Source* Author's own)

Luxembourg ($WL_{TTi} = 0.008$), Bhutan ($WL_{TTi} = 0.007$), Montenegro ($WL_{TTi} = 0.005$), Estonia ($WL_{TTi} = 0.005$), Mongolia ($WL_{TTi} = 0.005$), Latvia ($WL_{TTi} = 0.004$), Slovenia ($WL_{TTi} = 0.002$), Lithuania ($WL_{TTi} = 0.002$), Norway ($WL_{TTi} = 0.002$), Kyrgyzstan ($WL_{TTi} = 0.002$), Finland ($WL_{TTi} = 0.002$), Armenia ($WL_{TTi} = 0.002$), Macedonia ($WL_{TTi} = 0.002$), Georgia ($WL_{TTi} = 0.002$), Tajikistan ($WL_{TTi} = 0.001$) and Albania ($WL_{TTi} = 0.001$).

6.3.2.2 Climate Change Losers

Graph 6.12 (author's own) outlines the 111 countries with the least time to have a favorable climate for GDP production. The countries with the lowest time prospect to an economically unfavorable climate are Qatar ($WL_{TT} = -6005.46$), Bahrain ($WL_{TT} = -5577.83$), Brunei ($WL_{TT} = -5352.13$), United Arab Emirates ($WL_{TT} = -4965.38$), Kiribati ($WL_{TT} = -4687.02$), Mauritania ($WL_{TT} = -4541.19$), Tuvalu ($WL_{TT} = -4427.38$), Djibouti ($WL_{TT} = -4269.75$), Senegal ($WL_{TT} = -4099.99$), Burkina Faso ($WL_{TT} = -4007.08$), Maldives ($WL_{TT} =$

−3848.67), Trinidad and Tobago (WL_{TT} = 3847.55), Mali (WL_{TT} = −3826.13), Equatorial Guinea (WL_{TT} = −3683.81), Sri Lanka (WL_{TT} = −3593.57), Palau (WL_{TT} = −3554.02), Seychelles (WL_{TT} = −3522.63), Saint Vincent and the Grenadines (WL_{TT} = −3401.84), Samoa (WL_{TT} = −3395.72) and Guinea (WL_{TT} = −3373.02).

Global warming losers per inhabitants are outlined in Graph 6.13 (author's own), which ranks 110 countries around the world on the time of extinction of a favorable climate for GDP production per inhabitant. The countries worse off per inhabitant regarding climate change are Tuvalu (WL_{TTi} = −0.445), Palau (WL_{TTi} = −0.0165), Marshall Islands (WL_{TTi} = −0.055), Kiribati (WL_{TTi} = −0.041), Seychelles (WL_{TTi} = −0.036), Antigua and Barbuda (WL_{TTi} = −0.033), Saint Vincent and the Grenadines (WL_{TTi} = −0.031), Saint Kitts and Nevis (WL_{TTi} = −0.029), Grenada (WL_{TTi} = −0.026), Micronesia (WL_{TTi} = −0.018), Samoa (WL_{TTi} = −0.017), Tonga (WL_{TTi} = −0.016), Brunei (WL_{TTi} = −0.012), Mauritania (WL_{TTi} = −0.011), Saint Lucia (WL_{TTi} = −0.011), Maldives (WL_{TTi} = −0.010), Barbados (WL_{TTi} = −0.008),

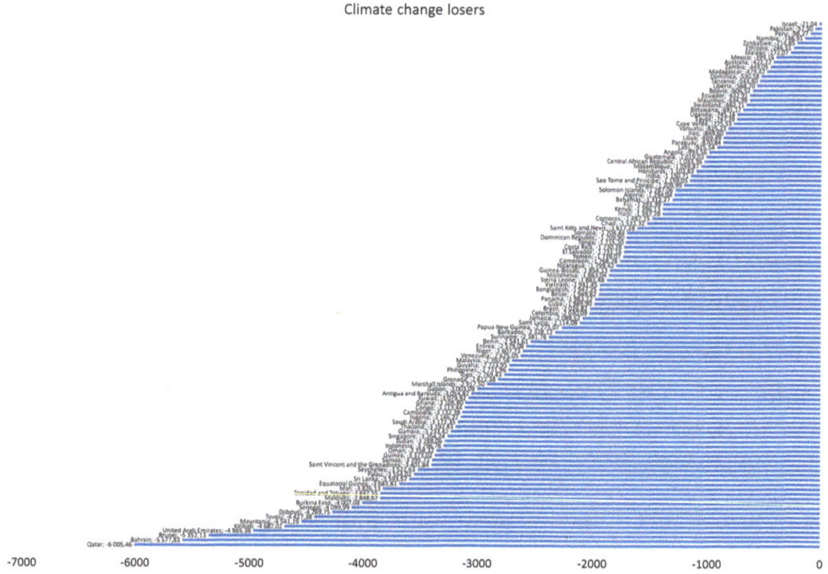

Graph 6.12 Climate change losers (*Source* Author's own)

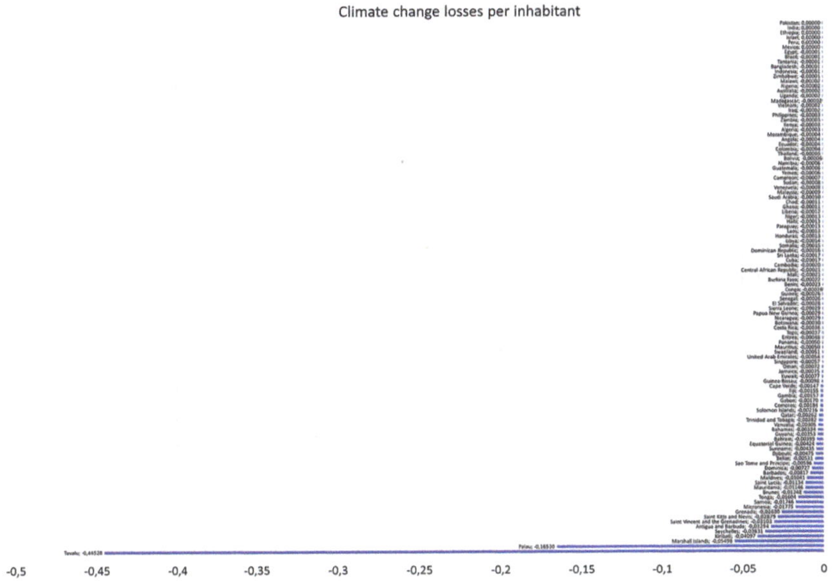

Graph 6.13 Climate change losses per capita (*Source* Author's own)

Dominica ($WL_{TTi} = -0.007$), Sao Tome and Principe ($WL_{TTi} = -0.006$) and Belize ($WL_{TTi} = -0.005$).

6.3.3 Gain and Emissions Connection

The relation between GDP growth prospects in light of climate change and percentage of GHG for ratification was investigated based on the total and percent of greenhouse gas emissions communicated by the Paris COP 21 Parties to the Convention retrieved in their national communications, GHG inventory reports as of December 2015.[5]

Over a sample of 181 countries of the world, a highly significant correlation of $r_{\text{Pearson}}(181) = 0.215$, $p < 0.004$, between the *WL* index over all models and the self-reported percentage of GHG emissions for ratification

[5] http://unfccc.int/resource/docs/2015/cop21/eng/10.pdf#page=30

was found. As a cross-validation check, the percentage of GHG emissions for ratification was significantly positively correlated ($r_{\text{Pearson}}(181) = 0.178$, $p < 0.016$) with self-reported GHG emissions per country.

This result leads to the conclusion that those countries that emit more GHG are the ones with a positive GDP prospect on the warming earth until 2100. The more time countries seem to have in a favorable climate for production, the more they are also likely to emit GHG and hence contribute to global warming. All these results outline the need for attention to climate justice.

6.4 Ethical Groundwork on the Metaphysics of the Gains of Global Warming

Uncertainty arising in assessing economic growth in relation to climate change creates an unprecedented predicament for scientists and global governance technocrats. Regarding climate justice, current generations are called upon to make sacrifices today for future generations to cut carbon emissions to avert global warming (Sachs, 2014). Climate change mitigation at the expense of lowered economic growth seems to pit the current generation against future ones. Costly climate change abatement prospects are thus hindering currently necessary action on climate change given a shrinking time window prior to reaching tipping points that make global warming irreversible (Oppenheimer, O'Neill, Webster, & Agrawal, 2011).

As a novel alternative, the results yield at funding today's climate mitigation and adaptation through an international benefits transfer backed by climate bonds and carbon tax (Marron & Morris, 2016; Sachs, 2014). Bonds are debt investment in which investors loan money to an entity, which borrows the funds for a defined period of time at a variable or fixed interest rate. Bonds are primarily used by companies, municipalities, states and sovereign governments to raise money and finance a variety of future-oriented long-term projects and activities. This solution appears as real-world relevant means to tap into the worldwide USD 80 trillion bond market in order to fund the incentives to a transition to a sustainable path (Flaherty et al., 2017; Puaschunder, 2016a; World Bank, 2015). Carbon tax will also be introduced. Sharing the costs of climate change aversion between and across generations appears as important strategy to instigate immediate climate change mitigation through incentivizing emission reduction and provide funding for climate change adaptation. Overall, this turns climate change burden sharing into a Pareto-improving option over time (Puaschunder, 2016a).

Recent IPCC research, international conferences on climate change and fund raising activities to combat global warming stress that it is advisable to pursue both mitigation as well as adaptation policies immediately (Puaschunder, 2016b). While climate justice will require both, climate change mitigation and adaptation, concurrently, no macroeconomic model exists to date that considers funding both approaches at once by allocating economic global warming benefits among nations. In addition, we lack information on the possible interdependencies, trade-offs and reciprocal influences between climate change benefits among nations and in-between generations. Another necessary extension of the model will be to phase in variegated climate risks. Therefore, the climate change mitigation and adaptation practices that are already in place—or are planned to be established—that help to buffer climate risk arising from weather extremes—such as sea level rise, flooding, severe droughts, desert formation, storms and hurricanes—are thus advised to be captured and analyzed in order to prepare for real-world relevant public policy recommendations on climate justice implementation. Concurrent knowledge of the benefits of a warming earth and the contemporary state of the world on climate instability will help retrieve information on the source of funding for climate change mitigation, alleviation and adaptation strategies and what areas of the world should become net beneficiaries of transfer aid.

While climate change benefits transfer is a novel economically and socio-psychologically superior strategy to nudge people into action and real-world relevant emergent risk prevention means (Centeno et al., 2013), we currently lack information on the impact of climate change. Investigating the concrete benefits a warming earth will hold crucial insights for modelling sustainability over time. Adding onto contemporary climate fund raising strategies ranging from emissions trading schemes (ETS) and carbon tax policies as well as financing climate justice through bonds as viable mitigation and adaptation strategies, the chapter sheds unprecedented light to the economic revenue from a warming globe which might then be used for pointed mitigation and adaptation efforts in specific crises areas and those parts of the world that will suffer from the greatest burden.

Overall, the chapter provided the first attempt to find a behavioral economics nudge into climate action based on positive incentives. Positive signals and incentives of gain prospects may help engage the many we need for climate justice. This novel, feasible and easily implementable way to steer many stakeholders toward climate action is only presented in order to help derive mitigation policies and communication strategies for a fairer climate solution. The revolutionary focus on the benefits of climate change is meant to open a gate for future research on how to distribute the gains equally among all world inhabitants in a Pareto-improving manner. All these endeavors may lead to a fairer response to global warming and hopefully provide a real-world relevant means on how to implement climate justice across the globe.

6.5 Climatorial Imperative

Society as a whole outlasts individual nation states' welfare. Pareto optimality for society differs from the individual nations' optimal conditions. As the sum of individual nations' optimality does not necessarily lead to societally favorable outcomes (Bürgenmeier, 1994; Klaassen & Opschoor, 1991), discounting future benefits and burdens for world nations on an individual level helps generate information on how to distribute unequal climate change advantages to offset global warming losses (Rawls, 1971). In general, international balance is accomplished through individual distributions around the world (Bauer, 1957). Policies curbing preferences and taxes distributing welfare between world inhabitants help alleviate the inequality arising from climate change (Chancel & Piketty, 2015).

In order to avoid a negative connotation and loss aversion distracting from the need to care for the impact of climate change around the globe, the presented results introduced the gains of a warming earth in order to find ways on how to share the expected benefits around the world or conserve the financial assets for future generations. Thereby, the results offer a novel contribution to contemporary carbon taxes and green bond solution strategies to help raise substantial revenue to subsidize countries for decarbonization (Marron & Morris, 2016; Puaschunder, 2016a).

The legal argumentation about climate stability touches on different fields of law, ranging from common goods to private property to human rights. While climate was historically understood as a common good, which is non-excludable and non-rivalrously shared and beneficial for all world members, this perception is currently challenged by climate change. If climate becomes less stable, climate may not be considered as common good anymore. For instance, if climate instability impacts certain world areas by weather extremes such as sea level rise, flooding, droughts, desert formation, storms and hurricanes more severe than other parts of the world that may even benefit from a warmer climate—for example, consider Russia's and Canada's access to natural resources when ice is melting in their Northern territories—a quest for living in a beneficial area of the world may leverage a favorable climate as a quasi-luxury. Living in a part of the world with a beneficial climate may become an exclusive privilege that is rivalrously contested. Under these circumstances, climate would not be considered as classical common good. Take also the example of rising sea levels. If climate gets warmer, sea levels drown landscape under water, and therefore the common good of climate directly impacts private property. Predictions of Venice disappearing and almost half of Manhattan being under water in 200 years[6] if we continue on a business-as-usual path underline the direct impact of the common good climate on private property rights. If temperature rises, private property will be destroyed or more expensive to maintain than before. Climate justice concerns thereby directly touch on human rights.

The greater goal of ensuring climate justice is that not one generation creates irreversible lock-ins for future generations. The claim for a human right of access to a stable climate stems from intergenerational fairness demands, which is a natural behavioral law or a human-imbued cue that has been practiced ever since (Puaschunder, 2016c). Climate justice is thereby traced in human rights and sustainable development as a quest to safeguard the rights of those affected by climate change. Connecting a stable climate to human rights has become a blatant demand in the light of changing sea levels (Ferreira, 2017). Yet to this day, climate refugees are not treated under the Geneva Convention[7] as long as their rights are not directly connected to a human right (Ferreira, 2017).

From the Arctic to the Indian Ocean to the South Pacific, small island states and coastal lines have therefore nowadays become home to the most vulnerable communities, whose rights are not fully protected by

[6] http://www.businessinsider.com/map-shows-how-much-of-nyc-could-be-underwater-in-200-years-2015-7

[7] https://www.icrc.org/en/war-and-law/treaties-customary-law/geneva-conventions

contemporary legal frameworks. The threat of rising sea levels in the wake of climate change pushes populations to relocate to safer areas. Contemporary legal frameworks do not recognize and protect the rights of these novel climate migrants as climate refugees do not fall under Geneva Convention[8] protectorate as long as access to a stable climate is not considered an eternal human right and international law. Subsequent legislation should set out to ensure a decent standard of living of those affected by global climate warming. Climate justice may thereby be leveraged as a human right in the years to follow.

As the destabilization of climate will directly destroy, damage or intensify maintenance costs of private property and a decent living, the climate stabilization efforts therefore also have entered private property and standard of living financialization considerations. In the financing of climate stabilization, private market rational has been proposed following World Bank and United Nations approaches to price natural resources. These financialization and commodification of nature efforts, however, have just started. In the commodification of climate, economic rational should be applied but with the caution of legal oversight. The basic economic rules of supply and demand suggest that an over-demand of climate elevates the price of a stable climate. Following the quest for justice between countries leads to the demand that countries that are going to win economically from a warming globe have a higher responsibility to protect and conserve a stable and decent climate. Therefore, the rising price of a stable climate should be proportionally paid based on the benefits a country has for a warming temperature. Fair climate change burden sharing between countries could therefore comprise two argumentations: First, those countries who benefit more from a warming climate, hence those with a higher ***WL*** index, should also bear a higher degree of the burden of climate change mitigation and adaptation efforts. Second, those countries who benefit more from global warming and reap benefits from a warmer earth temperature should redistribute some of the wealth accumulation due to climate change to offset the costs arising from global warming in other countries of the world that suffer from a decline in living conditions due to rising temperatures.

The legal foundation for different-country approaches toward climate mitigation and adaptation cost sharing can be found at the heart of

[8] https://www.icrc.org/en/war-and-law/treaties-customary-law/geneva-conventions

sustainability having been declared as one in which countries have 'common but differentiated responsibilities,' which was first discussed in Principle 7 of the Rio Declaration at the first Rio Earth Summit in 1992. New to this argument, however, is that the benefits of global warming—which are real and exist, for instance, in melting ice allowing unprecedented access to below surface resources and larger arable landscape—should be shared globally to offset costs and harms produced from a higher temperature in other parts of the world.

In addition, building on common and international law, those countries that have better means of protection or conservation of the common climate should also face a greater responsibility to protect the earth. The legal basis for this argument stems from an inverted subsumption of the argumentation whether climate stability is a common good or impacts on private property and draws on historical cases of legally justified expropriation. Private property rights are some of the starkest legal claims existent around the world. Private property rights hold through time, distance and space. If a neighbor goes on vacation, one cannot simply move in his or her home and claim oneself as new owner. But there is one interesting case in history, where private property rights could be neglected for the sake of common goods.

In history, the private property claim of a country was legally justified negligible if (1) those who owned a good were not alive anymore and direct ownership attribution was non-existent; (2) the former private property was turned into a common good and (3) the new owner had better means of protection and conservation for the good than before. Take historical examples such as the Stone of Rosetta but also other cases such as the 'Elgin' Parthenon Marbles as part of the Athenian Acropolis (Downs, 2008). Historically, the stone of Rosetta was discovered by Napoleon Bonaparte in 1799 in the small town of Rosetta in Egypt, and is now exhibited in the British Museum for more than 200 years.

There are several arguments for not restituting former private property to the countries of their origin as in the case of the Stone of Rosetta: (1) The *actual former owners* are not alive anymore and unknown, which naturally transposes any direct individual property claim to the international law domain, hence the country Egypt contests the country Great Britain over ownership and possession. Great Britain claims (2) to turn the former private property to a common good through granting a *large and diverse group of people access*, hence creating a larger summed up utility. London—as a metropolitan hub with a vast array of visitors—is a vibrant stage to explain the meaning of the stone to a broader public audience. The display of the stone in a museum adds additional value: (3) The British Museum has better *means of protection, preservation and conservation* of the stone than if being displayed by itself in a desert town benefiting not only from the political stability of Great Britain but also the strong culture of heritage awareness and valuation as a Royal mandate. (4) The surrounding in a museum grants the stone meaning and a *Gestalt* bringing to life the mysteries of the ancient Egypt and the stone's importance as a historical artifact in deciphering Greek, Hieroglyphs and demotic Egyptian (Downs, 2008). The stone's connection to history truly becomes apparent in the wealth of other displayed objects that grant the stone a certain collective Gestalt, which bestows upon the stone additional value. The sum of objects in a museum exhibition is worth more in its entirety than its sole pieces standing alone.

What can we learn from the historic case of expropriation for the future of climate justice? The standard ruling arguments for expropriation—in which the justification for these items to remain in former colonial powers' territories to this day lies—could be subsumed to the novel, contemporary pressing case of climate justice. In the arguments whether the first and the second world should bear the same burden of climate change mitigation and adaptation efforts, one has to acknowledge the arguments of developing nations and transitioning economies for economic growth by the same—unfortunately climate change causing—means of economic transition as the first world had in previous centuries.[9] In distinguishing now various responsibility levels, one could subsume from the above case of expropriation for the sake of common goods creation: (1) That climate change will potentially infringe on private property rights of *future owners* who are currently unknown. (2) Those countries should bear a higher burden of climate change mitigation and adaptation efforts that have more *access to a favorable and economically beneficial climate*; (3) but also those countries that have better means of *protection, preservation and conservation*, hence the first world may have to take on a greater responsibility in averting climate change. (4) There is a natural *Gestalt* over time regarding climate. The sum of a stable climate over time is more precious than the individual

[9] http://dotearth.blogs.nytimes.com/2009/12/23/views-on-china-and-copenhagen/?_r=0

country's or generation's costs incurred to maintain a favorable climate. Shedding light at these deficiencies underlines the need for considering climate justice a natural law over time that unites us around the globe and connects our common humankind's past to our future (Pope Francis, 2015; Puaschunder, 2016d).

The novel introduction of attention to climate change benefits leads to the demand of distributing the gains from a warming earth equally around the globe. In the distribution allocation decision, a climatorial imperative may lay foundation to the demand for sharing the positive externalities of climate change. With reference to Immanuel Kant's Categorical Imperative proposing to 'not impose on other what you do not wish for yourself' and suggesting to 'treat others how you wish to be treated,' the climatorial imperative should fortify the common but differentiated responsibility to ensure a stable climate and share the benefits of global warming among the world to alleviate those parts of the world that have run out of favorable climate time already (Kant, 1785/1993). Since the birth into a nation on earth is involuntary, by birth one owes to the population of the world a share of the windfall gains acquired simply by the fate of where one was born. As outlined by the correlation between GHG emissions and time to favorable climate extinction, birth in a climatorial prosperous region by fate naturally obliges to distribute some of the benefits due to the higher contribution to climate change onset and the unfair competitive advantage due to the unequally distributed time to climate extinction.

6.6 Discussion and Future Outlook

The preliminary findings of beneficial climate change are meant to pave the way to more solid investigations on the optimal policy mix to combat climate change financed through the distribution of global warming gains. The established fact of short-term benefits of climate change should thereby orchestrate the financialization of climate stability. Climate change mitigation and adaptation infrastructure should be derived from balancing the global warming gains.

In the wealth of literature on the variegated risks and losses in the wake of climate change, this chapter attempted to creatively address in particular the funding of burden sharing of climate policies by innovatively pointing out the short-term benefits of a warming earth, which turned out to be geographically-unequally distributed and related to the GHG emissions per country. Future research should build on the preliminary findings to complement the already existing knowledge, experiences and practices of issuing climate bonds and its relation to carbon tax to alleviate losses.

Future studies should introduce a time series perspective on economic growth for better specifying microeconomic behavior of economic agents in order to forecast growth and derive development policies. In addition, country-specific temperature forecasts including population density parameters could provide more accurate growth predictions in order to aid crafting the most valid development policies. Future endeavors should also directly relate contemporary growth models with legal theories on justice and public economics of distribution of benefits. More nuanced models should also take into consideration other options to promote growth including investing in infrastructure, education, research and development, and other activities that expand the productive capacity of the economy (Marron & Morris, 2016). In addition, climate risk beta could be phased into the macroeconomic model—for instance, one could extend the existing Winner-Loser index framework which measures GDP sectoral proximity to climate-optimal temperatures by incorporating systematic climate-related financial risk—analogous to how beta is used in the Capital Asset Pricing Model (CAPM).

Optimal climate change policy recommendations with respect for country-specificities could help design and calibrate an optimal financing of a mitigation and adaptation policy mix with real-world relevance for climate protection. Time-sensitive and country-specific models could elucidate the development of various world regions in the mitigation versus adaptation policy mix. Using macro- and microeconomic modeling and building on the DICE Model, the outlined benefits of mitigation and adaptation strategies are key in determining security strategies for vulnerable cities, communities and countries and protecting them from the variegated climate change risks (Nordhaus, 1994).

The results achieved help multivariate stakeholders for shaping economic growth and sustainable development. The described models will become the basis to be used for modeling how to smooth uneven distributional effects and thereby ensure climate justice. All these endeavors could help retrieve efficient climate modeling strategies leading to important contributions for the international climate negotiations.

Overall, our knowledge of climate change international interdependencies remains an open research gap important to be investigated in the eye of climate threats with enormous global impact. As we think about this topic, no single vantage point is sufficient by itself, and a genuine understanding of the problems and the possible solutions will require knowledge, expertise and experience from multiple fields. Therefore, future research must analyze climate change risks inherent in global environmental conditions and climate injustices in the larger global compound. International academic and practitioners' literature on climate change risk, climate mitigation and adaptation as well as climate justice should be reviewed. Future research endeavors should survey the current scholarship on contemporary climate policies and their funding (e.g., cap & trade, carbon tax, green energy). Here also climate change mitigation and adaptation strategies should be gathered in order to prepare the modeling and methodologies of systemic climate risk and climate analysis in the following. Insights on the connection of climate and political unrest and conflict should be added to the model since the warming temperature is captured in Burke et al. (2015) to increase the likelihood for economic risk, social and political conflict. In addition, the current measurement does not address interdependencies and cumulative effects for all climate measurements. Attention to the geographical context should be sophisticated by widening the analysis for cultural, historical, political and institutional differences of locations and local risks in regards to a warming earth.

A thorough literature analysis will form a foundation of knowledge on climate change, climate mitigation and adaptation as well as climate justice approaches on an international scale. A stakeholder-nuanced literature review should cover public and private, organizational and societal stakeholders to retrieve notions on global warming risks and climate change mitigation and adaptation in the international arena with a special focus on climate funding. The results should guide a descriptive analysis of climate change mitigation and adaptation strategies. The theoretical insights gained could lead to an interview guide to explore climate justice notions on an international basis. These activities should be oriented towards producing an interdisciplinary consensus on global climate justice and formulate guidelines for future-concerted climate justice implementation strategies.

Subsequent exploratory descriptive and qualitative data collection on climate change benefits should clarify how global warming gains are perceived, defined and reaped by various stakeholder groups in order to retrieve stakeholder-specific climate benefits with focus on climate justice implementation and gain distribution strategies. Contemporary notions of a warming climate and strategies of climate change risk mitigation and adaptation will draw a real-economy relevant climate justice picture. What factors contribute to the success of climate adaptation should be clarified in country-and-sector-specific approaches. Through the acknowledgment of favorable climate impacts and our common but differentiated responsibilities, potentially unknown international interdependencies on climate justice will be highlighted.

Future research should therefore aim at better understanding the structure, nature and challenges of these complex interaction and feedback systems of climate, climate change mitigation and adaptation choices and elaborate how it can be funded through right, just and fair global warming benefits distribution. The complexity and number of interactions will also require a qualitative analysis on the challenges of climate policy funding. In this context, it is also important to capture and map what regulatory and policy solutions exist and have been developed in response to climate crises. At the same time, also an informed understanding of legal and governmental benefits distribution strategies should become the basis for further socio-economic modeling of sustainable financing methods of climate stabilization.

In deriving information on climate justice implementation and management strategies, the question should be answered what institutions and regimes could guide benefits distribution. This information is essential in order to craft institutional climate justice management strategies and define feasible market structures to issue and policies to support climate justice on a global scale. In addition, the fiscal sustainability of climate justice distribution over time should be evaluated in order to estimate real-world relevant climate change mitigation and adaptation market strategies in the finance sector based on future global climate prospects.

The novel angle of climate gains gives rise to ethical questions concerning benefits distribution within a global society but also in-between generations. With the hope that the benefits attract attention of multiple stakeholders and incentivize action in order to avert future environmental lock-ins and irreversible global warmingtipping points at the expense of reversible overindebtedness (Kahneman & Tversky, 1979), benefits distribution is argued to be the Pareto-improving dominant strategy. Since financial assets are replaceable, an irreversible global temperature rise and climate imbalances would result in unforeseeable threats to humankind. Therefore, imposing the financial restructuring of the benefits of climate change in-between countries and partly onto future world inhabitants for the trade-off of a decent world temperature may thus be justified in the light of the complete replaceability of capital and its non-perishable nature in contrast to natural tipping points and an irreversibility of climate change (Oppenheimer et al., 2011). Redistribution of the benefits of a warming earth will therefore not only help avoiding to pit one generation against the other, but also lead to a quasi-positive multiplier effect, in which one monetary unit redistributed may lead to the aversion of worse conditions and multiplied better environmental conditions for all contemporary world inhabitants plus future generations. Future research may compile contemporary evidence on the benefits of climate change being unevenly heavy on current generations but also uncover new creative ways how to distribute and conserve the benefits for a wide variety of current and future stakeholders.

Building on contemporary modeling and empirical studies (e.g., Puaschunder, 2016a) on global warming as well as sensitivity and cost-benefit analyses of emission-constrained technological growth under uncertainty in natural emission (Rovenskaya, 2008) but also adding novel insights derived from Sachs (2014), future model extensions should also employ a DICE Model Nordhaus type IAM or RICE model (Nordhaus, 2008; Orlov et al., 2024). Solving the fair economic benefits distribution mathematically and economically will help implement sustainable development sustainability from a novel angle toward social environmental justice.

In addition, a clear methodological weakness of the current model is the neglect of structural changes in GDP composition, which could be refined in future studies to come. The overall interdependencies of economic benefits and burdens, visible in such as migration and financial foreign direct investment flows, will be outlined in the following as an example to undermine the implicit climate change competitive advantage of certain parts of the world. In addition, the current estimates are obviously limited to economic factors that should be coupled with well-being, quality of life and other socio-ecological factors in future research. Global warming risks will be phased into the model in later parts of the book. But also climate change-induced losses may be integrated and damages described more qualitatively featuring coastal area sea-level rising threats, weather conditions, heat waves, the human tragedies of losses of lives and animal extinctions as well as food insecurity in the developing world. Future research should also study climate change benefits distribution for global warming mitigation and adaptation in order to derive fairer and socially more just climate solution across various social, international and intergenerational strata. While the current climate change winners suffer a small income loss through benefits redistribution, the overall world and posterity is ensured by instigating immediate necessary action now that improves the overall well-being.

A novel theoretical model on benefits distribution and solution techniques will lead to an innovative and feasibly implementable climate change growth model that hopefully helps instigate action now for current climate change mitigation and future irreversible environmental damage reduction. In addition, the welfare effects of the enlarged model variants should be computed and compared between countries and over time. In the case of benefits transfers, the optimal size of benefits transfers and when and where it is optimal to produce in what sector should be explored in order to derive recommendations when to switch to mitigation financed through tax rates and bond issuing (Puaschunder, 2017). Thereby, the dynamics of the transition process regarding a regime switch from fossil fuels to renewable energy but also when to phase in adaptation to mitigation policies or financing repayments should be discovered.

Intergenerational climate change benefits transfers through intergenerational fiscal policies and sustainable finance methods could be delineated in order to introduce this method as novel approach to implement intergenerational climate justice. Based on preliminary work by Sachs (2014), an overlapping-generations model could be extended and converted into a continuous time model paying attention to economic growth, externalities, climate change mitigation funded through carbon tax and bond issuing. This intergenerational equity implementation turns climate change mitigation into a Pareto-improving strategy. Future research should also explore international dimensions of climate change risk mitigation and adaptation efforts and their interdependencies following the greater goal to derive strategies how to implement climate justice through benefits sharing. Comparisons of climate change risk reduction means on the international level will help retrieve insights for global governance experts on how to implement climate justice.

Climate change mitigation and adaptation study efforts should investigate how climate change is alleviated on the international level in order to outline international climate change prevention strategies. The adaptation efforts should be scrutinized on a global level in order to unravel interdependencies of climate change mitigation and adaptation on a worldwide basis. This may be done by economic market analyses featuring externality predictions and cross-market comparisons coupled with social network analyses.

As for concrete operationalizations, future research may combine theoretical and empirical research featuring qualitative and quantitative methodology. After a literature review of climate change risk, climate justice and climate change mitigation and adaptation strategies, quantitative research should target at gaining an in-depth understanding of climate change risk mitigation on the global scale, climate change stability implementation and climate adaptation in the international arena. Quantitative market analyses aim at capturing international climate change mitigation and adaptation interdependencies and how a benefits redistribution could aid to facilitate equal climate stability. The field-specific perspectives include a nomenclature creation, literature reviews, quantitative and qualitative methods, and public policy information of experts and institutions.

Developing our understanding of climate change benefits through the analysis of specific climate-related advantages in cooler areas could be approached by case studies and expert interviews. All these endeavors should lead to tangible solutions as well as recommendations of harmonious climate change mitigation and climate adaptation strategies. Developing a multidisciplinary methodological analysis of global climate advantages will help distribute climate change gains to alleviate climate burdens around the globe.

Community research will allow to present field-specific perspectives on systemic climate risk mitigation and most urgent benefits transfer beneficiary areas for climate crisis alleviation. Expert interviews will allow understanding aspect of climate change alleviation strategies that stakeholders find most relevant. Case studies on global climate risk mitigation will portray climate change abatement with attention to particular stakeholder perspectives in order to retrieve a real-world relevant climate strategy. Overall, all these endeavors will strengthen the research and design of climate, encourage interdisciplinary exchange on the contemporary complex climate agenda in strategic partnerships, as well as raise awareness and engage the broader international public on multiple climate regimes.

Expert interviews will reveal contemporary notions of climate change, climate change mitigation and adaptation efforts as well as their interdependencies. Shedding light on the common sense, but also stakeholder-specific nuances of climate change risk perceptions with a special focus on climate change mitigation solutions of the finance sector offers an invaluable opportunity to highlight unknown climate implementation strategies. Expert opinions will also help derive optimum climate change adaptation strategies featuring stakeholder-nuanced solutions. This working part should also include a meta-analysis of global warming benefits and urgent climate instability risk and its various meanings held by different constituency groups in order to provide the basis for global governance and public policy recommendations on how to mitigate and adapt to global warming.

Network analysis data should be gathered to be quantitatively analyzed by descriptive and multivariate methods in order to scrutinize the international climate gain and connected climate responsibility in order to foster climate risk mitigation and adaptation means. In order to unravel climate change gains in relation to risk mitigation and adaptation success factors, economic market data should be analyzed by descriptive and multivariate methods following the greater goal to outline prescriptive public policies to enhance climate justice and help develop recommendations on regulatory schemes. Multimodal approaches to capture climate mitigation and adaptation differences will help derive multifaceted climate justice implementation recommendations. Coupled with the study of climate change adaptation strategies by institutions, industry actors and policy makers, the results will lead to practical guidelines on how to implement environmental sustainability and climate justice for this generation and the following. The gained insight on climate mitigation and adaptation as well as the expert discussions will hopefully lead to scholarly exchange on how to distribute the gains of a global earth in order to prevent systemic risks through climate change.

A vital research exchange and scholar transfer between various stakeholder groups—featuring external quality control and results presentations—will help discuss benefits of a warming earth and craft risk definitions with colleagues prior to continuing to develop ideas and combining the lessons learned in a community on common endeavors. The information retrieved will also help create a coherent set of papers on systemic global warming gains in connection to climate change risks as well as policy briefings reflecting the different academic disciplines and viewpoints on the climate mitigation and adaptation agenda. The dialogue should be facilitated and captured for posterity by leading it over an open access interactiveonline climate change simulation as an attempt to map the contemporary climate efforts and regimes on a global scale to provide a possible application for future work. In addition, future research avenues open in the argumentation of the incentives of countries to share their benefits. Success factors in climate benefits distribution should be unraveled alongside providing direct policy implementation strategies to ensure a fair climate benefits distribution (Puaschunder, 2017).

Overall, these outlined research avenues should innovatively develop new interpretations, understandings and concepts of global warming gains and climate risks but also help deriving balanced approaches to implement climate justice and adapt to global warming. In compiling scholarship and theories on the benefits of a warming earth but also most pressing risk mitigation strategies in the climate action domain as well as by bringing together experts on climate from Europe and North America coupled with the financial sector insights on how to finance climate justice, the planned research will help create a central reference point and resources on aggregate information on the implementation and sophistication of climate justice.

The planned endeavors will elevate the importance of climate justice scholarship while deriving implications for climate stability. Emphasizing areas where to apply climate mitigation and where to promote climate adaptation strategies will help outlining practical implications for the global governance, private industry and public policy sectors. Understanding the manifold global warming gains in relation to the different climate risk attitudes but also shedding light on previously unknown climate mitigation and adaptation interdependencies on an international scale will aid environmental sustainability to ensure a future humankind. For practitioners, the prospective results will open eyes on how to profit from global warming. Public policy specialists can learn how to lower inequality and reduce potential institutional downfalls of increasingly interconnected and fragile global networks. For academia, the endeavors will spearhead interdisciplinary research on climate justice and lead to historically invaluable insights on the decision making under high uncertainty in the eye of systemic risk with short-term innovative benefit and long-term historic conservation values. On a grand level, the novel angle in the light of climate change will hopefully create a more right, just and fair climate for a sustainable society of the contemporary humankind and posterity.

References

Allais, M. (1947). *Economie et intérét*. Imprimerie Nationale.

Bauer, P. T. (1957). *Economic analysis and policy in underdeveloped countries*. Duke University Press.

Bürgenmeier, B. (1994). Environmental policy: Beyond the economic dimension. In B. Bürgenmeier (Ed.), *Economy, environment, and technology: A socio-economic approach* (pp. 166–175). Armonk Sharpe.

Burke, M., Hsiang, S. M., & Miguel, E. (2015). Climate and conflict. *Annual Review of Economics, 7*, 577–617. https://doi.org/10.1146/annurev-economics-080614-115430

Centeno, M. A., Creager, A. N., Elga, A., Felton, E., Katz, St. N., Massey, W. A., & Shapiro, J. N. (2013). *Global systemic risk: Proposal for a research community*. Working paper. Princeton Institute for International and Regional Studies, Princeton University.

Centeno, M. A., & Tham, A. (2012). *The emergence of risk in the global system*. Working paper. Princeton University.

Chancel, L., & Piketty, T. (2015). *Carbon and inequality: From Kyoto to Paris*. Paris School of Economics.

Downs, J. (2008). *Discovery of Rosetta: The stone that unlocked the mysteries of ancient Egypt*. Constable & Robinson.

Ferreira, V. (2017). Climate induced migration: Legal challenges. In J. M. Puaschunder (Ed.), *Intergenerational responsibility in the 21st century* (pp. 107–121). Vernon.

Flaherty, M., Gevorkyan, A., Radpour, S., & Semmler, W. (2017). Financing climate policies through climate bonds. *Research in International Business and Finance, 42*, 468-479.

Harrod, R. F. (1948). *Towards a dynamic economics*. Macmillan.

Kahneman, D., & Tversky, A. (1979). Prospect theory: An analysis of decision under risk. *Econometrica, 47*(2), 263–291. https://doi.org/10.2307/1914185

Kant, I. (1785/1993). *Grounding for the metaphysics of morals*. Hackett.

Klaassen, G. A. J., & Opschoor, J. B. (1991). Economics of sustainability or the sustainability of economics: Different paradigms. *Ecological Economics, 4*, 93–115. https://doi.org/10.1016/0921-8009(91)90024-9

Marron, D. B., & Morris, A. C. (2016). *How to use carbon tax revenues*. Tax Policy Center Urban Institute & Brookings Institution. https://doi.org/10.2139/ssrn.2737990

Nordhaus, W. D. (1994). *Mapping the global commons: The economics of climate change*. MIT Press.

Nordhaus, W. D. (2008). *A question of balance: Weighting the options on global warming*. Yale University Press.

Oppenheimer, M., O'Neill, B. C., Webster, M., & Agrawal, S. (2011). Climate change: The limits of consensus. *Science, 317*(5844), 1505–1506. https://doi.org/10.1126/science.1144831

Orlov, S., Rovenskaya, E., Puaschunder, J. M., & Semmler, W. (2024). Green bonds, transition to a low-carbon economy, and intertemporal welfare allocation: Evidence from an extended DICE model. AIMS Environmental Science, 11, 4, 628-648. https://doi.org/10.3934/environsci.2024031. International Institute for Applied Systems Analysis (IIASA) Working paper.

Pope Francis. (2015). *Encyclical Letter Laudato Si of the Holy Father Francis on care for our common home*. http://w2.vatican.va/content/dam/francesco/pdf/encyclicals/documents/papa-francesco_20150524_enciclica-laudato-si_en.pdf

Puaschunder, J. M. (2016a). Intergenerational climate change burden sharing: An economics of climate stability research agenda proposal. *Global Journal of Management and Business Research: Economics and Commerce, 16*(3), 31–38.

Puaschunder, J. M. (2016b). Mapping climate justice. In *Proceedings of the 2016 Young Scientists Summer Program Conference*, International Institute for Applied Systems Analysis (IIASA), Laxenburg, Austria.

Puaschunder, J. M. (2016c). On eternal equity in the fin-de-millénaire: Rethinking capitalism for intergenerational justice. *Journal of Leadership, Accountability and Ethics, 13*(2), 11–24.

Puaschunder, J. M. (2016d). The call for global responsible intergenerational leadership in the corporate world: The quest for an integration of intergenerational equity in contemporary Corporate Social Responsibility (CSR) models. In D. Jamali (Ed.), *Comparative perspectives in global corporate social responsibility* (pp. 275–288). Hershey: IGI Global Advances in Business Strategy and Competitive Advantage Book Series.

Puaschunder, J. M. (2017). Macroeconomic approach towards climate justice: The fair distribution of climate change gains and losses in the United States, Europe and the World. https://papers.ssrn.com/sol3/papers.cfm?abstract_id=2939007

Ramsey, F. P. (1928). A mathematical theory of saving. *Economic Journal, 38*(152), 543–559. https://doi.org/10.2307/2224098

Rawls, J. (1971). *A theory of justice*. Harvard University Press. https://doi.org/10.4159/9780674042605

Rovenskaya, E. (2008, January). Optimal economic growth under stochastic environmental impact: Sensitivity analysis. *International Institute for Applied Systems Analysis Interim Report*. Laxenburg, Austria.

Russell, J. A. (1957). *Industrial operations under extremes of weather* (Meteorological Monographs). American Meteorological Society.

Sachs, J. D. (2014). Climate change and intergenerational well-being. In L. Bernard & W. Semmler (Eds.), *The Oxford handbook of the macroeconomics of global warming* (pp. 248–259). Oxford University Press.

Sachs, J. D. (2021, November 17). Fixing climate finance. *Social Europe: Politics, Economy and Employment & Labor.*

Somanathan, E., Somanathan, R., Sudarshan, A., & Tewari, M. (2014). *The impact of temperature on productivity and labor supply: Evidence from Indian manufacturing*. Discussion Paper in Economics, 14–10. Indian Statistical Institute, Economics and Planning Unit.

The World Bank. (2015). Green bonds attract private sector climate finance, World Bank Brief.

The World Economic Forum Report. (2015). Davos, Switzerland: World Economic Forum.

CHAPTER 7

Climate in the Twenty-First Century

7.1 Introduction

Climate justice has been discussed in the focal point of law, economics and governance (Puaschunder, 2016c). The implementation of climate stability accounts for the most challenging contemporary global governance predicament that seems to open an abyss of world inequality regarding differing times and degrees to enjoy benefits of a warming earth around the globe. As a novel angle toward climate justice, this chapter proposes a well-balanced climate gains global governance distribution based on micro-, meso- and macroeconomic analyses results.

Overall, the following chapter investigates the nature of climate justice imbalances from an economic and a legal perspective in order to ensure economic justice solutions for advancing global climate stability. The structure of increasingly fragile environmental conditions will be captured in order to derive real-world relevant implications on how to improve the overall environmental conditions for humankind on a global scale but also over time inbetween overlapping generations. Through the understanding of climate change gains and losses, climate gain redistribution strategies will be presented. Shedding light on fair global warming gains distribution is meant to aid market economies to be brought to a path consistent with prosperity and sustainability.

A preliminary literature review revealed a missing focus on climate change gains. Holistic global systemic risk analysis centered around

J. Puaschunder, *Governance & Climate Justice*,
https://doi.org/10.1007/978-3-031-87971-5_7

fragility of the global environmental systems dominates the focus of contemporary climate justice discussions. Addressing these biases helps to gain an accurate climate change estimation. Innovatively the gains of a warming globe were recently measured in order to prepare for well-tempered climate stability policy mix recommendations financed through a right, just and fair global warming gainsdistribution strategy.

The following work combines theoretical and empirical research to review climate change gains in order to propose a right, just and fair strategy to implement climate justice around the globe. Outlining the benefits of a warming earth raises the demand for a fair asset allocation in order to offset the costs arising from climate change mitigation and adaptation strategies around the world. The chapter thereby targets at gaining an in-depth understanding of how to alleviate climate change risks and implement climate change stability in the international arena by the help of fair global warming benefits distribution. Theoretical legal arguments capturing international climate stability mandateinterdependencies are thereby innovatively coupled with quantitative global warming asset analyses in order to retrieve a fair and feasible climate stability implementation proposal. The overall goal of this work is to develop our understanding of climate stabilization through the analysis of global warming gains distribution. Legal studies as well as economic calculus enhanced will produce public policy recommendations following the greater goal of developing a multidisciplinary analysis of global climate alleviation.

The chapter is organized as follows: After a climate stability risk overview (Sect. 7.2), a qualitative legal analysis outlines climate change benefits and burden sharing strategies. Quantitative modeling of global warming gains (Sect. 7.3) focuses on finding a fair distribution of expected global warming benefits based on international climate interdependencies (Sect. 7.3.2). Global warming economic gains transfers will be proposed on a global and temporal scale in the discussion of an unprecedented tax-and-bonds-transfer strategy. In its entirety, the chapter offers a unique combination of the law and economics of global warming gains including a nomenclature creation, literature reviews and quantitative modeling in order to derive public policy directions for global governance experts and institutions pursuing a real-world relevant worldwide fair climate stability strategy.

7.2 Climate Change

Climate change accounts for one of the most pressing problems in the age of globalization as for exacerbating more complex risks than ever before. As never before in history since the birth of the earth, there is an environmental sensitivity to economic growth (Centeno & Tham, 2012; The World Economic Forum Report, 2015). While classic economics portrayed balancing the interests of different generations as ethical problem of competitive markets requiring governance for intergenerational transfers and some economists even opposed discounting of future utilities (Allais, 1947; Harrod, 1948; Puaschunder, 2015a; Ramsey, 1928), climate change has leveraged intergenerational equity as contemporary challenge of modern democracy and temporal justice an ethical obligation for posterity (Puaschunder, 2016e).

Global warming has become reality in temperature anomalies, extreme weather events, unprecedented hurricane seasons and up to 50 inches sea level rise predicted until the end of the century. History has also been made in reaching an iconic agreement on global warming mitigation at the UN Paris climate change conference and subsequent United Nations Conferences of the Parties (COP) meetings that debriefed on how to alleviate climate injustice. Literature on the topic centers around the economics of climate change, on causes of climate change, mitigation policies—such as cap and trade and carbon taxation—regulatory measures and adaptation. Substantive climate coverage through the Intergovernmental Panel on Climate Change (IPCC) research but also international conferences on climate change and global warming abatement stress the currently most urgent need for climate change mitigation and adaptation policies. The most recent research attempts address in particular innovative funding of a burden sharing of climate policies in the international compound raising challenging questions about the burden sharing in between countries and over generations.

7.2.1 Climate Justice Between Countries

The legal argumentation about climate stability touches on different fields of law, ranging from common goods to private property to human rights. While climate was historically understood as a common good, which is non-excludable and non-rivalrously shared and beneficial for all world members, this original perception is currently challenged by climate change. If climate becomes less stable, climate may not be considered as common good anymore. For instance, if climate instability impacts certain world areas by weather extremes—such as sea level rise, flooding, droughts, desert formation, storms and hurricanes—more severe than other parts of the world that may even benefit from a warmer climate (e.g., consider Russia's and Canada's access to natural resources when ice is melting in their Northern territories), a quest for living in a beneficial area of the world may leverage a favorable climate as a quasi-luxury. Living in a part of the world with a beneficial climate may become an exclusive privilege that is rivalrously contested. Under these circumstances, climate would not be considered as classical common good. But also take the example of rising sea levels. If climate gets warmer, sea levels rise and drown landscape under water, the common good of climate directly impacts on private property. Predictions of Venice disappearing and half of Manhattan being under water in 200 years[1] if we continue on a business-as-usual path underline the direct impact of the common good climate on private property rights. If temperature rises, private property will be destroyed or more expensive to maintain than before.

Climate justice concerns thereby directly touch on human rights (Puaschunder, 2016c). The greater goal of ensuring climate justice is that not one generation creates irreversible lock-ins for future generations. The claim for a human right of access to a stable climate stems from intergenerational fairness demands, which is a natural behavioral law or a human-imbued cue that has been practiced ever since (Puaschunder, 2011). Climate justice is thereby traced in human rights, and sustainable development is a quest to safeguard the rights of those affected by climate change. Connecting a stable climate to human rights has become a blatant demand in light of a changing sea level. Yet to this day, climate refugees are not treated under the Geneva Convention[2] as long as their rights are not directly connected to wartime displacement and human rights. From the Arctic to the Indian Ocean to the South Pacific, small island states and coastal lines have therefore become home to the most vulnerable communities, whose rights are not fully protected by contemporary legal frameworks. The threat of rising sea levels in the wake of

[1] http://www.businessinsider.com/map-shows-how-much-of-nyc-could-be-underwater-in-200-years-2015-7

[2] https://www.icrc.org/en/war-and-law/treaties-customary-law/geneva-conventions

climate change pushes populations to relocate to safer areas. Contemporary legal frameworks do not recognize and protect the rights of these novel climate migrants as climate refugees do not fall under the Geneva Convention[3] protectorate as long as climate justice is not considered an eternal law and human right. Climate justice may thus be leveraged as a human right in the years to follow. Subsequent legislation should set out to ensure a decent standard of living for those affected by global climate change. As the destabilization of climate will directly destroy, damage or intensify maintenance costs of private property and a decent living, the climate stabilization efforts therefore also have entered private property and standard of living financialization considerations.

In the financing of climate stabilization, private market rational has been proposed following World Bank and United Nations approaches to price natural resources. These financialization and commodification of nature efforts, however, have just started. In the commodification of climate, economic rational should be applied but with the caution of legal oversight. The basic economic rules of supply and demand suggest that an over-demand of climate elevates the price of a stable climate. Fair climate change burden sharing between countries could therefore comprise two argumentations: First, those countries that benefit more from a stable climate and have more access to climate than others and hence a greater summed up utility derived from a stable climate, should also bear a higher degree of the burden of climate change mitigation and adaptation efforts. Second, those countries that benefit more from global warming and reap benefits from a warmer earth temperature should redistribute some of the wealth accumulation due to climate change to offset the costs arising from global warming at other countries of the world that suffer from a decline in living conditions due to rising temperatures.

The legal foundation for different country approaches toward climate mitigation and adaptation cost sharing can be found at the heart of sustainability having been declared as one in which countries have 'common but differentiated responsibilities,' which was first discussed in Principle 7 of the Rio Declaration at the first Rio Earth Summit in 1992. New to this argument, however, is that the benefits of global warming—which are real and exist, for instance, in melting ice allowing unprecedented access to below surface resources and creating larger arable

[3] https://www.icrc.org/en/war-and-law/treaties-customary-law/geneva-conventions

landscape—should be shared globally to offset costs and harms produced from a higher temperature in other parts of the world. In addition, building on case and international law, those countries that have better means of protection or conservation of the common climate should also face a greater responsibility to protect the earth. The legal basis for this argument stems from an inverted subsumption of the argumentation whether climate stability is a common good or impacts on private property by drawing from historical cases of legally justified expropriation.

Private property rights are some of the starkest legal claims existent around the world. Private property rights hold through time, distance and space. If a neighbor goes on vacation, one cannot simply move in her or his home and claim oneself as new owner. But there is one interesting case in history, where private property rights could be neglected for the sake of the creation of common goods. In history, the private property claim of a country was legally justified negligible if (1) those who owned a good were not alive anymore and direct attribution of the owner non-existent; (2) the former private property was turned into a common good and (3) the new owner had better means of protection and conservation than the good had experienced before.

Take historical examples such as the Stone of Rosetta but also other cases such as the 'Elgin' Parthenon Marbles as part of the Athenian Acropolis (Downs, 2008). Historically, the stone of Rosetta was discovered by Napoleon Bonaparte in 1799 in the small town of Rosetta in Egypt, and is now exhibited in the British Museum for more than 200 years. There are several arguments for not restituting former private property to the countries of their origin as in the case of the Stone of Rosetta: (1) The actual former owners are not alive anymore and their heirs unknown, which naturally transposes any direct individual property claim to the international law domain, hence the country Egypt contests the country Great Britain over ownership and possession. Great Britain claims (2) to turn the former private property to a common good through granting a large and diverse group of people access, hence creating a larger summed up utility. London—as a vibrant metropolitan hub with a vast array of visitors—offers a vibrant stage to explain the meaning of the stone to a broader public. The display of the stone in a museum adds additional value: (3) The British Museum has better means of protection, preservation and conservation of the stone than if it were displayed by itself. In addition, the stone's exhibition in the United Kingdom benefits not only from the political stability of Great Britain but also the strong culture of heritage awareness and valuation as a Royal mandate. (4) The surrounding in a museum grants the stone meaning and a Gestalt bringing to life the mysteries of the ancient Egypt and the stone's importance as a historical artifact in deciphering Greek, Hieroglyphs and demotic Egyptian (Downs, 2008). The stone's connection to history truly becomes apparent in the wealth of other displayed objects that grant the stone a certain collective Gestalt, bestowing upon it an additional value. The sum of objects in a museum is worth more in its entirety than its sole pieces standing alone.

What can we learn from the historic case of expropriation for the future of climate justice? While raising critical questions of the right to hold onto cultural heritage goods in foreign territories, these standard argumentations for expropriation could be subsumed to the novel and contemporary case of climate justice. Already now developing nations and transitioning economies argue to have a right to economic growth by the same—unfortunately climate-change causing—means of economic transition as the first world had in previous centuries.[4]To balance between those calls, one could subsume from the above case of expropriation for the sake of common goods creation: (1) That climate change will potentially infringe on private property rights of future owners who are currently unknown. (2) Those countries should bear a higher burden of climate change mitigation and adaptation efforts that have more access to climate, hence those with larger gains from a warming earth; (3) but also those countries that have better means of protection, preservation and conservation have to take on a greater responsibility in averting climate change (Puaschunder, 2016c). (4) Overall, there is a natural Gestalt over time regarding climate. Over time, the sum of a stable climate is more precious than the individual generation's costs incurred to maintain a favorable climate. Shedding light at these deficiencies underlines the need for considering climate justice as a natural law over time that connects our common humankind's past to our future (Puaschunder, 2016d).

7.2.2 Climate Justice Between Generations

Society as a whole outlasts individual generations. Pareto optimality for society over time differs from the aggregated individual generations'

[4] http://dotearth.blogs.nytimes.com/2009/12/23/views-on-china-and-copenhagen/?_r=0

preferences. As the sum of individual generations' preferences does not necessarily lead to societally favorable outcomes over time (Bürgenmeier, 1994; Klaassen & Opschoor, 1991), discounting based on individual generations' preferences can lead to an unjust advantage of living generations determining future living conditions. In the climate domain, intertemporal questions arise whether to invest in abatement today—in order to prevent negative effects of global warming—or to delay investment until more information on climate change is gained (Rawls, 1971). In general, resources are balanced across generations by social discounting to weight the well-being of future generations relative to those alive today. Regarding climate justice, current generations are called upon to make sacrifices today for future generations to cut carbon emissions to avert global warming (Sachs, 2014). In general, intergenerational balance is therefore accomplished through individual saving decisions of the present generation (Bauer, 1957). Policies curbing preferences and taxes distributing welfare between the present and future generation, however, decrease economic growth. Climate change mitigation at the expense of lowered economic growth creates intergenerational predicaments. Costly climate change abatement prospects are currently hindering necessary action on climate change given a shrinking time window prior to reaching tipping points that make global warming irreversible (Oppenheimer, O'Neill, Webster, & Agrawal, 2011; Puaschunder, 2016a).

As an innovative angle in this debate of economic growth versus sustainability that pits the current generation against the future, a novel climate change mitigation approach with bonds funded through debt and taxation imposed on future generations is proposed. In order to avoid governmental expenditure on climate change hindering economic growth but also to instigate immediate action on climate change abatement (Barro, 1990), Sachs (2014) introduced to fund today's climate mitigation through an intertemporal fiscal policy mix backed by climate bonds and carbon tax (Marron & Morris, 2016). Bonds are debt investment in which investors loan money to an entity, which borrows the funds for a defined period of time at a variable or fixed interest rate. Bonds are primarily used by companies, municipalities, states and sovereign governments to raise money and finance a variety of future-oriented long-term projects and activities. In this debt investment strategy, investors loan money to an entity, which borrows the funds for a defined period of time at a variable or fixed interest rate (Puaschunder, 2016c). Inflation-adjusted climate bonds financing could subsidize the current world industry for transitioning to green solutions. Bonds are thereby a real-world relevant means to tap into the worldwide USD 80 trillion-bond market in order to incentivize a transition to a sustainable economic growth path (Puaschunder, 2016b; World Bank, 2015).

Carbon tax can also help financing a well-tempered climate change mitigation and adaptation mix. Carbon taxes can raise substantial revenue until the economy is largely decarbonized (Marron & Morris, 2016). In Sachs' (2014) two-period model, climate change mitigation is financed by debt to be repaid by tax revenues on labor income in the future. Leaving the current generation with unchanged disposable income allocates the burdens of climate change mitigation across generations without the need to trade off one generation's well-being for another's. While today's generation is left unharmed, the second period generation is made better off ecologically. Taxes on later generations are justified as for the assumed willingness of future generations to avoid higher costs of climate change prevention and environmental irreversible lock-ins. Shifting the ultimate costs of climate change aversion to later generations leverages climate stability into a Pareto-improving strategy for humankind.

Overall, in this tax-and-transfer mitigation policy, all generations are better off with mitigation through climate bonds as compared to the business-as-usual (BAU) non-mitigation scenario (Sachs, 2014). While future generations enjoy a favorable climate and averted environmental lock-ins, the current populace does not face economic growth drawbacks. Sharing the costs of climate change aversion between and across generations appears as powerful strategy to instigate immediate climate change mitigation through incentivizing emission reduction. Although intergenerational burden sharing on climate change appears as viable real-world relevant emergent risk prevention, we currently lack an analytical understanding of the impact of climate mitigation through bonds on economic growth, the coordinated implementation of climate change burden sharing bonds as well as the model's long-term effects.

A literature review and preliminary studies have to be undertaken on the current discussion on sustainable finance and the diverse methods of funding of mitigation and adaptation policies. Particular emphasis was given to the already existing literature, experiences and practices of issuing climate bonds and its relation to carbon tax. The issue of sustainable financing of climate policies was found to be less developed. Regarding creative financing strategies, a focus group was staged during August 2016 at the International Institute for Applied Systems

Analysis (IIASA) in Laxenburg, Austria, with thirty-nine young scientists representing diverse disciplines, well-balanced gender composition and differing nationalities from around the globe (Puaschunder, 2016c). The focus group revealed primary interests in the novel climate change bonds' financing strategy (Puaschunder, 2016c). Questions were raised about the legal and policy frameworks to enact the bonds; feasibility, efficiency and pricing of bonds; multidimensionality and incentive concerns of the general bond financing strategy idea; international differences in climate change mitigation and adaptation attempts as well as justice between countries to share the benefits and burdens of global warming equally. In addition, several sustainable development financing approaches were discussed during the 2016 Alpbach Laxenburg Group Retreat comprising of open source investment platforms, innovative public-private partnership plans as well as self-financing tools to create constant revenue streams to settle long-term expenses (Puaschunder, 2016a).

Uncertainty arising in assessing economic growth in relation to climate change creates an unprecedented predicament for scientists and global governance technocrats. International questions are posed as to which countries should be paying to invest in abatement today—in order to prevent negative effects of global warming around the globe (Rovenskaya, 2008). In general, resources are balanced across the globe by social discounting to weight the overall well-being. Regarding climate justice, current climate change beneficiaries are called upon to make sacrifices today to cut carbon emissions to avert global warming (Sachs, 2014). The implementation of climate change avoidance and the adaptation against the coming climate risk were previously described as to pit today's climate change winners and losers against each other in the trade-off of economic growth versus sustainability (Puaschunder, 2016b; Sachs, 2014).

In this framing, the problem of climate change is therefore mainly associated with risks and burden sharing costs, which may have caused a lethargy on action (Puaschunder, 2016a). Climate change burden sharing strategies have been thematized alongside mitigation and adaptation policies against climate risks (Puaschunder, 2016b). Recent IPCC research, international conferences on climate change and fund-raising activities to combat global warming stress that it is advisable to pursue both climate change mitigation as well as adaptation strategies concurrently. As climate justice will require both, climate change gains and losses should be analyzed. Yet, no macroeconomic model exists to date on the potential transfers of global warming benefits to areas that need funding to offset for climate change losses.

7.2.3 Research Questions

The empirical part features a stylized model to introduce intergenerational burden sharing financed through distributing the expected economic global warming gains around the world. The discussion will then focus on the actual implementation of climate justice through the idea of issuing climate bonds coupled with a taxation solution. Empirical analyses will help revealing the model's viability in order to derive suggestions for global governance policy makers to efficiently herald climate justice in the twenty-first century.

Shedding unprecedented light on the advantages of global warming will help to retrieve real-world relevant climate justice implementation recommendations. Contrary to negative connotations of burden sharing on climate change, outlining the benefits of global warming promises to draw attention from agnostic market actors. As a positive incentive, gains raise awareness for the issue at stake and ensure that the positive advantages of a warming earth can be distributed based on right, just and fair ethical principles. Without knowledge and quantification of the gains of climate change, climate inequality may grow unnoticed (Chancel & Piketty, 2015). Only by the sound understanding of who will gain what on a warming earth, justice can be established—as to the entire world benefitting from the gains of a warming globe in a just way. Knowledge of the concrete benefits based on contemporary finance and growth models maximizing utility over the world can thereby lead the implementation of climate justice between countries but also over time.

This measurement and description of the short-term winners and losers of a warming earth is an innovative and novel angle toward accomplishing climate justice that is introduced in order to find a viable solution to steer action through positive incentives. The following empirical part therefore elucidates climate change gains around the world in order to find right, just and fair benefit sharing strategies and mechanisms, which will be proposed in the following discussion section based on a tax-and-bonds diversification strategy.

As the very first preliminary attempt in the benefits distributions direction, the chapter provides real-world relevant means on how to implement climate justice on a global and long-term scale. Outlining the distribution of benefits of climate change is key in determining redistribution strategies for vulnerable cities, communities and countries to protect them from variegated climate change risks (Nordhaus, 1994). The prospective results of a climate change gains analysis will therefore help multivariate stakeholders achieving compensation for climate change mitigation, adaptation and a transition into renewable energy for sustainable development. The climate model helps analyze how global governance experts can distribute the gains of a warming earth around the globe. Winking with insights on the benefits of climate change appears as novel, feasible and easily-implementable solution to gain interest from the very many contemporary stakeholders involved in the causes and consequences of a warming earth. In addition, nudging overlapping generations toward future-oriented sustainability follows the greater goal to make the world a fairer place in the access to a stable climate for this generation and the following.

7.3 Method

7.3.1 Modeling Climate Change Gains and Losses Distribution

Building on subsequent chapters, economic output was measured under projected conditions of a warming earth. In order to offset the losses from climate change based on the overall WL_{TT} index, global warming benefits are redistributed in a fair way to climate change loser countries for climate change mitigation and adaptation efforts and to instigate a transition into renewable energy.

To ensure a fair benefit transfer strategy, the difference of $\Delta WL =$ 121,426,447 should be distributed based on following criteria: A fair benefit transfer strategy in between countries entails per-country contributions to alleviate the losses caused by global warming to come from those countries that win from a warming earth, hence seventy-eight nation states. The climate change economic winning countries' contribution should directly be guided by the economic gains that the country has from a warming temperature.

In order to account for country differences in offsetting global warming through GDP growth (especially on the winners' side) and the country differences in their ability to transfer into renewable energy, the overall GDP per inhabitant was factored into the transfer equation as outlined in formula (7.1):

$$R^W = \left(\frac{P_W}{WL_G}\right) \times \left(\frac{GDP_c}{I_c}\right) \tag{7.1}$$

whereby R^W denotes the total transfer of climate change wins obligation per winning country, and P_W is the country-specific GDP win divided by all wins WL_G = 3,540,396,345 from a warming earth for the year 2100 business-as-usual projection in order to derive the percentage fraction of wins per country. GDP_C equals the per country GDP estimate per inhabitant I_c per country for the year 2016.

Losses were calculated based on formula (7.2):

$$R^L = \left(\frac{P_L}{P_{LT}}\right) \times \left(\frac{GDP_c}{I_c}\right) \tag{7.2}$$

whereby R^L denotes the total transfer of climate change compensation, and P_L is the country-specific GDP loss divided by all losses WL_L = −232,613,188 from climate change for the year 2100 business-as-usual projection in order to derive the relative fraction of losses per country. GDP_C equals the per country GDP estimate per inhabitant I_c per country for the year 2016.

The overall transfers R^T equate with the sum of contributions from the countries benefiting from global warming and the losses incurred by the countries that lose from global warming until 2100. Overall, all transfers R^T derive from R^W and R^L, which must be equal based on the following formula (7.3):

$$R^T = R^W + R^L \tag{7.3}$$

whereby R^T denotes the total transfer of wins for compensation of losses.

7.3.2 Climate Change Transfers

Based on the WL_{TT} index, which measures the distance to cardinal temperature per GDP sector in 188 countries of the world, all 77 climate change transfer grantor countries and 109 climate change transfer beneficiary countries are outlined in Graph 7.1 (author's own).

Graph 7.2 (author's own) features the countries with most time ahead to the most favorable climate weighted by overall GDP and

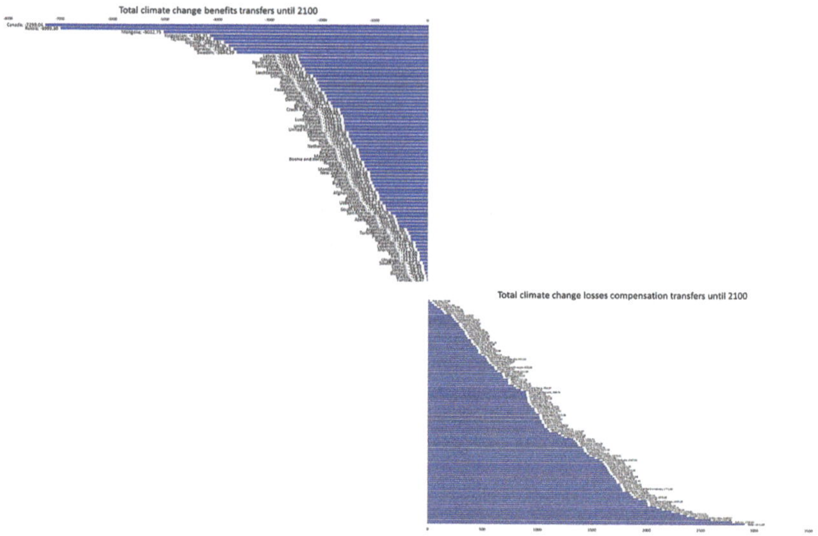

Graph 7.1 Total climate change transfers until 2100 (*Source* Author's own)

their overall transfer burden. The top 19 transfer grantor countries are Canada ($WL_{TTB} = -7299.04$), Russia ($WL_{TTB} = -6999.3$), Mongolia ($WL_{TTB} = -5032.75$), Kyrgyzstan ($WL_{TTB} = -4154.31$), Tajikistan ($WL_{TTB} = -4089.42$), Iceland $WL_{TTB} = -3887.91$), Finland ($WL_{TTB} = -3798.36$), Norway ($WL_{TTB} = -3708.43$), Sweden ($WL_{TTB} = -3644.39$), Latvia ($WL_{TTB} = -2465.58$), Georgia ($WL_{TTB} = -2452.19$), North Korea ($WL_{TTB} = -2428.53$), Switzerland ($WL_{TTB} = -2394.05$), Estonia ($WL_{TTB} = -2342.77$), Liechtenstein ($WL_{TTB} = -2318.88$), Lithuania ($WL_{TTB} = -2188.99$), Nepal ($WL_{TTB} = -2131.26$), Austria ($WL_{TTB} = -2103.02$), Belarus ($WL_{TTB} = -2077.12$) and Kazakhstan ($WL_{TTB} = -2039.07$).

All global warming transfer beneficiary countries weighted per inhabitant are outlined in Graph 7.3 (author's own). The major transfer beneficiaries are Qatar ($WL_{TT} = 2911.69$), Bahrain ($WL_{TT} = 2793.87$), Brunei ($WL_{TT} = 2591.98$), United Arab Emirates ($WL_{TT} = 2446.67$), Kiribati ($WL_{TT} = 2370.55$), Mauritania ($WL_{TT} = 2311.14$), Tuvalu ($WL_{TT} = 2228.85$), Djibouti ($WL_{TT} = 2140.24$), Senegal ($WL_{TT} = 2091.74$), Burkina Faso ($WL_{TT} = 2009.04$), Maldives ($WL_{TT} = 2008.46$), Trinidad and Tobago ($WL_{TT} = 1997.28$), Mali ($WL_{TT} =$

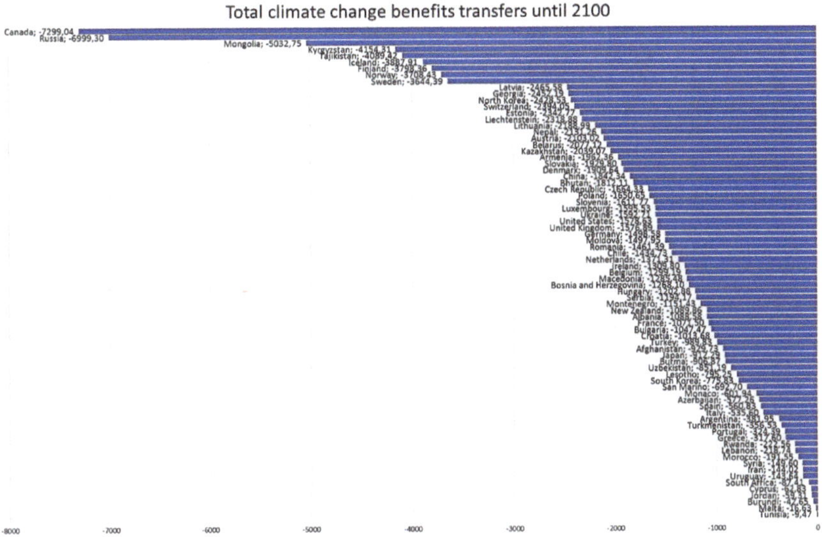

Graph 7.2 Total climate change benefits transfers until 2100 (*Source* Author's own)

1922.99), Equatorial Guinea (WL_{TT} = 1875.88), Sri Lanka (WL_{TT} = 1855.24), Palau (WL_{TT} = 1838.85), Seychelles (WL_{TT} = 1775.79), Saint Vincent and the Grenadines (WL_{TT} = 1772.60), Samoa (WL_{TT} = 1760.75) and Guinea (WL_{TT} = 1753.07).

Based on the WL_{TT} index weighted per GDP and per inhabitant, which measures the distance to cardinal temperature per GDP sector in 186 countries of the world, all 76 climate change transfer grantor countries and 110 climate change transfer beneficiary countries are outlined in Graph 7.4 (author's own).

Graph 7.5 (author's own) features the highest grantor countries weighted by GDP and per inhabitant. The top 20 transfer grantor countries are Liechtenstein (WL_{TTB} = −10.34), Canada (WL_{TTB} = −8.88), Norway (WL_{TTB} = −7.65), Iceland (WL_{TTB} = −6.35), Switzerland (WL_{TTB} = −5.67), Luxembourg (WL_{TTB} = −5.52), Finland (WL_{TTB} = −3.77), Monaco (WL_{TTB} = −3.41), Denmark (WL_{TTB} = −3.19), Austria (WL_{TTB} = −3.03), United States (WL_{TTB} = −2.87), Ireland (WL_{TTB} = −2.18), Germany (WL_{TTB} = −2.15), United Kingdom (WL_{TTB} = −1.80), Belgium (WL_{TTB} = −1.44), New Zealand (WL_{TTB}

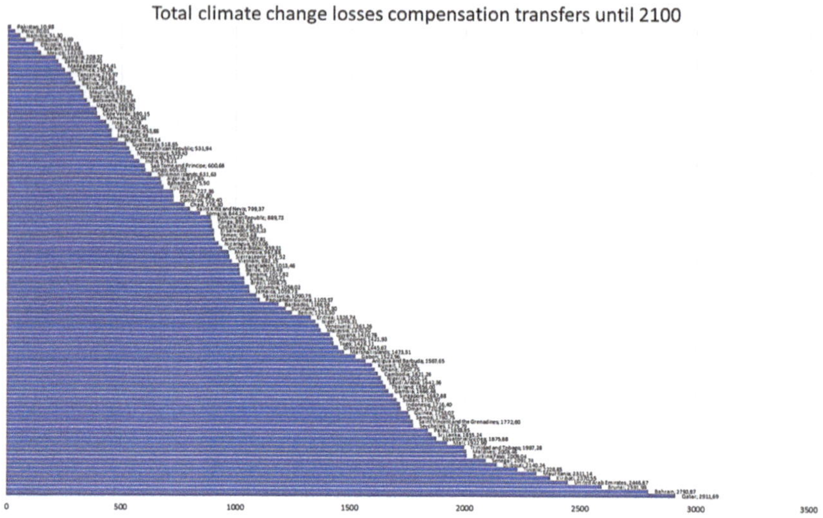

Graph 7.3 Climate change losses compensation transfers (*Source* Author's own)

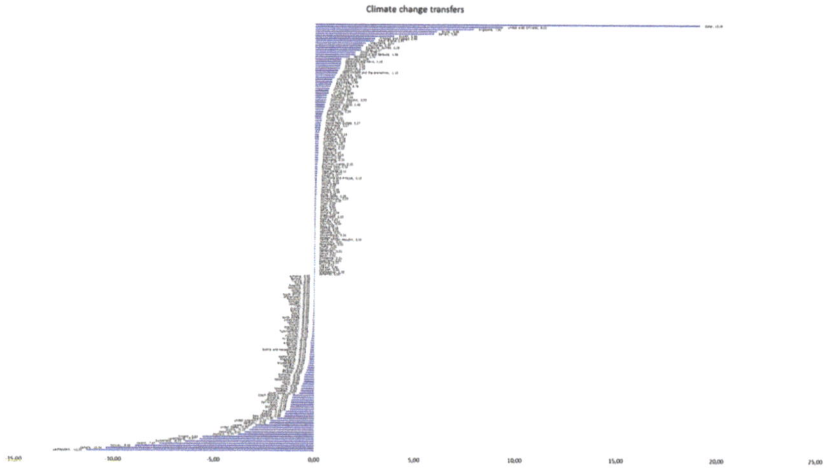

Graph 7.4 Climate change transfers weighted by GDP per inhabitant (*Source* Author's own)

$= -1.41$), Estonia ($WL_{TTB} = -1.38$), France ($WL_{TTB} = -1.18$), Latvia ($WL_{TTB} = -1.15$) and Slovenia ($WL_{TTB} = -1.15$).

All global warming transfer beneficiary countries are outlined in Graph 7.6 (author's own). The major transfer beneficiaries are Qatar ($WL_{TT} = 19.06$), United Arab Emirates ($WL_{TT} = 9.33$), Singapore ($WL_{TT} = 7.90$), Brunei ($WL_{TT} = 6.06$), Bahrain ($WL_{TT} = 5.90$), Kuwait ($WL_{TT} = 3.94$), Trinidad and Tobago ($WL_{TT} = 2.98$), Saudi Arabia ($WL_{TT} = 2.92$), Mauritania ($WL_{TT} = 2.51$), Seychelles ($WL_{TT} = 2.39$), Equatorial Guinea ($WL_{TT} = 2.29$), Palau ($WL_{TT} = 2.27$), Oman ($WL_{TT} = 2.00$), Antigua and Barbuda ($WL_{TT} = 1.99$), Barbados ($WL_{TT} = 1.70$), Venezuela ($WL_{TT} = 1.29$), Saint Kitts and Nevis ($WL_{TT} = 1.28$), Panama ($WL_{TT} = 1.27$), Grenada ($WL_{TT} = 1.26$) and Malaysia ($WL_{TT} = 1.24$).

Graph 7.7 (author's own) provides an overview of all climate change winning transfer grantors in green and yellow colors and all climate change loser country transfer beneficiaries in red colors.

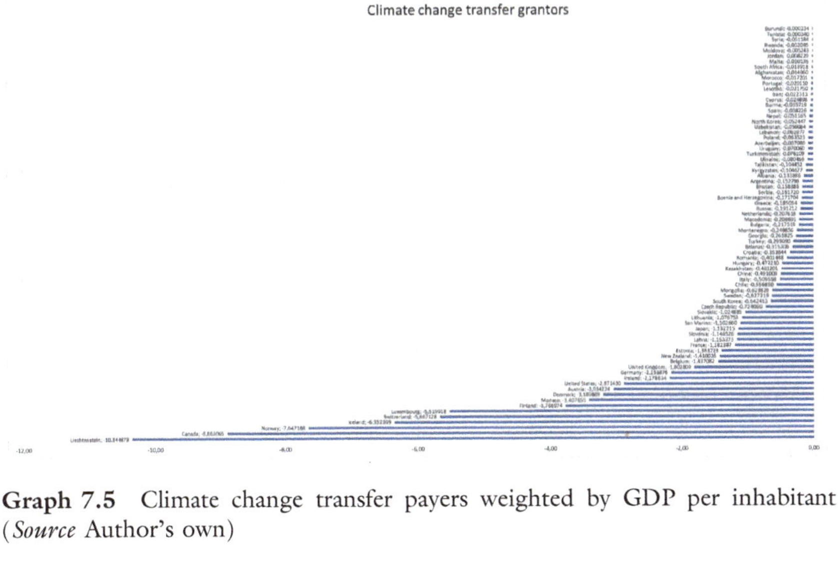

Graph 7.5 Climate change transfer payers weighted by GDP per inhabitant (*Source* Author's own)

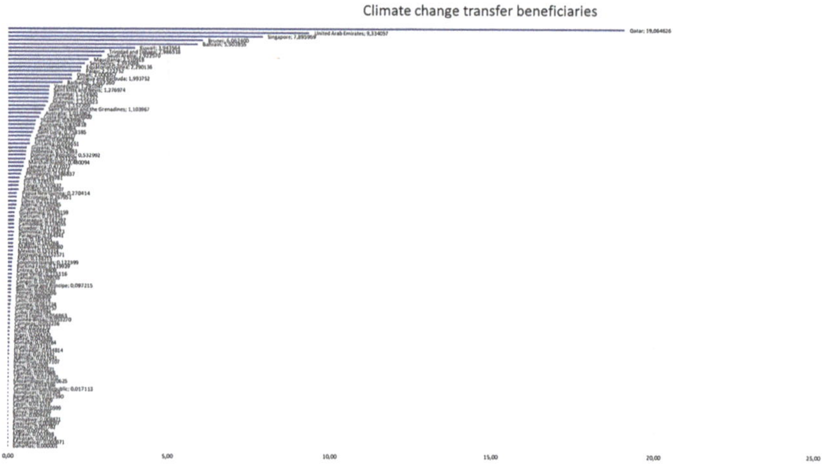

Graph 7.6 Climate change transfer beneficiaries weighted by GDP per inhabitant (*Source* Author's own)

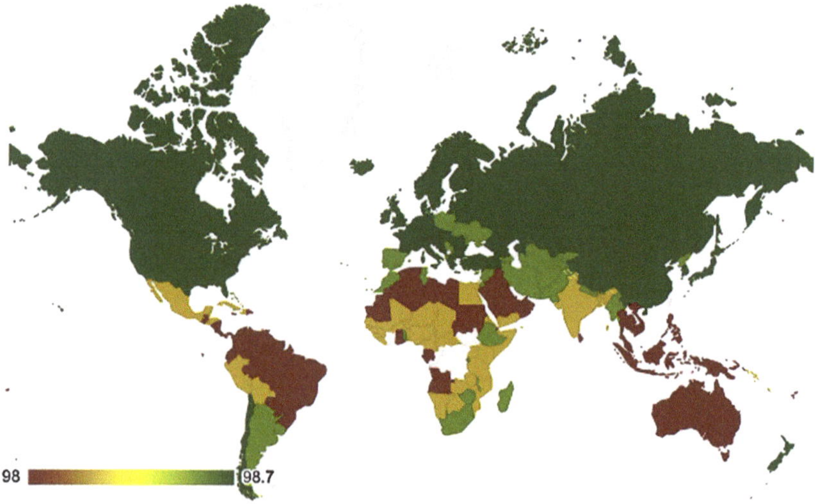

Graph 7.7 World climate change benefit transfers, green = highest benefit transfer countries, yellow = medium benefit transfer countries, light red = medium beneficiary countries, red = highest beneficiary countries (*Source* Author's own)

7.4 Discussion

The implementation of climate stability accounts for the most challenging contemporary global governance predicament that seems to pit world countries against each other but also today's generation against future world inhabitants. For enacting climate justice, current world nation states are called upon to make sacrifices today for future generations to cut carbon emissions to avert global warming (Sachs, 2014). Climate change mitigation at the expense of lowered economic growth seems to infringe on the current generation's welfare. Costly climate change abatement prospects are thus hindering currently necessary action on climate change given a shrinking time window prior to reaching tipping points that make global warming irreversible (Oppenheimer et al., 2011).

In a trade-off of economic growth versus sustainability, a broad-based international coalition could establish climate stability. As a novel angle toward climate justice, the attention to global warming gains and losses being distributed unequally around the globe allows proposing a well-balanced climate mitigation and adaptation public policy mix guided by micro- and macroeconomic analysis results. This chapter offers a new way of funding climate change mitigation and adaptation policies but also the transition to renewable energy through a broad-based climate stability bonds-and-tax-transfer-mix that also involves future generations (World Bank, 2015). Contemporary climate stability financing strategies are discussed in order to derive recommendations on how market economies can be brought to a path consistent with prosperity and sustainability. Finding innovative ways how to finance climate abatement over time coupled with future risk prevention as well as adaptation to higher temperatures appears as an innovative and easily implementable solution to nudge overlapping generations toward climate justice in the sustainability domain (Rawls, 1971).

Having shed light on the gains of a warming earth demands for the redistribution of climate change benefits to those areas of the world that will be losing from a warming earth. In the implementation, a climate change bonds but also taxation strategies are recommended. Having found that there are gains from a warming earth demands to partially transfer benefits into areas of the world that will be primarily losing from climate change. In order to avoid governmental expenditure on climate change hindering economic growth (Barro, 1990), the climate transfers should be enacted through bonds and taxes.

First Jeffrey Sachs (2014) proposed an intergenerational burden sharing idea by presenting a three-model climate change burden sharing through fiscal policy with bond issuing. This model reflects the implementation regarding contemporary finance and growth models with respect for maximizing utility. In an overlapping-generations type model, climate change abatement and mitigation policies financed through climate gains redistribution lead to a fairer solution across the globe and over generations. Thereby, the current generation mitigates climate change and provides infrastructure against climate risk financed through climate bonds to be paid by future generations. Since for future generations the currently created externalities from economic activities—the effects of CO_2 emissions—are removed, this entails that the current generation remains financially as well off as without mitigation while improving the environmental well-being of future generations. As Sachs (2014) shows, this intergenerational tax-and-transfer policy turns climate change mitigation and adaptation policy into a Pareto-improving strategy. Shifting the costs for climate abatement to the recipients of the benefits of climate stability appears as novel, feasible and easily implementable solution to nudge overlapping generations toward future-oriented loss aversion in the sustainability domain (Puaschunder, 2016b).

One of the most prominent forms to create revenues for public long-term investment causes are taxes. Taxation is codified in all major societies and a hallmark of democracy. Aimed at redistributing assets to provide public goods and ensure societal harmony, taxation improves societal welfare and fairness notions within society. Tax compliance is a universal phenomenon based on cooperation in the wish for improving the social compound. Taxpayers voluntarily decide to what extent to pay or avoid tax that limits the personal freedom. In a social dilemma, individual interests are in conflict with collective goals. From a myopic economic perspective, the optimal strategy of rational individuals would be to not cooperate and thus evade tax. Short-term single civilian tax contribution does not make a significant difference in the overall maintenance of public goods—if only a few taxpayers evade taxes, public goods will not disappear or be reduced. But if a considerable number of taxpayers do not contribute to tax over time, common goods are not guaranteed and ultimately everyone will suffer from suboptimal societal conditions (Dawes, 1980; Stroebe & Frey, 1982; Puaschunder, 2015b).

Contemporary economic research has focused on costs and risks of tax evasion (Tyler & De Cremer, 2006). Coercive means—such as audits and fines—were found to crowd out tax morale and ultimately result in greater non-compliance as people feel controlled and not being trusted (Cialdini, 1996; Feld & Frey, 2002; Frey, 1992; Hasseldine, 1998). In the last decade, researchers have started to recognize the importance of incorporating morals and social dynamics in economic theory on tax behavior (Andreoni, Erard, & Feinstein, 1998). When analyzing tax behavior, recently behavioral economics insights have drawn attention to social influences (Puaschunder, 2015b).

Behavioral economists widen the lens of incorporating sociological and socio-psychological notions of fairness into standard neoclassical theories. In this sense, social comparisons regarding tax were found as positive drivers of tax compliance to overcome the 'burden of taxes' and associations of losses (Puaschunder, 2015b). The cases of voluntary, self-chosen tax ethics and situational influences on social tax compliance norms have just recently been covered by behavioral approaches toward public administration. In general, social comparisons determine social norms that define internalized standards on how to behave. Yet internalized social norms are based on comparisons with others that may determine tax morale (Frey, 1997; Mumford, 2001; Schmölders, 1960). Social norms elicit concurring behavior when taxpayers identify with the goals of a group but also if they feel being treated in a fair manner by that group. Social fairness considerations in a tax reference group may further taxpayer compliance. Fairness is believed to decrease egoistic utility maximization leveraging trust and reciprocity as interesting social norms building factors (Kirchler, 2007). Social perceptions of fairness breeding trust and reciprocity are powerful tax ethics nudges (Puaschunder, 2015b). But psychological facets of fairness for the formation of social norms have been left out. If taxpayers believe that non-compliance is widespread and socially accepted, then they will more likely not comply as well. Non-compliance may stem from the notion of unfairness in how the tax burden is weighted heavier on some parts of society.

The respective bonds-and-tax climate stability financing strategy therefore proposes to bear the burden of climate in a right, just and fair way around the globe. In the climate change winner countries, taxation should become the main driver over financing climate stability strategies. Foremost, the industries winning from a warming climate should be taxed. The Winner-Loser index is based on the cardinal temperatures for all GDP contributing sectors. Based on the cardinal temperatures for the three GDP components agriculture, industry and service, the taxation should be enacted for those sectors having most time ahead. The underlying rational thereby is that the sectors with the most time ahead will be gaining the most and flourishing over time on a warming earth.

Regarding concrete climate taxation strategies, a carbon tax on top of the existing tax system should be used to reduce the burden of climate change and encourage economic growth through subsidies (Chancel & Piketty, 2015). Within a country, high- and low-income households should face the same burden of climate stabilization adjusted for their disposable income. First, climate justice within a country should pay tribute to the fact that low- and high-income households share the same burden proportional to their income, for instance enabled through a progressive carbon taxation. Finding the optimum balance between consumption tax adjusted for income through a progressive tax scheme will aid to unravel drivers of tax compliance in the sustainability domain. Those who caused climate change could be regulated to bear a higher cost through carbon tax in combination with retroactive billing through a corporate inheritance tax. Besides progressive taxation schemes to imbue a sense of fairness in climate change burden sharing, a corporate inheritance taxation whenever there is a board change is also a flexible means to reap past wealth accumulation, which potentially caused environmental damage. The burden of climate change mitigation and adaptation could also be allocated in a fair way within society through contemporary inheritance tax in order to reap benefits of past wealth accumulation.

If climate taxation is perceived as fair and just allocation of the climate burden, this could convince tax payers to pay one's share. A novel 'service-and-client' atmosphere could promote taxpayers as cooperative citizens who are willing to comply if they feel their share is a fair contribution to the environment. Taxpayers as cooperative citizens would then voluntarily give by following the greater goal to promote taxpayer collaboration and enhance tax morale in the environmental domain.

But also developed and underdeveloped countries as well as various overlapping generations are affected differently. International comparisons of tax behavior also reveal tax norms being related to different stages of institutional development of the government, which is an essential consideration in sharing the climate change burden in a fair manner between countries. A completely novel approach is to shed light on the benefits of a warming earth in order to derive fair climate gains distribution strategies around the world.

Introducing financing climate change mitigation through bonds to be paid back by future generations through taxation is an additional means to raise funds for offsetting the losses of global warming. As a novel way to amend individual saving preferences in favor of future generations, Sachs (2014) proposes to mitigate climate change by debt to be repaid by tax revenues on labor income in the future. In a two-period model, one generation works in period 1 and retires in period 2. Part of the disposable wage income is saved for consumption in the second period. CO_2 emission mitigation imposes immediate costs onto current generations and reduces wages. Greenhouse gas concentrations in period 2 are determined by the emissions in period 1. Wages of the young in the second period are reduced by climate change dependent on greenhouse gas levels. Disposable labor income of the young equals market wage net of taxes. Leaving the current generation with unchanged disposable income allocates the burdens of climate change mitigation across generations without the need to trade off one generation's well-being for another's. While today's young generation is left unharmed, the second period young generation is made better off ecologically.

The bonds solution should primarily be pursued in climate change loser countries, in order to offset the costs for climate change in a more intergenerationally harmonious way. Since there are no expected climate change gains in the losing countries, future generations should be serving as last resort to pay for climate stability. All generations are better off with mitigation through climate bonds as compared to the business-as-usual (BAU) non-mitigation scenario (Sachs, 2014). While future generations enjoy a favorable climate and averted environmental lock-ins, the current populace does not face drawbacks on economic growth. Governments in loser countries should receive tax transfers in the present from the winning countries. Since here borrowing equals loans or issuing of bonds to be paid back by future generations, the government must pay back debt plus interest payments by raising taxes on later generations. This strategy is justified as for the assumed willingness of future generations to avoid higher costs of climate change prevention and environmental irreversible lock-ins. Overall, this tax-and-transfer mitigation policy is thus Pareto-improving and an overall fair solution across the world and generations.

7.5 Conclusion

As a novel alternative to raise funds to instigate climate change mitigation and adaptation, the results yield at funding today's climate stabilization efforts through an international benefits transfer backed by climate bonds and carbon tax. Sharing the costs of climate change aversion between countries and across generations appears as important strategy to instigate immediate climate change mitigation through incentivizing emission reduction and provide adaptation funding opportunities.

While the proposed climate change benefits transfer strategy through bonds and taxation is a novel strategy to nudge people into action and real-world relevant emergent risk prevention means, we currently lack information on the concrete impact of financing climate change mitigation and adaptation as well as transition to renewable energy. The concrete benefits of a warming earth and their interconnectedness and interdependencies with climate change losses is yet unknown.

Overall, the chapter thereby provided the first attempt to find a solution to nudge people into necessary climate action based on positive incentives. Positive signals of gain prospects may help engage the many we need for instigating action now on climate justice. This novel, feasible and easily implementable solution to steer many stakeholders toward climate action is only presented in order to help derive mitigation policies and communication strategies for a fairer climate stability solution. All these endeavors may hopefully lead to real-world relevant means on how to implement climate justice across the globe and over time. The variegated climate risks as well as lurking tipping points and irreversible lock-ins are not understood completely, which lowers the sustainability of the presented solution over time.

In deriving information on climate justice implementation and management strategies, future research should investigate what multilateral institutions and regimes could guide benefits distribution. This information is essential in order to craft institutional climate justice management strategies and define feasible market structures to issue and policies to support climate justice on a global scale. In addition, the fiscal sustainability of climate justice distribution over time should be evaluated in order to estimate real-world relevant climate change mitigation and adaptation market strategies in the finance sector based on future global climate prospects. In this line of research, a more nuanced investigation should follow on what sector-specific gains are expected to flourish as well as what GDP sectors will likely see growth and which ones will likely see a reduction due to changing climate conditions. Further, climate instigated migration should become prospected based on the findings of global climate change winners and losers as well as prospective findings of sector and industry-specific climate change-induced growth. Further insights gained will lead to mapping climate-induced migration streams and a climate refugee asylum strategy in countries winning from a warming earth. These insights on climate-induced migration could become the backbone of acknowledging climate refugees under the Geneva Convention (Ferreira, 2017).

The preliminary finding of beneficial climate change is aimed at paving the way to more solid investigations on the optimal policy mix to combat climate change financed through the distribution of global warming benefits. The established fact of short-term benefits of climate change should thereby orchestrate the financialization of climate stability. Climate change mitigation and adaptation infrastructure should be derived from balancing the global warming gains around the globe and over time. Future research endeavors should help multiple stakeholders shape economic growth with respect for sustainable development on the basis of climate change burden sharing through global warming benefit transfers enacted through bonds and taxation in between countries and in between generations. All these endeavors target at serving the greater goal of enacting climate justice in the twenty-first century.

References

Allais, M. (1947). *Economie et intérêt*. Imprimerie Nationale.

Andreoni, J., Erard, B., & Feinstein, J. S. (1998). Tax compliance. *Journal of Economic Literature, 36*(2), 818–860.

Barro, R. (1990). Government spending in a simple model of endogenous growth. *Journal of Political Economy, 98*, 103–125. https://doi.org/10.1086/261726

Bauer, P. T. (1957). *Economic analysis and policy in underdeveloped countries*. Duke University Press.

Bürgenmeier, B. (1994). Environmental policy: Beyond the economic dimension. In B. Bürgenmeier (Ed.), *Economy, environment, and technology: A socio-economic approach* (pp. 166–175). Armonk Sharpe.

Centeno, M. A., & Tham, A. (2012). *The emergence of risk in the global system*. Working paper. Princeton University.

Chancel, L., & Piketty, T. (2015). *Carbon and inequality: From Kyoto to Paris*. Paris School of Economics.
Cialdini, R. B. (1996). Social influence and the triple tumor structure of organizational dishonesty. In D. M. Messick & A. E. Tenbrunsel (Eds.), *Codes of conduct* (pp. 44–58). Sage.
Dawes, R. M. (1980). Social dilemmas. *Annual Review of Psychology, 31*(1), 169–193. https://doi.org/10.1146/annurev.ps.31.020180.001125
Downs, J. (2008). *Discovery of Rosetta: The stone that unlocked the mysteries of ancient Egypt*. Constable & Robinson.
Feld, L. P., & Frey, B. S. (2002). Trust breeds trust: How taxpayers are treated. *Economics of Governance, 3*(2), 87–99. https://doi.org/10.1007/s101010100032
Ferreira, V. (2017). Climate induced migration: Legal challenges. In J. M. Puaschunder (Ed.), *Intergenerational responsibility in the 21st century* (pp. 107–121). Vernon.
Frey, B. S. (1992). Tertium datur: Pricing, regulation and intrinsic motivation. *Kyklos, 45*(2), 161–184. https://doi.org/10.1111/j.1467-6435.1992.tb02112.x
Frey, B. S. (1997). *Not just for the money: An economic theory of personal motivation*. Edward Elgar.
Harrod, R. F. (1948). *Towards a dynamic economics*. Macmillan.
Hasseldine, J. D. (1998). Prospect theory and tax reporting decisions: Implications for tax administrators. *International Bureau of Fiscal Documentation, 52*(11), 501–505.
Kirchler, E. M. (2007). *The economic psychology of tax behaviour*. Cambridge University Press. https://doi.org/10.1017/CBO9780511628238
Klaassen, G. A. J., & Opschoor, J. B. (1991). Economics of sustainability or the sustainability of economics: Different paradigms. *Ecological Economics, 4*, 93–115. https://doi.org/10.1016/0921-8009(91)90024-9
Marron, D. B., & Morris, A. C. (2016). *How to use carbon tax revenues*. Tax Policy Center Urban Institute & Brookings Institution. https://doi.org/10.2139/ssrn.2737990
Mumford, A. (2001). *Taxing culture*. Ashgate.
Nordhaus, W. D. (1994). *Mapping the global commons: The economics of climate change*. MIT Press.
Oppenheimer, M., O'Neill, B. C., Webster, M., & Agrawal, S. (2011). Climate change: The limits of consensus. *Science, 317*(5844), 1505–1506. https://doi.org/10.1126/science.1144831
Puaschunder, J. M. (2011). *Intergenerational equity as a natural behavioral law*. Library of Congress United States of America Copyright Office Copyright Catalogue TXu001743422/2011-03-08.

Puaschunder, J. M. (2015a). On the social representations of intergenerational equity. *Oxford Journal of Finance and Risk Perspectives, 4*(4), 78–99.
Puaschunder, J. M. (2015b). Trust and reciprocity drive social common goods contribution norms. In *Proceedings of the Cambridge Business and Economics Conference 2015*. Academy of Management Annual Meeting Proceedings, 2016.
Puaschunder, J. M. (2016a, October 10). A world on the move. *Nexus: The Research Blog of IIASA*. https://www.google.com/#q=puaschunder+2016+nexus+a+world+on+the+move
Puaschunder, J. M. (2016b). Intergenerational climate change burden sharing: An economics of climate stability research agenda proposal. *Global Journal of Management and Business Research: Economics and Commerce, 16*(3), 31–38.
Puaschunder, J. M. (2016c). Mapping climate justice. In *Proceedings of the 2016 Young Scientists Summer Program Conference*, International Institute for Applied Systems Analysis (IIASA), Laxenburg, Austria.
Puaschunder, J. M. (2016d). On eternal equity in the fin-de-millénaire: Rethinking capitalism for intergenerational justice. *Journal of Leadership, Accountability and Ethics, 13*(2), 11–24.
Puaschunder, J. M. (2016e). The call for global responsible intergenerational leadership in the corporate world: The quest for an integration of intergenerational equity in contemporary Corporate Social Responsibility (CSR) models. In D. Jamali (Ed.), *Comparative perspectives in global corporate social responsibility* (pp. 275–288). IGI Global Advances in Business Strategy and Competitive Advantage Book Series.
Ramsey, F. P. (1928). A mathematical theory of saving. *Economic Journal, 38*(152), 543–559. https://doi.org/10.2307/2224098
Rawls, J. (1971). *A theory of justice*. Harvard University Press. https://doi.org/10.4159/9780674042605
Rovenskaya, E. (2008). Optimal economic growth under stochastic environmental impact: Sensitivity analysis. *International Institute for Applied Systems Analysis Interim Report*. Laxenburg, Austria, January.
Sachs, J. D. (2014). Climate change and intergenerational well-being. In L. Bernard & W. Semmler (Eds.), *The Oxford handbook of the macroeconomics of global warming* (pp. 248–259). Oxford University Press.
Schmölders, G. (1960). *Das Irrationale in der öffentlichen Finanzwirtschaft*. Suhrkamp.
Stroebe, W., & Frey, B. S. (1982). Self-interest and collective action: The economics and psychology of public goods. *British Journal of Social Psychology, 21*(2), 121–137. https://doi.org/10.1111/j.2044-8309.1982.tb00521.x
The World Bank. (2015). Green bonds attract private sector climate finance. World Bank Brief.

The World Economic Forum Report. (2015). Davos, Switzerland: World Economic Forum.

Tyler, T. R., & De Cremer, D. (2006). How do we promote cooperation in groups, organizations, and societies? In P. A. M. van Lange (Ed.), *Bridging social psychology. Benefits of transdisciplinary approaches* (pp. 427–433). Lawrence Erlbaum.

CHAPTER 8

Financial Redistribution to Implement Climate Justice

Climate change financialization demands for urgent action of the global community and an innovative solution. The economic impact of climate change on the world will have national differences in Gross Domestic Product (GDP) prospect under climate change around the world. Redistribution of some of the expected gains enacted by a taxation-and-bonds strategy are proposed in this chapter. A fair redistribution of relative short-term economic gains under global warming is envisioned in order to offset for economic losses based on economic, ecologic, historic, and political factors. The model determining redistribution patterns throughout the world is based on the geo-impact of climate change, the financial crisis resilience capabilities as well as the global connectivity and science diplomacy leadership of a country. Empirically, nine indices determine which countries should be using a taxation strategy and what countries should be granted climate bonds premiums in order to enact a fair redistribution between countries. The indices highlight different facets of climate justice as mentioned in the following: A country's starting ground on the climate gains and losses spectrum, a country's climate flexibility in terms of temperature zones, and a country's CO_2 emissions contributions in production and consumption levels as well as a country's CO_2 emissions levels changes and historically grown the bank lending rate as well as resilient finance strategies coupled with science diplomacy leadership and economic connectivity on the international

J. Puaschunder, *Governance & Climate Justice*,
https://doi.org/10.1007/978-3-031-87971-5_8

stage. These factors will determine whether a country will be on the taxation regime for funding mutual climate stabilization or whether a country will be on the receiving end of climate bonds solutions. Geopolitically, countries economically gaining from climate change and being climate flexible as well as countries with high CO_2 emissions and not changing CO_2 emissions levels as well as consuming goods and services from other countries will be argued to have a higher responsibility to protect from climate change. As for financial resilience, those countries that have favorable bank lending rates and a history of resilience finance and crisis intervention expertise should face a higher responsibility and likely be taxed to fund climate bonds. In terms of global connectivity, countries that are leaders on science diplomacy and trade leadership advantages could be taxed to transfer funds via climate bonds for regions of the world that are losing from global warming and are not climate flexible as well as countries with low CO_2 emissions and lowering CO_2 emissions levels that are producing goods and services that are consumed in other parts of the world as well as having unfavorable bank lending rates and missing resilience finance expertise as science diplomacy and trade followers. The proposed taxation-and-bonds strategy could aid a broad-based and long-term market incentivization of a transition to a clean energy economy. The discussion features the implications of the proposed strategy, implementation details, and the economic impetus of redistribution mechanisms. The monitoring, evaluation, and accountability control of the outlined plan are thematized in the estimation of the feasibility and limitations. Future research and outlooks on the redistribution via taxation-and-bonds strategy are provided.

8.1 Introduction

8.1.1 *Climate Stabilization Financialization Model*

The climate change problem exacerbated in the most recent decades. As outlined by the world surface temperature anomalies since the start of world surface temperature recordings in 1891, temperatures have a constant rising trend throughout the world (Tokyo Climate Center, 2022) as outlined in Fig. 8.1.

The New York Times most recently discussed the disparate impact of climate policies and climate protection attention disparities (Flavelle, 2021a, 2021b; Puaschunder, 2023). Today's economically advanced

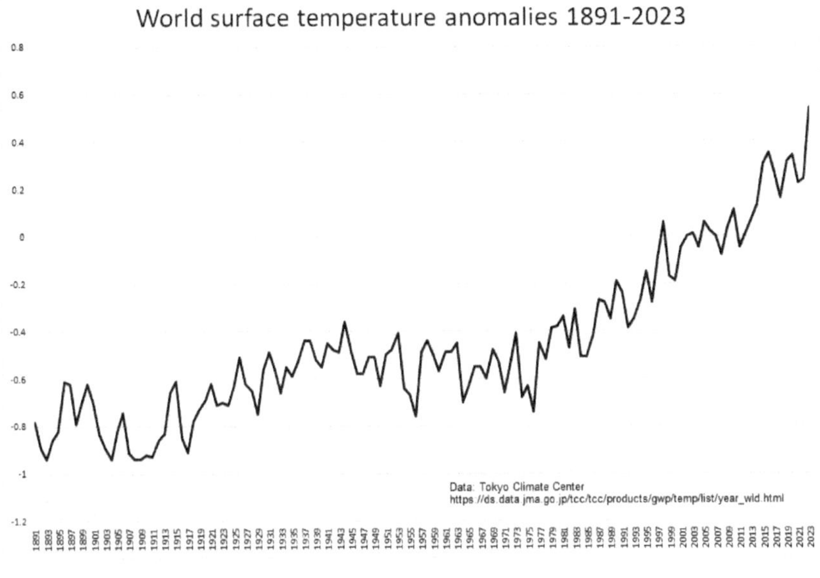

Fig. 8.1 World surface temperature anomalies 1891–2023

countries have gained economic growth and welfare improvements by the use of fossil fuel energy and intensive CO_2 emissions. The developing countries and transitioning economies appear now more burdened with climate disasters and have been assigned transfer payments to help sustainable growth. Geographically determined economic prospects in light of climate change reveal vast inequalities in the distribution of future climate-induced economic gain or loss prospects (Puaschunder, 2023). Ethical imperatives lead to the conclusion for redistribution of some of the expected gains of global warming into territories that are losing out from climate change the most and the fastest.

Political historical facts may cause hesitancy for redistribution, as was shown during the Copenhagen Intergovernmental Panel on Climate Change Conferences of the Parties (COP). In the aftermath of the 2021 United Nations Conference of the Parties (COP26) meeting on Climate Change, it has been argued that the advanced countries have an obligation and responsibility to help low-income countries' transition to sustainable growth through direct transfers and credit guarantees (Puaschunder,

2023; Sachs, 2021). The 2022 COP27 fortified the call for climate redistribution of economic gains and losses inbetween countries and over time (Moneer, 2022; Puaschunder, 2023).

In order to provide redistribution mechanism independent from political denominations, economic growth perspectives in light of climate change based on national temperature conditions and factors of productivity are taken into account (Hansen, 2014). Economic growth prospects under warming globe are being distributed unequally between countries, which can become the basis of climate justice redistribution schemes. In the aftermath of the COP26 annual climate meeting of the United Nations, Jeffrey Sachs (2021) put forward an idea of funds for climate change mitigation and adaptation that should be raised by taxation for transfer payments. Determining which countries should be grantors and which recipients can be based on macroeconomically informed criteria, such as expected economic gains and losses under climate change. Economic facts can quantify how much what countries could contribute to the collective burden-sharing strategy in the climate gains redistribution scheme. The following part will provide different indices that highlight quantified transfer strategies throughout the globe in order to distribute the gains and losses of climate change in a fair way.

Climate change heralded a call for a fair climate stabilization solution and burden sharing *within society*, *between countries*, and *over time* (Puaschunder, 2023). Innovative financialization strategies unleashing a sustainable economy in harmony with equity pledges for a healthy population have become the core of COVID-19 rescue and recovery aid as captured in the U.S. Green New Deal Program as well as the European Green Deal (Puaschunder, 2022, 2023). Green bonds are at the forefront to raise funds internationally and over time for a green transition as well as to fund climate change mitigation and adaptation (Orlov et al., 2024). Within the country level, environmental pricing—foremost enabled via ecotaxation—curbs harmful emissions and sets incentives to reduce energy or transition to a renewable solution (Braga et al., 2020; Puaschunder, 2023).

Based on the measured the Gross Domestic Product (GDP) prospect differences under climate change worldwide that outlines climate inequalities, the climate change winners and losers index represents relative economic climate change windfall gain reaper and victim countries, based on the economic prospects under climate change around the world and over time. Stressing climate-induced economic inequalities around the world, the model assumes that there are relative economic climate change reapers that have a windfall gain from a warming globe while other relative economic climate change victims face immediate disadvantages due to global warming (Puaschunder, 2023).

Climate justice addresses inequalities inherent in global warming with a mandate to fairness regarding climate benefits and burden sharing. To alleviate inequalities in climate change impacts between countries, ethical imperatives of Immanuel Kant's categorical imperative (1783/1993) and John Rawls' veil of ignorance (1971) but also economic calculus as put forward in Kaldor-Hicks' compensation criteria guide redistribution schemes (Kaldor, 1961; Law & Smullen, 2008; Puaschunder, 2023).

A novel policy recommendation for enacting climate justice entails a taxation funding strategy coupled with a climate bonds repayment based on the following influence factors: (1) The country's initial position on the relative economic climate change winners and losers spectrum as well as (2) a country's climate flexibility as a broad economic degree of freedom and future trade asset and (3) the country's human-made contribution to climate change as measured by CO_2 emissions but also the (4) ability and willingness of a country to lower its CO_2 emissions in relation to others and (5) the estimated bank lending rate of bonds in that country as well as (6) the consumption-based, trade-adjusted CO_2 emissions of a country (Puaschunder, 2023). A country's responsibility to act on climate change is also determined by (1) its past financial crisis intervention expertise and (2) its capability to administer future-oriented resilience finance, as well as (3) the country's science diplomacy as the ability to cooperate with other nations on scientific grounds and (4) its global connectivity in international trade and human capital flows (Puaschunder, 2023). These ten drivers are the basis of nine indices that target at determining either the burden on carbon taxes (abatement costs) to finance climate bonds to be repaid later or the transfer payments receivable by climate victim countries.

As a global financialization method employing taxation and bonds, countries could raise funds via taxation or become bond premium beneficiaries. Being a taxation country or a bonds transfer payment recipient be determined by a country's initial position on the climate change economic gains and losses spectrum, regular and consumption-based, trade-adjusted CO_2 emissions per country, climate flexibility as the range of temperatures within a national territory as future comparative trade advantage to other nations in the world, the willingness of countries to change CO_2

emissions, and the historically determined banking lending regimes of a country (Puaschunder, 2023).

The idea of diversified tax and bonds is also extendable to sector-specific bond yield interest rate regimes (Puaschunder, 2023). Within a country, the bonds could offer industry-specific diversified interest rate maturity bond yields based on the environmental sustainability of an industry, e.g., as measured by the European Sustainable Finance Taxonomy (Puaschunder, 2023). The bonds could also feature a long-time financialization via debt that can be repaid by future generations for inheriting a stable climate (Puaschunder, 2023). All these Law & Economics-informed recommendations aim at sharing the burden but also the benefits of climate change within society in an economically efficient, legally equitable, and practically feasible way now around the world and also between generations (Puaschunder, 2023).

A country's starting ground on the relative economic climate gains and losses spectrum, a country's climate flexibility, and a country's CO_2 emissions contributions to production and consumption levels, as well as a country's CO_2 emissions levels changes and the bank lending rate, will determine whether a country will be on the taxation regime for funding mutual climate stabilization or whether a country will be on the receiving end of climate bonds solutions (Puaschunder, 2023). Pegging the country's situation to the initial starting levels on rational grounds and relations to each other allows for turning the general price-cutting competitive race-to-the-bottom into a race-to-the-top for the mutual climate fund allocations (Puaschunder, 2023). Integrating the CO_2 emissions levels changes over time imbues incentives to change harmful behavior to this common climate fund solution (Puaschunder, 2023).

The more a country lacks financial crisis intervention expertise and the less likely a country is able to administer future-oriented resilience finance, as well as the less likely the country engages in science diplomacy and is globally connected in trade and human capital flows, the more the country should follow a climate resilience finance solution that features climate taxation-and-bonds redistribution schemes around the world (Puaschunder, 2023).

The more a country has historic financial crisis intervention expertise and is able to administer future-oriented resilience finance, the more a country should face responsibility to protect from climate change on a global level (Puaschunder, 2023). The more a country engages in science diplomacy and is globally connected in trade and human capital flows, the more the country should lead on finding a common climate resilience finance solution that features climate taxation-and-bonds redistribution schemes around the world (Puaschunder, 2023).

8.2 Redistribution Factors

8.2.1 *Theoretical Framework*

Redistribution of expected economic gains from climate change to territories that will lose from a warming globe enacts climate justice between countries and over time. The following model introduces fair redistribution of short-term economic gains under global warming in order to offset for economic losses based on economic, ecologic, historic, and political factors. Climate justice between countries and over time is proposed by nine indices to implement a climate taxation-and-bonds strategy:

(1) *Climate winners and losers CO_2 emission Index*: Climate justice between countries could be based on the idea that climate change winning countries that contribute to human-made global warming in CO_2 emissions should pay for the establishment and maintenance of climate bonds via carbon taxation, while climate change losing territories with low CO_2 emissions should be climate bond issuers with a higher interest rate premium (Puaschunder, 2023).

(2) *Climate winners and losers' climate flexibility and CO_2 emission Index*: Climate change winning countries that also feature relative climate flexibility in terms of temperature ranges on their territory and that contribute to human-made global warming in CO_2 emissions should pay for the establishment and maintenance of climate bonds via carbon taxation; while climate change losing territories with low CO_2 emissions and a narrow range of temperatures on their soil and thus low climate flexibility should be recipients of climate bonds with relative high interest rate premium and thus be relative beneficiaries in the common climate taxation-and-bonds transfer scheme (Puaschunder, 2023).

(3) *Climate winners and losers CO_2 emission change Index*: Climate justice over time could be fortified by climate change winning countries that contribute to human-made global warming in CO_2 emissions and have a rising trend of CO_2 emissions compared to

other countries should pay for the establishment and maintenance of climate bonds via carbon taxation; while climate change losing territories with low CO_2 emissions that have declining CO_2 emissions compared to other countries should be recipients of climate bonds with higher interest rates and thus be climate bond premium beneficiaries (Puaschunder, 2023). Creating incentives for countries to compete over CO_2 emissions reductions naturally leads toward a transition to renewable energy (Puaschunder, 2023). This solution is also fair in light of the positive correlation between being a climate change winner and CO_2 emissions, which indicates that climate winners have the propensity to have created the problem of global warming.

(4) *Climate winners and losers CO_2 emission Financial Crisis Intervention Index*: Climate justice between countries could be based on the idea that climate change winning countries that contribute to human-made global warming in CO_2 emissions and have a history of financial crisis interventions should pay for the establishment and maintenance of climate bonds via carbon taxation; while climate change losing territories with low CO_2 emissions that also have no crisis intervention expertise should be climate bond recipient beneficiaries being granted a relatively higher bond interest rate premium funded by taxation of CO_2 emitting industries in climate winning countries with well-established financial crisis intervention means (Puaschunder, 2023).

(5) *Climate winners and losers CO_2 emission Resilience Finance Index*: Climate justice between countries could be based on the idea that climate change winning countries that contribute to human-made global warming in CO_2 emissions and have expertise in resilience finance strategies should pay for the establishment and maintenance of climate bonds via carbon taxation; while climate change losing territories with low CO_2 emissions that lack resilience finance infrastructures should be climate bond recipient beneficiaries being granted a relatively higher bond interest rate premium funded by taxation of CO_2 emitting industries in climate winning countries with well-established resilience finance means (Puaschunder, 2023).

(6) *Climate winners and losers CO_2 emission bond lending rate Index*: Climate justice between countries could be based on the idea that climate change winning countries that contribute to human-made

global warming in CO_2 emissions and have a low bank lending rate should pay for the establishment and maintenance of climate bonds via carbon taxation; while climate change losing territories with low CO_2 emissions that also have a high bank lending rate should be climate bond recipient beneficiaries being granted a relatively higher bond interest rate premium funded by taxation of CO_2 emitting industries in climate winning countries with low bank lending rates (Puaschunder, 2023).

(7) *Climate winners and losers consumption-based, trade-adjusted CO_2 emission Index*: Climate justice between countries could be based on the idea that climate change winning countries that contribute to human-made global warming in CO_2 emissions goods and services consumption should pay for the establishment and maintenance of climate bonds via carbon taxation; while climate change losing territories with low CO_2 emissions goods and services consumption should be climate bond issuers with a higher interest rate premium (Puaschunder, 2023).

(8) *Science Diplomacy Climate Responsibility Index*: The implementation of climate justice between countries could be based on the idea that climate change winning countries that contribute to human-made global warming in CO_2 emissions goods and services consumption and have an established science diplomacy expertise with networks around the world should have a heightened responsibility to protect from climate change and pay for the establishment and lead in implementing the maintenance of climate bonds via carbon taxation; while climate change losing territories with low CO_2 emissions goods and services consumption that are not so well versed in science diplomacy should follow the plan to accept climate bond issuers with a higher interest rate premium (Puaschunder, 2023).

(9) *Global Connectivity Climate Responsibility Index*: The implementation of climate justice between countries could be based on the idea that climate change winning countries that contribute to human-made global warming in CO_2 emissions goods and services consumption and have an established global connectivity through trade, finance, and human capital transfer with networks around the world should have a heightened responsibility to protect from climate change and pay for the establishment and lead the maintenance of climate bonds via carbon taxation; while climate change

losing territories with low CO_2 emissions goods and services consumption that are not so well versed in global networking should follow the plan to accept climate bond issuers with a higher interest rate premium (Puaschunder, 2023).

All mentioned influence factors in the determination of which countries have a responsibility to act on climate change and best starting positions to implement a planned climate change taxation-and-bonds strategy to redistribute funds to offset for climate change losses and transition to a renewable resilient finance economy are outlined in Table 8.1 (Puaschunder, 2023).

8.2.2 Operationalization

The operationalization of calculating indices to determine if a country falls on either climate taxation-bonds funders or climate bonds recipients in light of climate change depends on CO_2 emissions and consumption-based, trade-adjusted CO_2 emissions and carbon pollution changes as well as historic bank lending rates and resilient financing strategies coupled with science diplomacy leadership credentials. Empirically, nine indices were macroeconomically modeled based on the following data sources: Table 8.2 holds a summary of nine different indices' components' underlying data (Puaschunder, 2023).

In the following indices, the ten drivers are first translated into a scale featuring an actual numerical position on the scale for every country of the world included in the data (Puaschunder, 2023). The ten drivers are then integrated in the nine different indices by multiplying the respective country position on the created scales (Puaschunder, 2023).

8.2.3 Results

8.2.3.1 Climate Winners and Losers CO_2 Emission Index

Climate justice between countries could be based on the idea that countries economically relatively winning in the short run and countries that contribute to human-made global warming in CO_2 emissions should pay for the establishment and maintenance of climate bonds via carbon taxation, while territories that lose from climate change the fastest and the most with low CO_2 emissions should be climate bond issuers with a higher interest rate premium (Puaschunder, 2023). The $WL_{TT}CCC$

Table 8.1 Influence variables on climate taxation-and-bonds strategy (Puaschunder, 2023)

Climate taxation-bonds strategy		
Index description	*Low values*	*High values*
Geo-impact of climate change		
(1) Relative economic climate change winner and loser countries and CO_2 emission Index (integrating normalized values of F_1 and F_3)	Climate change economic loser countries with low CO_2 emissions → climate bonds strategy with high discount rates regime and longer-term maturity to spread the burden of climate stabilization financing between generations and a relatively high bond yield-at-issue with relatively high interest rates from subsidies funded by high index value countries	Relative climate change economic short-term winner countries with high CO_2 emissions → taxation strategy coupled with low discount rates regime and short-term maturity to act now to protect future generations from climate change impacts. The higher the index, the more the country should be obliged to fund the climate bond via taxation. The lower the relatively high values of the index, the more likely the country should have access to bonds with growing yield-at-issue instead of taxation
(2) Relative economic climate change winner and loser countries' climate flexibility and CO_2 emission Index (integrating normalized values of F_1, F_2, and F_3)	Climate change economic loser countries with low climate flexibility and low CO_2 emissions → climate bonds strategy with high discount rates regime and longer-term maturity to spread the burden of climate stabilization financing between generations and a relatively high bond yield-at-issue with relatively high interest rates from subsidies funded by high index value countries	Relative climate change economic short-term winner countries with high climate flexibility and high CO_2 emissions → taxation strategy coupled with low discount rates regime and short-term maturity to act now to protect future generations from climate change impacts. The higher the index, the more the country should be obliged to fund the climate bond via taxation. The lower the relatively high values of the index, the more likely the country should have access to bonds with growing yield-at-issue instead of taxation

(continued)

Table 8.1 (continued)

Climate taxation-bonds strategy		
Index description	*Low values*	*High values*
(3) Relative economic climate change winner and loser countries and CO_2 emission change Index (integrating of F_1, F_3, and F_4 by multiplication)	Climate change economic loser countries with low CO_2 emissions and lowering CO_2 emissions → climate bonds strategy with high discount rates regime and longer-term maturity to spread the burden of climate stabilization financing between generations and a relatively high bond yield-at-issue with relatively high interest rates from subsidies funded of high index value countries	Relative climate change economic short-term winner countries with high CO_2 emissions and increasing CO_2 emissions → taxation strategy coupled with low discount rates regime and short-term maturity to act now to protect future generations from climate change impacts. The lower the relatively high values of the index, the more likely the country should have access to bonds with growing yield-at-issue instead of taxation. The yield-to-maturity should depend on the country's CO_2 emission changing behavior insofar as a reduction over time improves the maturity over time
Financial crisis resilience		
(4) Relative economic climate change winner and loser countries and CO_2 emission and Financial Crisis Intervention Index (integrating normalized values of F_1, F_3, and F_7 by multiplication)	Climate change economic loser countries with low CO_2 emissions and low crisis intervention capability → climate bonds strategy with high discount rates regime and longer-term maturity to spread the burden of climate stabilization financing between generations and a relatively high bond yield-at-issue with relatively high interest rates from subsidies funded of high index value countries	Relative climate change economic short-term winner countries with high CO_2 emissions and high crisis intervention capability → taxation strategy coupled with low discount rates regime and short-term maturity to act now to protect future generations from climate change impacts. The lower the relatively high values of the index, the more likely the country should have access to bonds with growing yield-at-issue instead of taxation

(continued)

Table 8.1 (continued)

Climate taxation-bonds strategy

Index description	*Low values*	*High values*
(5) Relative economic climate change winner and loser countries and CO_2 emission and Resilience Finance Index (integrating normalized values of F_1, F_3, and F_8 by multiplication)	Climate change economic loser countries with low CO_2 emissions and low resilience finance → climate bonds strategy with high discount rates regime and longer-term maturity to spread the burden of climate stabilization financing between generations and a relatively high bond yield-at-issue with relatively high interest rates from subsidies funded of high index value countries	Relative climate change economic short-term winner countries with high CO_2 emissions and high resilience finance capability → taxation strategy coupled with low discount rates regime and short-term maturity to act now to protect future generations from climate change impacts. The lower the relatively high values of the index, the more likely the country should have access to bonds with growing yield-at-issue instead of taxation
(6) Relative economic climate change winner and loser countries and CO_2 emission and bank lending rate Index (integrating normalized values of F_1, F_3, and F_5 by multiplication)	Climate change economic loser countries with low CO_2 emissions and high bond lending rate → climate bonds strategy with high discount rates regime and longer-term maturity to spread the burden of climate stabilization financing between generations and a relatively high bond yield-at-issue with relatively high interest rates from subsidies funded of high index value countries	Relative climate change economic short-term winner countries with high CO_2 emissions and low bank lending rate → taxation strategy coupled with low discount rates regime and short-term maturity to act now to protect future generations from climate change impacts. The lower the relatively high values of the index, the more likely the country should have access to bonds with growing yield-at-issue instead of taxation

(continued)

Table 8.1 (continued)

Climate taxation-bonds strategy		
Index description	*Low values*	*High values*
Global connectivity & leadership		
(7) Relative economic climate change winner and loser countries and CO_2 emission and consumption-based, trade-adjusted CO_2 emission Index (integrating normalized values of F_1 and F_6 by multiplication)	Climate change economic loser countries with low consumption-based, trade-adjusted CO_2 emissions and high bond lending rate → climate bonds strategy with high discount rates regime and longer-term maturity to spread the burden of climate stabilization financing between generations and a relatively high bond yield-at-issue with relatively high interest rates from subsidies funded of high index value countries	Relative climate change economic short-term winner countries with high consumption-based, trade-adjusted CO_2 emissions → taxation strategy coupled with low discount rates regime and short-term maturity to act now to protect future generations from climate change impacts. The lower the relatively high values of the index, the more likely the country should have access to bonds with growing yield-at-issue instead of taxation
(8) Relative economic climate change winner and loser countries and CO_2 emission and Science Diplomacy Index (integrating normalized values of F_1, F_3, and F_9 by multiplication)	Climate change economic loser countries with low CO_2 emissions and low science diplomacy → climate bonds strategy with high discount rates regime and longer-term maturity to spread the burden of climate stabilization financing between generations and a relatively high bond yield-at-issue with relatively high interest rates from subsidies funded of high index value countries	Relative climate change economic short-term winner countries with high CO_2 emissions and high science diplomacy → taxation strategy coupled with low discount rates regime and short-term maturity to act now to protect future generations from climate change impacts. The lower the relatively high values of the index, the more likely the country should have access to bonds with growing yield-at-issue instead of taxation

(continued)

Table 8.1 (continued)

Climate taxation-bonds strategy		
Index description	*Low values*	*High values*
(9) Relative economic climate change winner and loser countries and CO_2 emission and Global Economic Connectivity (integrating normalized values of F_1, F_3, and F_{10} by multiplication)	Climate change economic loser countries with low CO_2 emissions and low science diplomacy → climate bonds strategy with high discount rates regime and longer-term maturity to spread the burden of climate stabilization financing between generations and a relatively high bond yield-at-issue with relatively high interest rates from subsidies funded of high index value countries	Relative climate change economic short-term winner countries with high CO_2 emissions and high science diplomacy → taxation strategy coupled with low discount rates regime and short-term maturity to act now to protect future generations from climate change impacts. The lower the relatively high values of the index, the more likely the country should have access to bonds with growing yield-at-issue instead of taxation

index therefore integrated whether a country is a climate change winner or loser and the relative CO_2 emissions per country in relation to other countries (Puaschunder, 2023).

Figure 8.2 displays a worldmap that colors those countries in red that are advised to fund climate bonds with a taxation strategy (Puaschunder, 2023). Countries colored in green are climate change losers and have less CO_2 emissions per country (Puaschunder, 2023). The greener the country is colored, the higher bonds premium are advised to be paid out in this country as for being a higher climate loser and having less CO_2 emissions, hence contributing less to the climate problem (Puaschunder, 2023). Countries that range in the middle of the index are displayed in yellow in the display spectrum from countries ranked on taxation strategy in red to the highest bonds premium recipient country colored in green (Puaschunder, 2023).

Figure 8.3 ranks 185 countries based on their relative $WL_{TT}CCC$ index status (Puaschunder, 2023).

Table 8.2 Operationalization data sources

Climate taxation-bonds strategy			
Index description	*Input factor*	*Input factor*	*Input factor*
(1) Climate winners and losers CO_2 emission Index $WL_{TT}CCC$ (integrating normalized values of F_1 and F_3 by multiplication) for 185 countries of the world	Factor 1 (F_1) called $WL_{TT}CCC$ for 185 countries of the world, standing for being a climate change winner or loser as measured by a newly created normalized $PRWL_{TT}$ Index based on the original WL_{TT} index as retrieved from Puaschunder (2020b) including GDP growth prospects in light of climate change, GDP sector composition per country, and mean temperature per country as well as optimal temperature for production by GDP sector		Factor 3 (F_3) human-made self-reported country's Climate Change Causing (CCC) CO_2 emissions data for 185 countries of the world as retrieved from *Our World in Data*[1] and presented by the Global Carbon Project

(continued)

[1] https://github.com/owid/co2-data.

Table 8.2 (continued)

Climate taxation-bonds strategy			
Index description	*Input factor*	*Input factor*	*Input factor*
(2) Climate winners and losers' climate flexibility and CO_2 emission Index $CFCCC$ (integrating normalized values of F_1, F_2, and F_3 by multiplication) for 84 countries of the world	Factor 1 (F_1) called $WL_{TT}CCC$ for 185 countries of the world, standing for being a climate change winner or loser as measured by a newly created normalized $PRWL_{TT}$ Index based on the original WL_{TT} index as retrieved from Puaschunder (2020b) including GDP growth prospects in light of climate change, GDP sector composition per country, and mean temperature per country as well as optimal temperature for production by GDP sector	Factor 2 (F_2) climate flexibility index (Wx) paying attention to the GDP prospects of countries around the globe in a changing temperature based on climate flexibility as measured by the individual country's latitude and altitude naturally determining extreme temperatures as retrieved from the *List of Countries by Extreme Temperatures*[2]	Factor 3 (F_3) human-made self-reported country's Climate Change Causing (CCC) CO_2 emissions data for 185 countries of the world as retrieved from *Our World in Data*[3] and presented by the Global Carbon Project

(continued)

[2] https://en.wikipedia.org/wiki/List_of_countries_by_extreme_temperatures.

[3] https://github.com/owid/co2-data.

Table 8.2 (continued)

Climate taxation-bonds strategy			
Index description	*Input factor*	*Input factor*	*Input factor*
(3) Climate winners and losers CO_2 emission change Index $WL_{TT}CCCG$ (integrating of F_1, F_3, and F_4 by multiplication) for 185 countries of the world	Factor 1 (F_1) called $WL_{TT}CCC$ for 185 countries of the world, standing for being a climate change winner or loser as measured by a newly created normalized $PRWL_{TT}$ Index based on the original WL_{TT} index as retrieved from Puaschunder (2020b) including GDP growth prospects in light of climate change, GDP sector composition per country, and mean temperature per country as well as optimal temperature for production by GDP sector	Factor 3 (F_3) human-made self-reported country's Climate Change Causing (CCC) CO_2 emissions data for 185 countries of the world as retrieved from *Our World in Data*[4] and presented by the Global Carbon Project	Factor 4 (F_4) as the percentage of CO_2 emissions CCC as indicated for the year 2019 data retrieved from *Our World in Data*[5] and presented by the Global Carbon Project

(continued)

[4] https://github.com/owid/co2-data.

[5] https://github.com/owid/co2-data.

Table 8.2 (continued)

Climate taxation-bonds strategy			
Index description	*Input factor*	*Input factor*	*Input factor*
(4) Climate winners and losers CO_2 emission Financial Crisis Intervention Index *CCFCI* (integrating normalized values of F_1, F_3, and F_7 by multiplication) for 112 countries of the world	Factor 1 (F_1) called $WL_{TT}CCC$ for 185 countries of the world, standing for being a climate change winner or loser as measured by a newly created normalized $PRWL_{TT}$ Index based on the original WL_{TT} index as retrieved from Puaschunder (2020b) including GDP growth prospects in light of climate change, GDP sector composition per country, and mean temperature per country as well as optimal temperature for production by GDP sector	Factor 3 (F_3) human-made self-reported country's Climate Change Causing (*CCC*) CO_2 emissions data for 101 countries of the world as retrieved from *Our World in Data*[6] and presented by the Global Carbon Project	Factor 7 (F_7) called Financial Crisis Intervention Index *FCI*, included historical cross-sectional country data on crisis intervention measures of guarantees (including account guarantee, blanket guarantee, other liability guarantee, asset guarantee), lending (including ad hoc liquidity assistance, broad-based liquidity assistance, market liquidity assistance), capital injection (including ad hoc capital injection, broad-based capital injection), resolutions (including restructuring or resolution, stakeholder bail-in), rules (including suspension or bank holiday, debt or payment moratorium, credit rules, other rules), asset management (including ad hoc asset management, broad-based asset management), and other financial crisis intervention tools (including major communication, stress test, other) retrieved from the *Metrick-Schmelzing Banking-Crisis Intervention Database* for the years 1960–2019[7]

(continued)

[6] https://github.com/owid/co2-data.

[7] https://som.yale.edu/centers/program-on-financial-stability/metrick-schmelzing-paper-and-database.

Table 8.2 (continued)

Climate taxation-bonds strategy			
Index description	*Input factor*	*Input factor*	*Input factor*
(5) Climate winners and losers CO_2 emission Resilience Finance Index *CCRFI* (integrating normalized values of F_1, F_3, and F_8 by multiplication) for 130 countries of the world	Factor 1 (F_1) called $WL_{TT}CCC$ for 185 countries of the world, standing for being a climate change winner or loser as measured by a newly created normalized $PRWL_{TT}$ Index based on the original WL_{TT} index as retrieved from Puaschunder (2020b) including GDP growth prospects in light of climate change, GDP sector composition per country, and mean temperature per country as well as optimal temperature for production by GDP sector	Factor 3 (F_3) human-made self-reported country's Climate Change Causing (*CCC*) CO_2 emissions data for 101 countries of the world as retrieved from *Our World in Data*[8] and presented by the Global Carbon Project	Factor 8 (F_8) called future-oriented Resilience Finance *RF*, included historical cross-sectional country data on future-oriented resilience finance measures of guarantees (including account guarantee, blanket guarantee, other liability guarantee, asset guarantee), lending (including ad hoc liquidity assistance, broad-based liquidity assistance, market liquidity assistance), capital injection (including ad hoc capital injection, broad-based capital injection), retrieved from the *Metrick-Schmelzing Banking-Crisis Intervention Database* for the years 1960–2019[9]

(continued)

[8] https://github.com/owid/co2-data.

[9] https://som.yale.edu/centers/program-on-financial-stability/metrick-schmelzing-paper-and-database.

Table 8.2 (continued)

Climate taxation-bonds strategy			
Index description	*Input factor*	*Input factor*	*Input factor*
(6) Climate winners and losers CO_2 emission bank lending rate Index $WL_{TT}CCCL$ (integrating normalized values of F_1, F_3, and F_5 by multiplication) for 101 countries of the world	Factor 1 (F_1) called $WL_{TT}CCC$ for 185 countries of the world, standing for being a climate change winner or loser as measured by a newly created normalized $PRWL_{TT}$ Index based on the original WL_{TT} index as retrieved from Puaschunder (2020b) including GDP growth prospects in light of climate change, GDP sector composition per country, and mean temperature per country as well as optimal temperature for production by GDP sector	Factor 3 (F_3) human-made self-reported country's Climate Change Causing (CCC) CO_2 emissions data for 101 countries of the world as retrieved from *Our World in Data*[10] and presented by the Global Carbon Project	Factor 5 (F_5) called NL, the inverse normalized Lending rate L_n in percent retrieved from the World Bank recorded International Monetary Fund *Lending interest rate in percent* for the year 2019[11]

(continued)

[10] https://github.com/owid/co2-data.

[11] https://data.worldbank.org/indicator/FR.INR.LEND.

Table 8.2 (continued)

Climate taxation-bonds strategy			
Index description	*Input factor*	*Input factor*	*Input factor*
(7) Climate winners and losers consumption-based, trade-adjusted CO_2 emission Index $WL_{TT}CBTAE$ (integrating normalized values of F_1 and F_6 by multiplication) for 116 countries of the world	Factor 1 (F_1) called $WL_{TT}CCC$ for 185 countries of the world, standing for being a climate change winner or loser as measured by a newly created normalized $PRWL_{TT}$ Index based on the original WL_{TT} index as retrieved from Puaschunder (2020b) including GDP growth prospects in light of climate change, GDP sector composition per country, and mean temperature per country as well as optimal temperature for production by GDP sector	Factor 6 (F_6) country consumption-based, trade-adjusted CO_2 emissions data for 116 countries of the world as retrieved from Our World in Data[12]	

(continued)

12 https://ourworldindata.org/co2-emissions#consumption-based-trade-adjusted-emissions.

Table 8.2 (continued)

Climate taxation-bonds strategy			
Index description	*Input factor*	*Input factor*	*Input factor*
(8) Climate winners and losers CO_2 Science Diplomacy Index $CCSDI$ (integrating normalized values of F_1, F_3, and F_9 by multiplication) for 48 countries of the world	Factor 1 (F_1) called $WL_{TT}CCC$ for 185 countries of the world, standing for being a climate change winner or loser as measured by a newly created normalized $PRWL_{TT}$ Index based on the original WL_{TT} index as retrieved from Puaschunder (2020b) including GDP growth prospects in light of climate change, GDP sector composition per country, and mean temperature per country as well as optimal temperature for production by GDP sector	Factor 3 (F_3) human-made self-reported country's Climate Change Causing (CCC) CO_2 emissions data for 101 countries of the world as retrieved from *Our World in Data*[13] and presented by the Global Carbon Project	Factor 9 (F_9) called Science Diplomacy Index SDI, , integrating the Academia quota per country as retrieved from the World Bank, Educational Attainment, at least Bachelor's or equivalent, population 25 + , total (%) cumulative,[14] World Ranking of academic institutions based on the Ranking Web of Universities weighted by relevance,[15] and the diplomatic relations in embassies, consulates, and other representations in total based on the *Lowy Global Diplomacy Index* for the year 2021[16]

(continued)

[13] https://github.com/owid/co2-data.

[14] https://data.worldbank.org/indicator/SE.TER.CUAT.BA.ZS.

[15] https://www.webometrics.info/en/distribution_by_country.

[16] https://globaldiplomacyindex.lowyinstitute.org/.

Table 8.2 (continued)

Climate taxation-bonds strategy			
Index description	*Input factor*	*Input factor*	*Input factor*
(9) Climate winners and losers CO_2 emission Global Connectivity Index *CCGC* (integrating normalized values of F_1, F_3, and F_{10} by multiplication) for 158 countries of the world	Factor 1 (F_1) called $WL_{TT}CCC$ for 185 countries of the world, standing for being a climate change winner or loser as measured by a newly created normalized $PRWL_{TT}$ Index based on the original WL_{TT} index as retrieved from Puaschunder (2020b) including GDP growth prospects in light of climate change, GDP sector composition per country, and mean temperature per country as well as optimal temperature for production by GDP sector	Factor 3 (F_3) human-made self-reported country's Climate Change Causing (*CCC*) CO_2 emissions data for 101 countries of the world as retrieved from *Our World in Data*[17] and presented by the Global Carbon Project	Factor 10 (F_{10}) called Global Connectivity Index *GCI*, integrates Trade Freedom in The Global Economy Ranking Trade Freedom Index of 2022,[18] Remittances in Personal Remittances paid (current US$) as derived from the World Bank[19] normalized by Gross Domestic Product as derived from PopulationU,[20] Personal Remittances received (% of GDP),[21] Foreign Direct Investments as derived from the World Bank, Foreign direct investment, net inflows (% of GDP),[22] World Bank, Foreign direct investment, net outflows (% of GDP),[23] as well as migration derived from Our world in data, Migration % of population for the year 2020[24]

17 https://github.com/owid/co2-data.

18 https://www.theglobaleconomy.com/rankings/herit_trade_freedom/#:~:text=Trade%20freedom%20index%20(0%2D100,countries%20where%20data%20are%20available.

19 https://data.worldbank.org/indicator/BM.TRF.PWKR.CD.DT.

20 https://www.populationu.com/gen/countries-by-gdp.

21 https://data.worldbank.org/indicator/BX.TRF.PWKR.DT.GD.ZS?name_desc=false.

22 https://data.worldbank.org/indicator/BX.KLT.DINV.WD.GD.ZS.

23 https://data.worldbank.org/indicator/BM.KLT.DINV.WD.GD.ZS?end=2020&start=1970.

24 https://ourworldindata.org/explorers/migration.

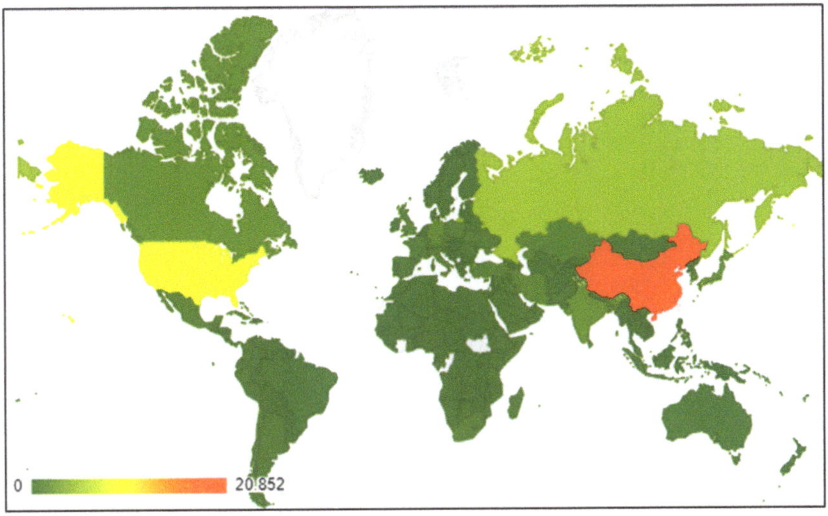

Fig. 8.2 Worldmap of $WL_{TT}CCC$ index per country for 185 countries

8.2.3.2 *Climate Winners and Losers' Climate Flexibility and CO_2 Emission Index*

Countries that relatively win economically from climate change in the short run and that benefit from climate flexibility in terms of temperature ranges on their territory but also contribute to human-made global warming in CO_2 emissions should pay for the establishment and maintenance of climate bonds via carbon taxation; while climate change losing territories with low CO_2 emissions and a narrow range of temperatures and thus low climate flexibility should be recipients of climate bonds with relative high interest rate premium and thus be relative beneficiaries in the common climate taxation-and-bonds transfer scheme (Puaschunder, 2023). The $CFCCC$ index therefore integrated whether a country is a climate change winner or loser and if the country has relatively high or low climate flexibility as well as the relative CO_2 emissions per country in relation to other countries (Puaschunder, 2023).

Figure 8.4 displays a worldmap that colors those countries with advised high taxation strategy in red, high bonds premium recipients in green, and countries that are in the middle of the index range in yellow (Puaschunder, 2023). In Graph 6.6 countries are greener if being climate

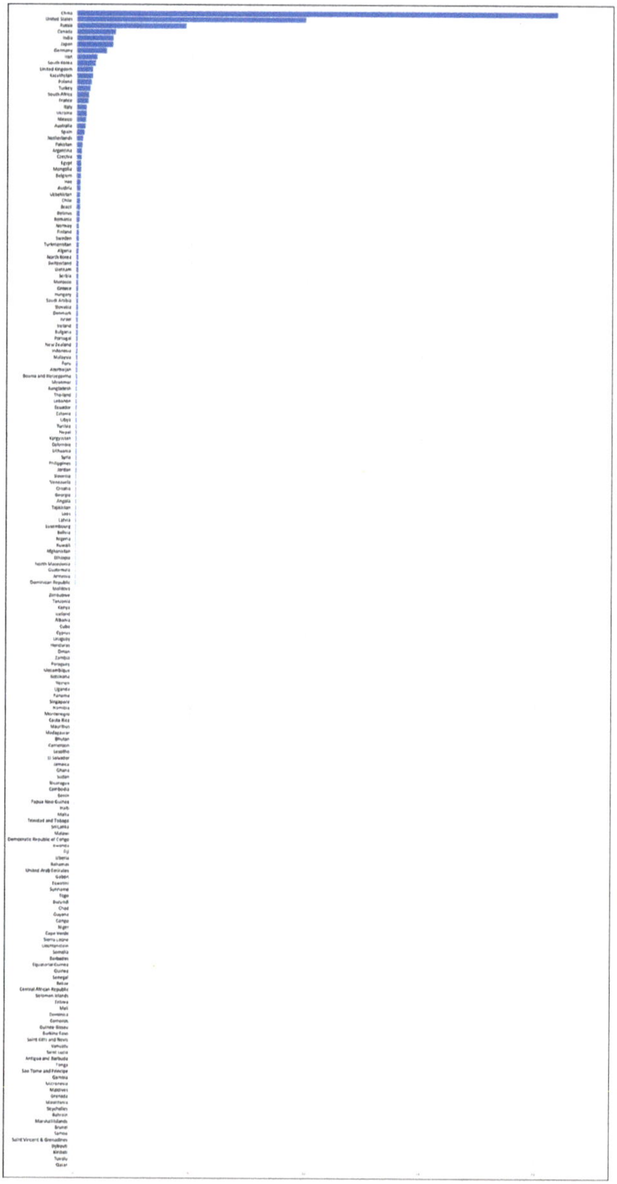

Fig. 8.3 Country ranking of $WL_{TT}CCC$ index for 185 countries

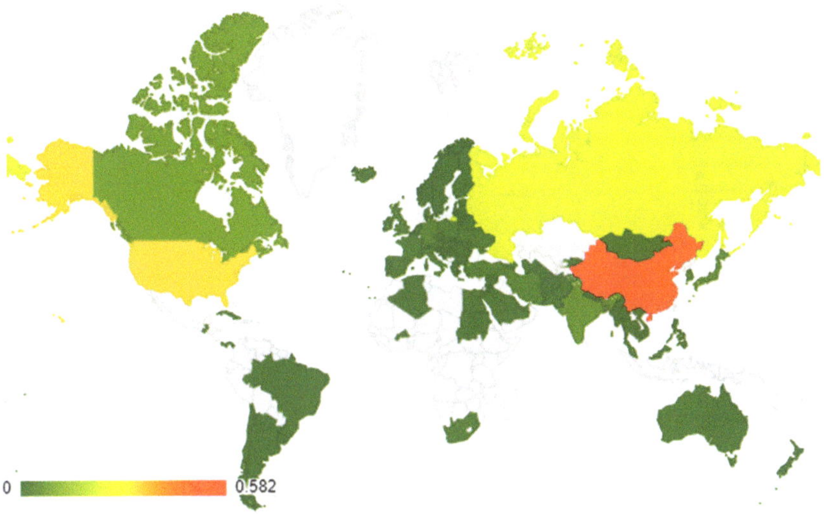

Fig. 8.4 Worldmap of $CFCCC_x$ index per country for 84 countries

change losers and having less climate flexibility as well as having less CO_2 emissions per country (Puaschunder, 2023). The redder a country is, the more likely the country is a climate change winner and benefits from more climate flexibility as well as is more likely to feature more CO_2 emissions (Puaschunder, 2023). Medium levels of climate change economic changes with medium climate flexibility and moderate levels of CO_2 emissions are colored in yellow (Puaschunder, 2023). Red countries are advised to having a climate taxation strategy, yellow countries a mixed climate taxation-and-bonds strategy, and green countries a climate bonds recipient standing among all countries of the world (Puaschunder, 2023).

Figure 8.5 ranks 84 countries based on their relative $CFCCC_x$ index status from the highest climate taxation country to the highest bonds premium recipient country (Puaschunder, 2023).

8.2.3.3 Climate Winners and Losers CO_2 Emission Change Index

Climate justice over time could be fortified by climate change winning countries that contribute to human-made global warming in CO_2 emissions and have a rising trend of CO_2 emissions compared to other countries paying for the establishment and maintenance of climate bonds

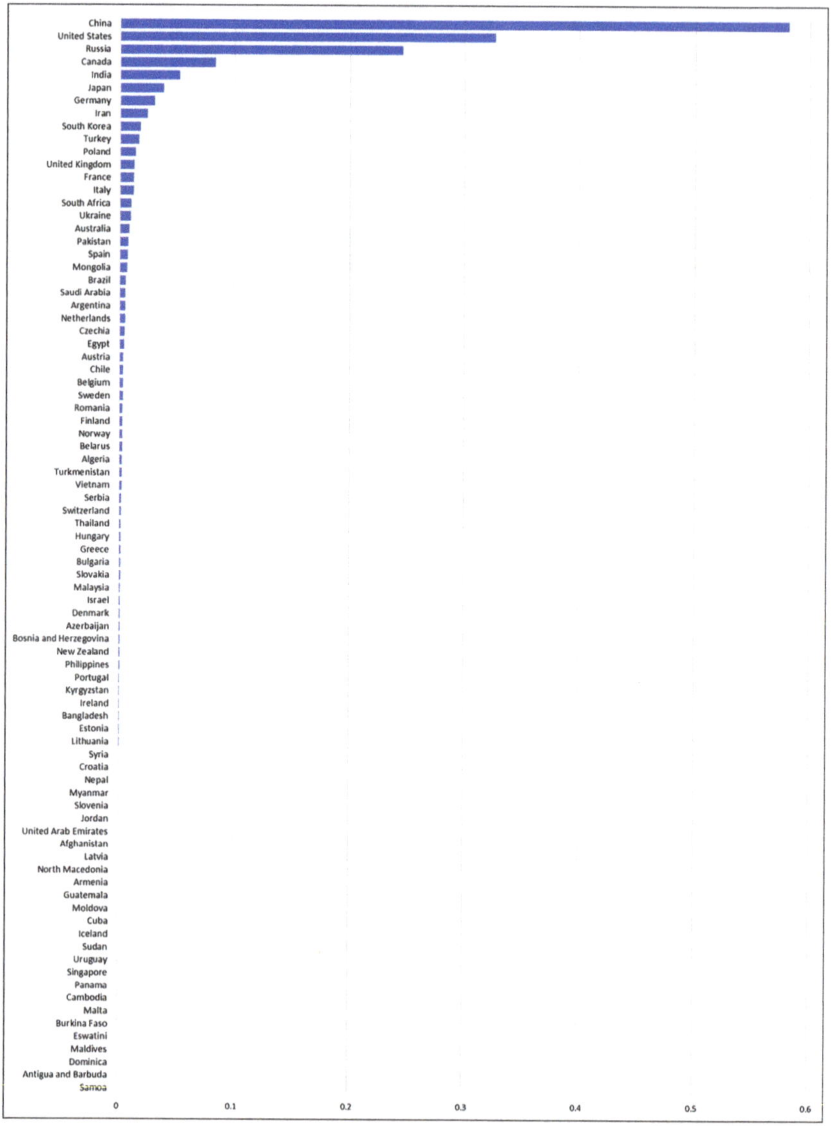

Fig. 8.5 Country ranking of *CFCCC* index for 84 countries

via carbon taxation; while climate change losing territories with low CO_2 emissions that have declining CO_2 emissions compared to other countries should be recipients of climate bonds with higher interest rate and thus be climate bond premium beneficiaries (Puaschunder, 2023). This would create market incentives for countries to compete over CO_2 emissions reductions and naturally lead toward a transition to renewable energy (Puaschunder, 2023). The $WL_{TT}CCCG$ index therefore integrated whether a country is a climate change winner or loser as well as the relative CO_2 emissions per country in relation to other countries and the relative CO_2 emissions changes per country compared to the rest of the world (Puaschunder, 2023). This index incentivizes positive changes in terms of CO_2 emissions reductions over time (Puaschunder, 2023).

Figure 8.6 displays a worldmap that colors those countries with recommended high taxation strategy in red, high bonds interest premium recipients in green, and countries that are in the middle of the index range in yellow (Puaschunder, 2023). In Fig. 8.6 the greener a country is colored, the more likely the country is a climate change loser as well as has less CO_2 emissions and decreasing CO_2 emissions in relation to other countries (Puaschunder, 2023). The redder a country is, the more likely the country is a climate change winner as well as is more likely to feature more CO_2 emissions and rising CO_2 emission levels (Puaschunder, 2023). Medium levels of climate change economic changes and moderate levels of CO_2 emissions that are not changing are colored in yellow (Puaschunder, 2023). Red countries are advised to having a climate taxation strategy, yellow countries a mixed climate taxation-and-bonds strategy, and green countries a climate bonds recipient standing among all countries of the world (Puaschunder, 2023).

Figure 8.7 ranks 66 countries based on their relative $WL_{TT}CCCG$ index status (Puaschunder, 2023).

Figure 8.7 and Fig. 8.8 are graphical displays of 120 countries in total ranked from the highest climate taxation country to the lowest based on the $WL_{TT}CCCG$ index accounting for climate change loser as well as having less CO_2 emissions per country and decreasing CO_2 emissions (Puaschunder, 2023).

8.2.3.4 Climate Winners and Losers CO_2 Emission Financial Crisis Intervention Index

The Climate winners and losers CO_2 emission Financial Crisis Intervention Index $CCFCI$ included the past performance of countries during

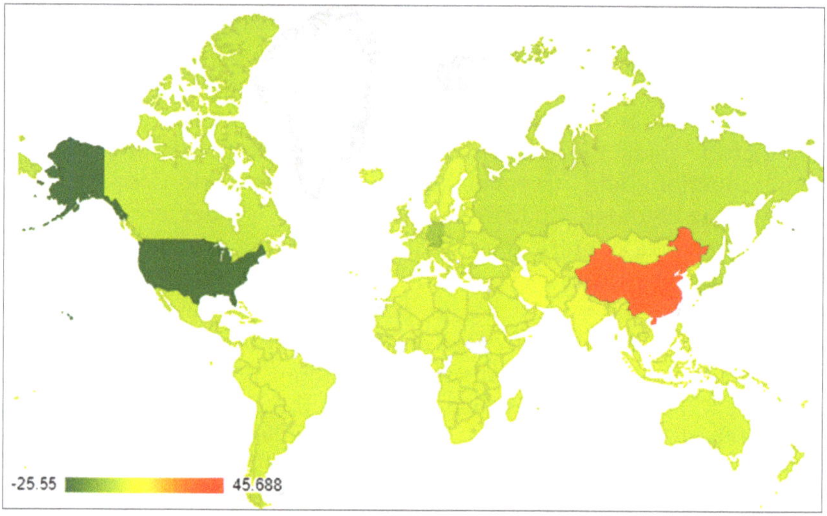

Fig. 8.6 Worldmap of $WL_{TT}CCCG$ index per country

financial crisis (Puaschunder, 2023). Based on historical data of past crisis responses around the world, the Financial Crisis Intervention Index $CCFCI$ included historical accounts of crisis intervention measures of guarantees (including account guarantee, blanket guarantee, other liability guarantee, asset guarantee), lending (including ad hoc liquidity assistance, broad-based liquidity assistance, market liquidity assistance), capital injection (including ad hoc capital injection, broad-based capital injection), resolutions (including restructuring or resolution, stakeholder bail-in), rules (including suspension or bank holiday, debt or payment moratorium, credit rules, other rules), asset management (including ad hoc asset management, broad-based asset management), and other financial crisis intervention tools (including major communication, stress test, other), countries were ranked based on their crisis intervention propensity (Puaschunder, 2023). The rationale is that countries that have better established means to combat crises financially are better equipped to lead the world in the proposed climate financialization taxation-and-bonds strategy (Puaschunder, 2023). Countries that have rising GDP prospects in light of climate change and countries that contribute to human-made global warming in CO_2 emissions and have historical

Fig. 8.7 Country ranking of $WL_{TT}CCCG$ index for CO_2 reducing countries

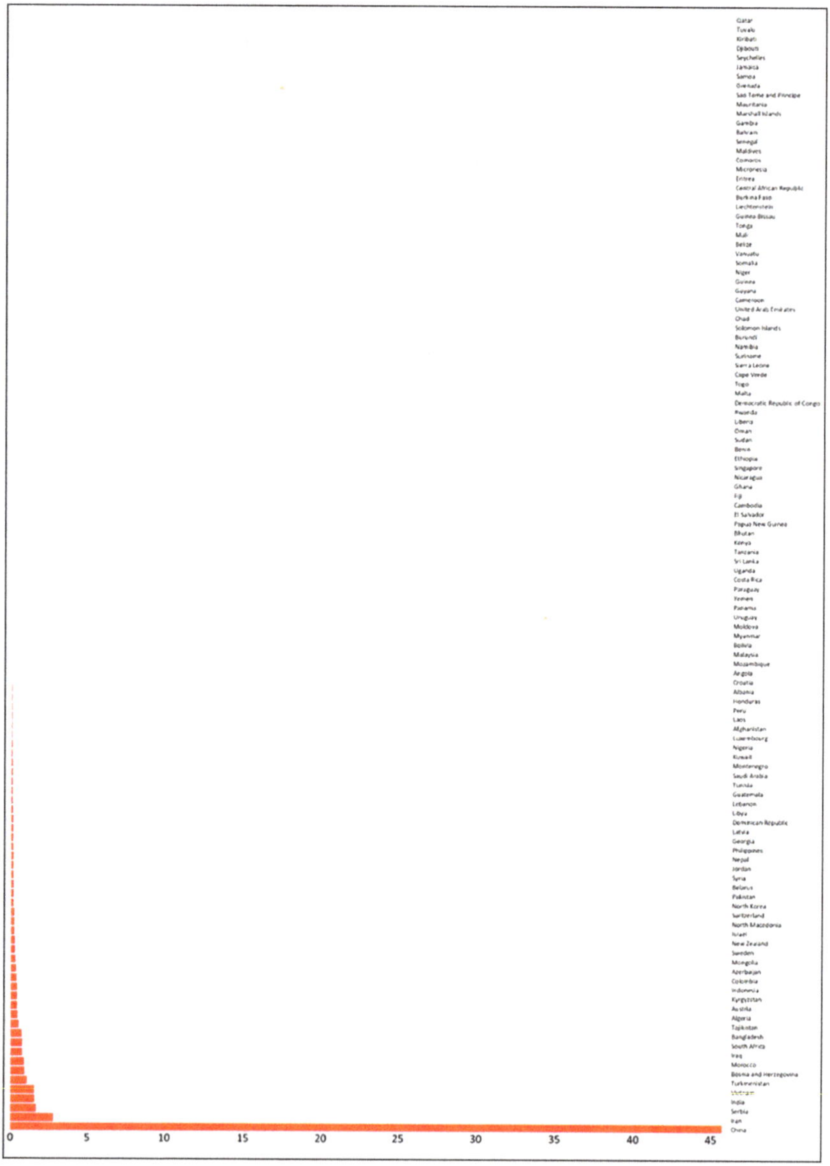

Fig. 8.8 Country ranking of $WL_{TT}CCCG$ index for CO_2 increasing countries

knowledge to combat crises with financial means, are argued to face a higher responsibility to act on climate change and lead the world in the proposed climate taxation-and-bonds solution (Puaschunder, 2023). In these countries the financial sector should pay for the establishment and maintenance of climate bonds via carbon taxation; while climate change losing territories with low CO_2 emissions that also have no experience in financial crisis intervention based on past historic records should be climate bond recipient beneficiaries being granted a relatively higher bond interest rate premium funded by taxation of CO_2 emitting industries in climate winning countries with financial crisis intervention expertise (Puaschunder, 2023). The $CCFCI$ index therefore integrated whether a country is a climate change winner or loser as well as the relative CO_2 emissions per country in relation to other countries and the historic crisis intervention propensity of the country (Puaschunder, 2023).

Figure 8.9 exhibits a worldmap that colors those countries with recommended high taxation strategy in red, high bonds premium recipients in green, and countries that are in the middle of the index range in yellow (Puaschunder, 2023). Figure 8.9 displays a worldmap that colors those countries in red that are advised to fund climate bonds with a taxation strategy (Puaschunder, 2023). Red countries tend to be more likely climate winners with high CO_2 emissions and crisis intervention propensity (Puaschunder, 2023). Countries are greener if being climate change losers as well as having less CO_2 emissions per country and crisis intervention followership (Puaschunder, 2023). The greener the country is colored, the higher bonds premium are advised to be paid out in this country as for being a higher climate loser and having less CO_2 emissions with less crisis intervention capabilities, hence contributing less to the climate problem and being advised to be climate bond followers (Puaschunder, 2023). Countries that range in the middle of the index are displayed in yellow in the display spectrum from countries ranked on taxation strategy in red to the highest bonds premium recipient country colored in green (Puaschunder, 2023). Figure 8.10 ranks 112 countries based on their relative $CCFCI$ index status from the highest climate taxation country to the highest bonds premium recipient country.

8.2.3.5 Climate Winners and Losers CO_2 Emission Resilience Finance Index

The Climate winners and losers CO_2 emission Resilience Finance Index $CCRFI$ included the resilience finance expertise of countries

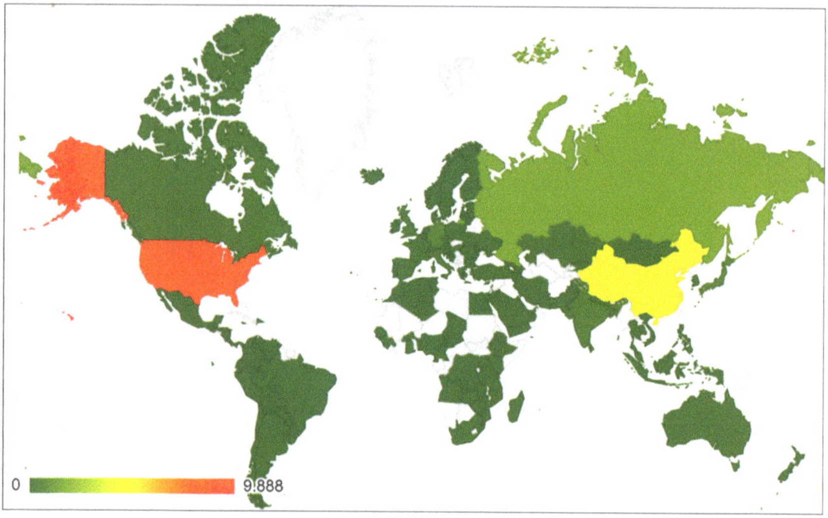

Fig. 8.9 Worldmap of *CCFCI* index per country

(Puaschunder, 2023). Based on historical data of resilience finance propensities around the world, that included historical accounts of resilience finance measures of guarantees (including account guarantee, blanket guarantee, other liability guarantee, asset guarantee), lending (including ad hoc liquidity assistance, broad-based liquidity assistance, market liquidity assistance), and capital injection (including ad hoc capital injection, broad-based capital injection), countries were ranked based on their resilience finance leadership potential (Puaschunder, 2023). The rationale is that countries that have better established resilience finance are better equipped to lead the world in the proposed climate financialization taxation-and-bonds strategy (Puaschunder, 2023). Countries that have rising GDP prospects in light of climate change and countries that contribute to human-made global warming in CO_2 emissions and have historical knowledge in resilience finance applications are argued to face a higher responsibility to act on climate change and lead the world in the proposed climate taxation-and-bonds solution (Puaschunder, 2023). In these countries the financial sector should pay for the establishment and maintenance of climate bonds via carbon taxation; while climate change losing territories with low CO_2 emissions that also have no experience

Fig. 8.10 Country ranking of *CCFCI* index for 112 countries

in resilience finance intervention based on past historic records should be climate bond recipient beneficiaries being granted a relatively higher bond interest rate premium funded by taxation of CO_2 emitting industries in climate winning countries with resilience finance expertise (Puaschunder, 2023). The $CCRFI$ index therefore integrated whether a country is a climate change winner or loser as well as the relative CO_2 emissions per country in relation to other countries and the historic resilient finance propensity of the country (Puaschunder, 2023).

Figure 8.11 exhibits a worldmap that colors those countries with recommended high taxation strategy in red, high bonds premium recipients in green, and countries that are in the middle of the index range in yellow (Puaschunder, 2023). Figure 8.11 displays a worldmap that colors those countries in red that are advised to fund climate bonds with a taxation strategy (Puaschunder, 2023). Red countries tend to be more likely climate winners with high CO_2 emissions and resilience finance propensity (Puaschunder, 2023). Graph 6.13 displays countries greener if being climate change losers as well as having less CO_2 emissions per country and resilience finance followership (Puaschunder, 2023). The greener the country is colored, the higher bonds premium are advised to be paid out in this country as for being a higher climate loser and having less CO_2 emissions with less resilience finance capabilities, hence contributing less to the climate problem and being advised to be climate bond followers (Puaschunder, 2023). Countries that range in the middle of the index are displayed in yellow in the display spectrum from countries ranked on taxation strategy in red to the highest bonds premium recipient country colored in green (Puaschunder, 2023). Figure 8.12 ranks 130 countries based on their relative $CCRFI$ index status from the highest climate taxation country to the highest bonds premium recipient country.

8.2.3.6 Climate Winners and Losers CO_2 Emission Bank Lending Rate Index

Climate justice between countries could be based on the idea that climate change winning countries that contribute to human-made global warming in CO_2 emissions and have a low bank lending rate based on historic financial sector performance should pay for the establishment and maintenance of climate bonds via carbon taxation; while climate change losing territories with low CO_2 emissions that also have a high bank lending rate based on past financial records should be climate bond recipient beneficiaries being granted a relatively higher bond interest rate premium funded

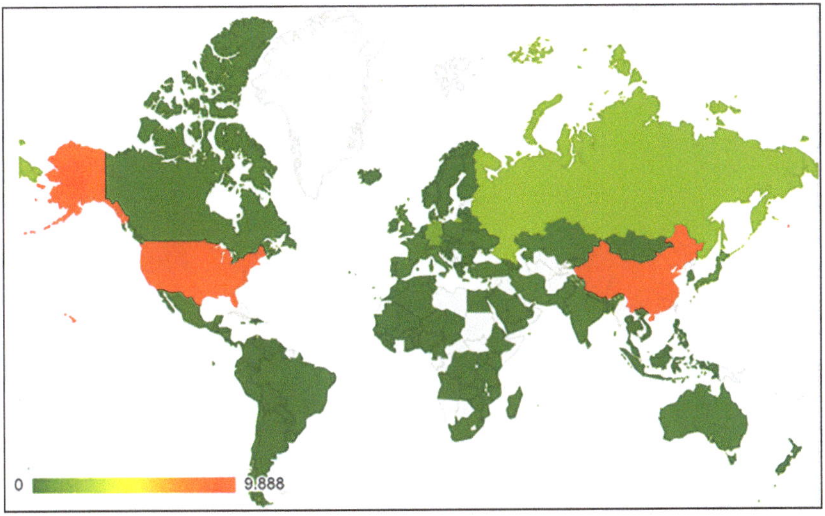

Fig. 8.11 Worldmap of $CCRFI$ index per country

by taxation of CO_2 emitting industries in climate winning countries with low bank lending rates (Puaschunder, 2023). The $WL_{TT}CCCL$ index therefore integrated whether a country is a climate change winner or loser as well as the relative CO_2 emissions per country in relation to other countries and the bank lending rate of the country (Puaschunder, 2023).

Figure 8.13 exhibits a worldmap that colors those countries with recommended high taxation strategy in red, high bonds premium recipients in green, and countries that are in the middle of the index range in yellow (Puaschunder, 2023). Figure 8.13 displays a worldmap that colors those countries in red that are advised to fund climate bonds with a taxation strategy (Puaschunder, 2023). Red countries tend to be more likely climate winners with high CO_2 emissions and low bank lending rates (Puaschunder, 2023). Figure 8.13 displays countries greener if being climate change losers as well as having less CO_2 emissions per country and high bank lending rates (Puaschunder, 2023). The greener the country is colored, the higher bonds premium are advised to be paid out in this country as for being a higher climate loser and having less CO_2 emissions, hence contributing less to the climate problem as well as high bank lending rates (Puaschunder, 2023). Countries that range in the middle

Fig. 8.12 Country ranking of *CCRFI* index for 130 countries

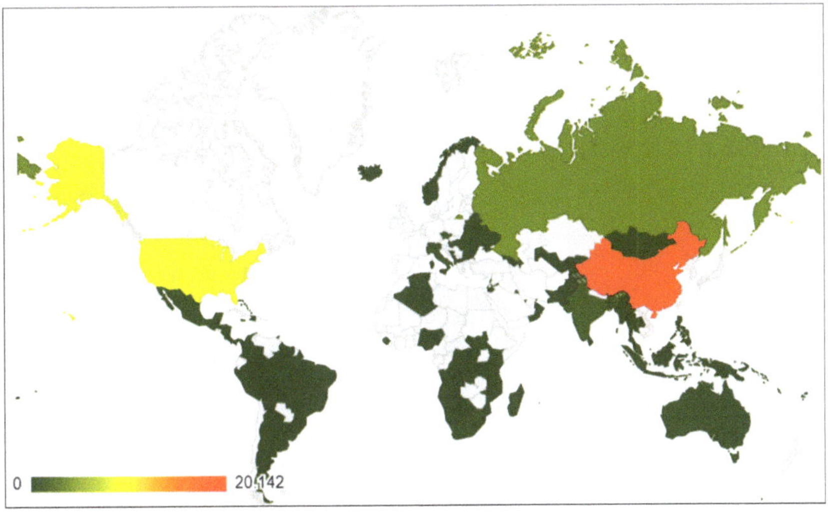

Fig. 8.13 Worldmap of $WL_{TT}CCCL$ index per country

of the index are shown in yellow in the display spectrum from countries ranked on taxation strategy in red to the highest bonds premium recipient country colored in green (Puaschunder, 2023). Figure 8.14 ranks 101 countries based on their relative $WL_{TT}CCCL$ index status from the highest climate taxation country to the highest bonds premium recipient country (Puaschunder, 2023).

8.2.3.7 Climate Winners and Losers Consumption-Based, Trade-Adjusted CO_2 Emission Index

Climate justice between countries could be based on the idea that countries that have relative short-term economic climate change gain prospects that contribute to human-made global warming in consuming CO_2 emissions should pay for the establishment and maintenance of climate bonds via carbon taxation, while climate change losing territories with low CO_2 emissions consumption should be climate bond issuers with a higher interest rate premium (Puaschunder, 2023). The $WL_{TT}CBTAE$ index therefore integrated whether a country is a climate change winner or loser and the relative consumption-based, trade-adjusted CO_2 emissions per country in relation to other countries (Puaschunder, 2023).

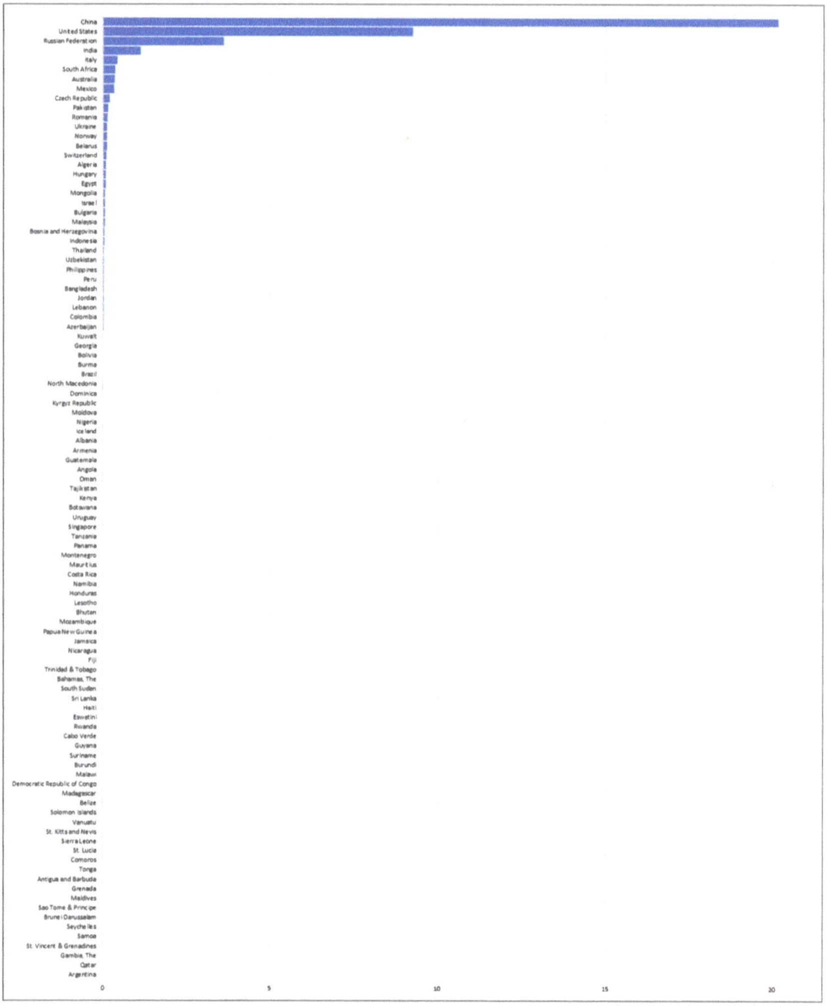

Fig. 8.14 Country ranking of $WL_{TT}CCCL$ index for 101 countries

Figure 8.15 displays a worldmap that colors those countries in red that are advised to fund climate bonds with a taxation strategy (Puaschunder, 2023). Countries colored in green are climate change losers and have less CO_2 emissions consumption per country (Puaschunder, 2023). The

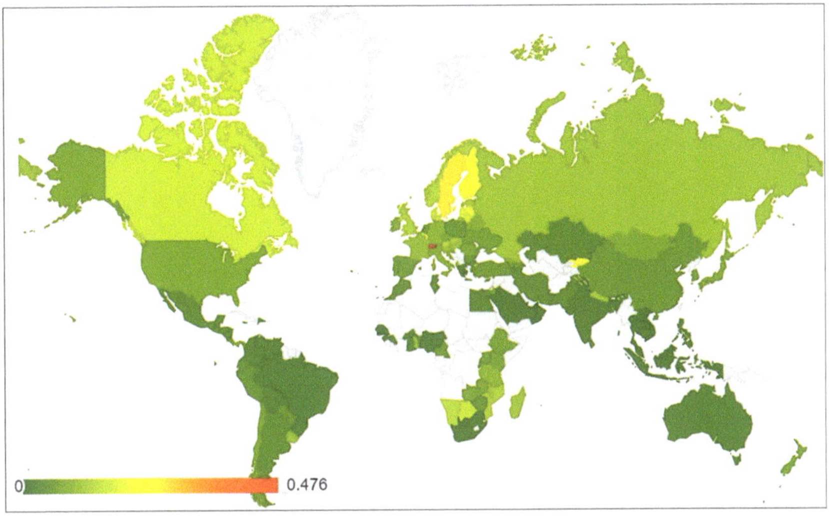

Fig. 8.15 Worldmap of $WL_{TT}CBTAE$ index per country for 116 countries

greener the country is colored, the higher bonds premium are advised to be paid out in this country as for being a higher climate loser and having less CO_2 emission consumption, hence contributing less to the climate problem (Puaschunder, 2023). Countries that range in the middle of the index are shown in yellow in the display spectrum from countries ranked on taxation strategy in red to the highest bonds premium recipient country colored in green (Puaschunder, 2023).

Figure 8.16 ranks 116 countries based on their relative $WL_{TT}CBTAE$ index status from the highest climate taxation country to the highest bonds premium recipient country.

8.2.3.8 Science Diplomacy Climate Responsibility Index

The Science Diplomacy Climate Responsibility Index is applied to the macroeconomic model on disparate economic impacts of climate change around the world and country-specific CO_2 emission levels for the year 2019 derived from *Our World in Data*, in order to determine what countries have excellent starting grounds on science diplomacy leadership but also a heightened responsibility to engage in science diplomacy to reverse

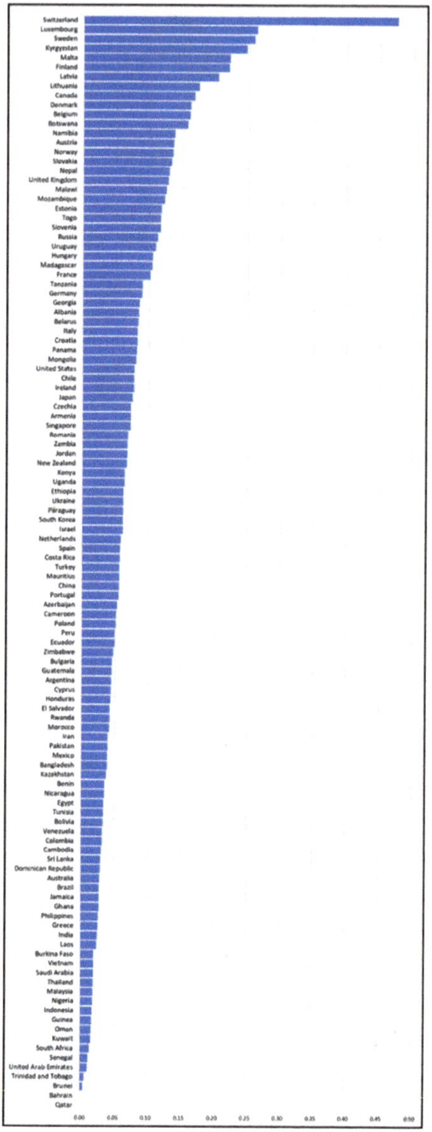

Fig. 8.16 Country ranking of $WL_{TT}CBTAE$ index for 116 countries

the negative impacts of global warming via redistribution of prospective short-term economic gains (Puaschunder, 2023).

Figure 8.17 exhibits a worldmap that colors those countries with recommended high taxation strategy in red, high bonds premium recipients in green, and countries that are in the middle of the index range in yellow (Puaschunder, 2023). Figure 8.17 displays a worldmap that colors those countries in red that are advised to fund climate bonds with a taxation strategy (Puaschunder, 2023). Red countries tend to be more likely climate winners with high CO_2 emissions and global connectivity (Puaschunder, 2023). Figure 8.17 features countries greener if being climate change losers as well as having less CO_2 emissions per country and global followership (Puaschunder, 2023). The greener the country is colored, the higher bonds premium are advised to be paid out in this country as for being a higher climate loser and having less CO_2 emissions with science diplomacy leadership, hence contributing less to the climate problem and being advised to be climate bond followers (Puaschunder, 2023). Countries that range in the middle of the index are displayed in yellow in the display spectrum from countries ranked on taxation strategy in red to the highest bonds premium recipient country colored in green (Puaschunder, 2023).

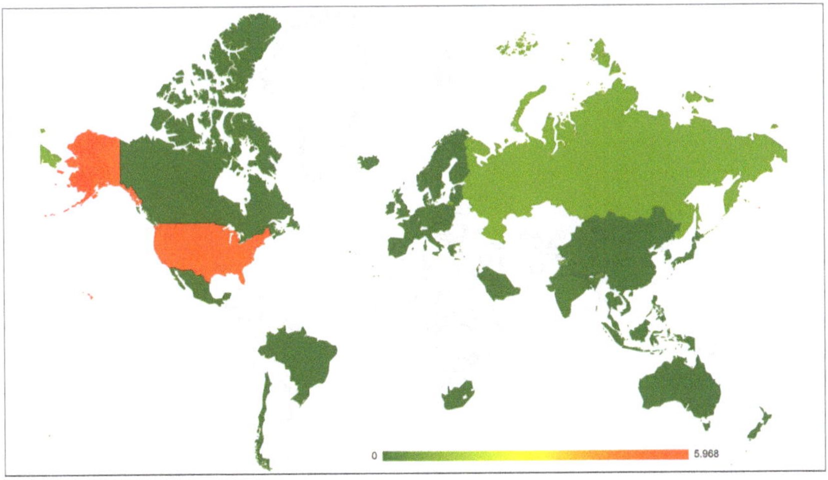

Fig. 8.17 Worldmap of *CCSDI* index per country for 48 countries

The Science Diplomacy Climate Responsibility Index for 48 countries of the world indicates that the United States offers best conditions to lead the world on science diplomacy as exhibited by Fig. 8.18 (Puaschunder, 2023). Russia, Japan, India, Germany, United Kingdom, Canada, China, South Korea, France, Poland, Mexico, and Brazil are additional countries that have best starting grounds on science diplomacy in the climate responsibility domain.

Overall, the results indicate that the United States and Russia are key players in science diplomacy with highest responsibility to act to avert climate change (Puaschunder, 2023). Africa offers science diplomacy climate stabilization leadership potential foremost in South Africa (Puaschunder, 2023). In Asia, Japan, India, and China but also South Korea, Indonesia as well as Malaysia, the Philippines, Vietnam, Thailand, Mongolia, Bangladesh, Singapore, and Nepal play a role in science diplomacy with responsibility for climate control (Puaschunder, 2023). Australia and New Zealand should take a responsible role in science diplomacy for global warming alleviation (Puaschunder, 2023). Within

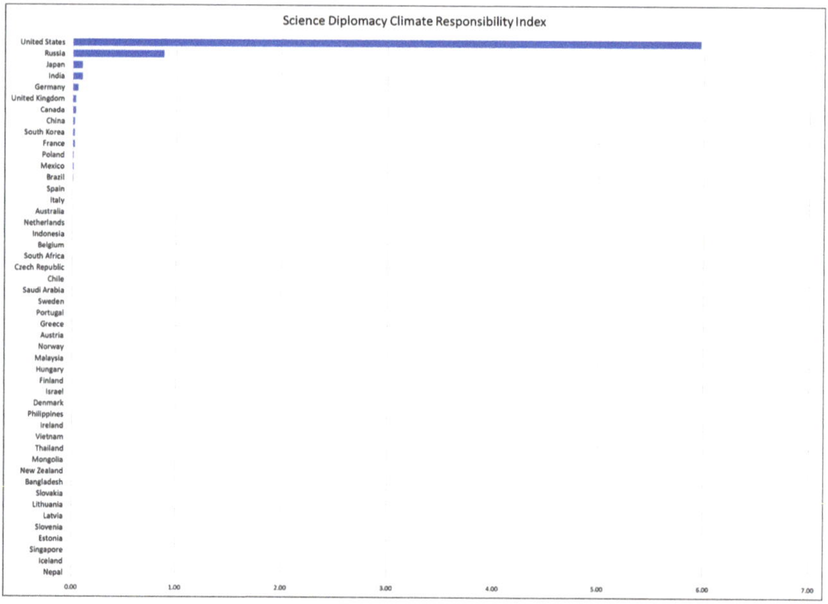

Fig. 8.18 Country ranking of *CCSDI* index for 48 countries

Eurasia, Saudi Arabia and Israel are key players on science diplomacy with respect for climate mitigation and adaptation (Puaschunder, 2023). In Europe, Germany, the United Kingdom, and France lead on science diplomacy for climate awareness followed by Poland, Spain, Italy, the Netherlands, Belgium, Czech Republic, Sweden, Portugal, Greece, Austria, Norway, Hungary, Finland, Denmark, Ireland, Slovakia, Lithuania, Latvia, Slovenia, Estonia, and Iceland (Puaschunder, 2023). In North America, the United States leads on science diplomacy but also Canada has potential for academic diplomacy and a responsibility to protect from the downsides of global warming (Puaschunder, 2023). In South America, Mexico, Brazil and Chile have good starting grounds on science diplomacy for a common climate solution (Puaschunder, 2023).

8.2.3.9 Climate Winners and Losers CO_2 Emission Global Connectivity Index

The Climate winners and losers CO_2 emission Global Connectivity Index *CCGC* included the global connectivity of countries in terms of their trade, finance, and human migration flows (Puaschunder, 2023). Based on data of Global Economic Trade Freedom and remittances normalized by Gross Domestic Product, Foreign Direct Investment flows as well as migration in percent of the population of countries around the world, countries were ranked based on their economic, financial as well as human capital flows (Puaschunder, 2023). The rationale is that countries that are better connected with other countries in terms of economics, finance, and human flows are better equipped to lead the world in the proposed climate financialization taxation-and-bonds strategy (Puaschunder, 2023).

Countries that have rising GDP prospects in light of climate change and countries that contribute to human-made global warming in CO_2 emissions and have a solid economics, finance, and human network around the world, are argued to face a higher responsibility to act on climate change and lead the world in the proposed climate taxation-and-bonds solution (Puaschunder, 2023). In these countries the financial sector should pay for the establishment and maintenance of climate bonds via carbon taxation; while climate change losing territories with low CO_2 emissions that also have less global connectivity based on economic, finance, and human migration flows should be climate bond recipient beneficiaries being granted a relatively higher bond interest rate premium funded by taxation of CO_2 emitting industries in climate winning countries with global connectivity (Puaschunder, 2023). The *CCGC* index therefore integrated whether a country is a climate change winner or loser as well as the relative CO_2 emissions per country in relation to other countries and the global connectivity in terms of trade, finance, and human migration of the country (Puaschunder, 2023).

Figure 8.19 exhibits a worldmap that colors those countries with recommended high taxation strategy in red, high bonds premium recipients in green, and countries that are in the middle of the index range in yellow (Puaschunder, 2023). Figure 8.19 displays a worldmap that colors those countries in red that are advised to fund climate bonds with a taxation strategy (Puaschunder, 2023). Red countries tend to be more likely climate winners with high CO_2 emissions and global connectivity (Puaschunder, 2023). Figure 8.19 displays countries greener if being climate change losers as well as having less CO_2 emissions per country and global followership (Puaschunder, 2023). The greener the country is colored, the higher bonds premium are advised to be paid out in this country as for being a higher climate loser and having less CO_2 emissions without global connectivity, hence contributing less to the climate problem and being advised to be climate bond followers (Puaschunder, 2023). Countries that range in the middle of the index are displayed in yellow in the display spectrum from countries ranked on taxation strategy in red to the highest bonds premium recipient country colored in green (Puaschunder, 2023).

Figure 8.20 ranks 158 countries based on their relative *CCGC* index status from the highest climate taxation country to the highest bonds premium recipient country (Puaschunder, 2023).

8.3 Discussion

Climate justice as one of the most challenging global governance goals requiring financialization of mitigation and adaptation efforts is a formidable task. The complex interplay of many actors that are bound by multiple constraints creates a multifaceted predicament for finding the funds needed for a transition to sustainability. From the economic standpoint, sustainability and a transitioning to a zero economy with renewable energy is an economic game theoretical problem on the international level but also over time. Free rider problems arise as removal efforts of global warming raises costs with outcomes that can be enjoyed by all nation-states since the world climate has no borders. This leads to a natural first mover-hesitancy and over time inter-temporally sub-optimal choices

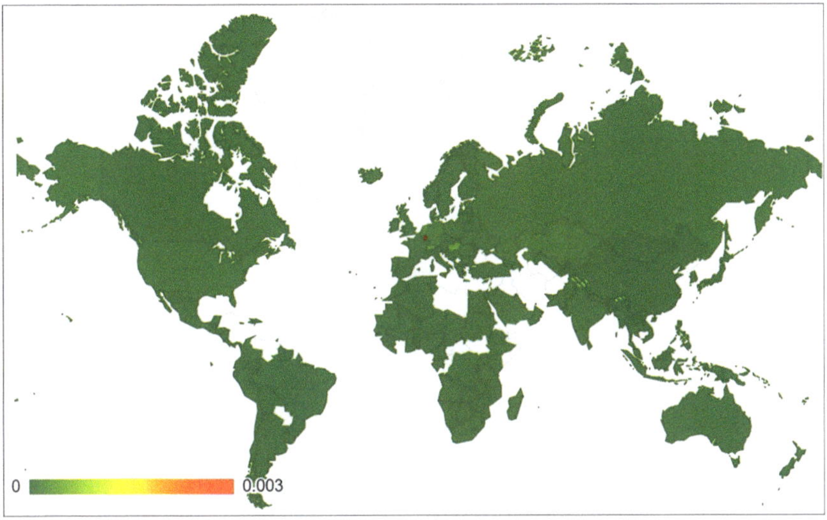

Fig. 8.19 Worldmap of *CCGC* index per country

(Nordhaus, 2008, 2013). Ignorance of climate problems and non-action of some countries can raise international political and diplomatic tensions.

International negotiations' success is often slowed by non-ratification of treaties and lack of commitment of the many parties and stakeholders involved (Chappe, 2014). The enforceability of environmental treaties is another challenged in international environmental law as—for instance—soft laws, such as policy declarations, are not legally binding (Chappe, 2014). Ambitious long-term aspirational climate goals are usually dependent on a wide range of multiple actors and often vague, which hinders their actual implementation (Chappe, 2021). All these influencing factors oftentimes lead to widespread non-compliance with climate goals (Chappe, 2021). Procedural obstacles in the length of negotiation processes, coordination of multiple actors for ratifications as well as missing treaty violations sanctions mechanisms can further delay necessary climate action (Chappe, 2021; Puaschunder, 2023). Fairness in climate negotiations is a formidable endeavor struck by a political and historical complexity in finding common ground (Haurie et al., 2014). The political reality of climate control is bound by policymakers facing short-term election cycles and an overall focus on serving concrete

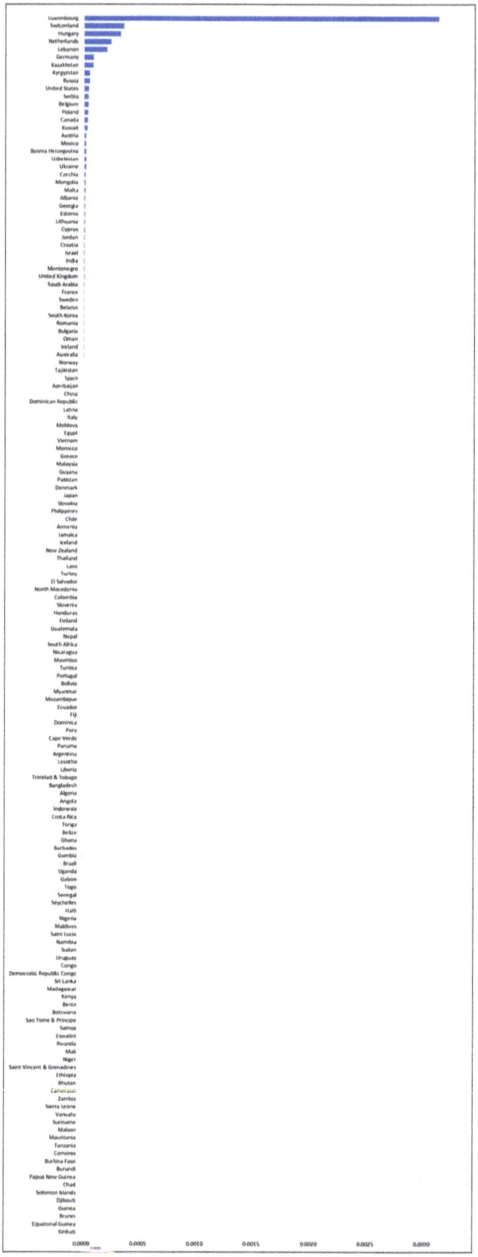

Fig. 8.20 Country ranking of *CCGC* index for 158 countries

voter demands, which deflects from long-term international environmental sustainability pledges. Human beings are also prone to myopic choice patterns that zone out long-run uncertainty and complexity with ignorance (Puaschunder, 2020a). This kind of short-termism may lead to a natural propensity to hyperbolic discounting, which materializes in an underappreciation of natural resource preservation over time (Heal, 2000; Puaschunder, 2020a).

In finding funds to transition to renewable energy and foster sustainability, the economic disparities under climate change have just recently been considered. Though measuring the short-term economic growth prospects that lie in global warming allows for innovative financialization strategies in redistribution but also to derive inferences how to cope with a warming globe. Capturing the gains and losses of global warming being distributed unequally throughout the world helps to map fair redistribution mechanisms. Employing climate bonds in combination with taxation can now also enact climate justice over time. Grounded in historical ethical imperatives and hallmarks of economic theories to guide on societal fairness predicaments, redistribution helps to aid with some of the expected gains from climate change to offset for some of the expected losses of global warming in territories that will be hit the hardest and fastest from climate change. All these endeavors breed hope for harmony with the environment and future generations (Jonas, 1979; Nikulin, 2021).

Based on the overall relative climate winners and losers index, global warming benefits are proposed to be redistributed in a fair way to offset climate change loser countries for climate change mitigation and adaptation efforts (Puaschunder, 2023). Climate justice comprises fairness within society, between countries but also over generations in a unique and unprecedented tax-and-bonds climate change gains and losses distribution strategy. Climate change winning countries are advised to use taxation to raise revenues to offset the losses incurred by climate change in climate change loser countries (Puaschunder, 2023). Within the winning countries, foremost the gaining GDP sectors and carbon-emitting industries should be taxed proportional to their CO_2 emissions. Climate change loser countries could raise revenues by issuing bonds that are subsidized by the relative climate winning territories and also have to be paid back by taxing future generations in order to shift the economic burden of climate stabilization financing in time (Puaschunder, 2016, 2023).

8.3.1 *Implications*

In this chapter a fair global warming benefits distribution was proposed that grants climate justice around the world and over time via a taxation-and-bonds transfer strategy. The novel policy recommendation proposed features climate bonds funded by taxation in some countries with differing bonds premium and maturity yield regimes based on different variables such as climate change gains and losses, climate flexibility, CO_2 emissions production and consumption as well as greenhouse gas emission changes (Puaschunder, 2023). The responsibility to protect from global warming was attributed to countries that have favorable conditions to draw from public finance in bank lending rates, crisis and resilient finance capabilities as well as science diplomacy and economic ties, which all offer best conditions to enact a universal climate justice scheme (Puaschunder, 2023).

The proposed taxation and bonds would feature some countries' financing green bonds via carbon taxation revenues, while other countries are climate bonds premium payment recipients. Based on (1) climate change-related economic prospects, (2) climate flexibility as the range of temperatures a country has, (3) CO_2 emission production and (4) CO_2 emission consumption levels, (5) CO_2 emission changes as well as (6) bank lending rates as indicator for access to capital, (7) a country's financial crisis intervention propensity, (8) a country's resilience finance capabilities and (9) a country's science diplomacy ranking and (10) a country's global connectivity ranking; countries could either give the guarantee for climate bonds paid by their taxes or get a revenue from the bonds to invest in a large-scale transition of markets and de-risk their economics.

Nine indices would determine which country should be using a taxation strategy and what country should be granted climate bonds. Countries that are having a rising GDP prospect and are polluting and consuming goods and do not change their polluting behavior as well as benefit from low bank lending rates, should repay bonds with their taxes as a transfer to countries that face declining GDP prospect, are not emitting CO_2 in production and consumption or have declining CO_2 levels and face higher bank lending rates (Puaschunder, 2023). Those countries should face a higher responsibility to protect from global warming that have a financial crisis intervention propensity and resilience finance capabilities as well as are well connected in science diplomacy and economic terms (Puaschunder, 2023).

The bonds could be used to create market incentives to transition economies via competition for the economic benefits of climate change. Countries would be nudged to change policies and implement sustainability and international cooperation on averting climate change in order to gain a favorable position on a climate-taxation-obligation to climate-benefits-recipient spectrum.

8.3.2 Economic Impetus

The proposed climate tax-and-bonds solution based on different responsibility levels throughout the world is an innovative way to enact climate justice. The idea to determine different levels of responsibility to protect the earth around the globe based on various geo-political influence factors—such as climate change contribution, climate flexibility, financial resilience and international leadership capabilities—is a novel approach to find fair solutions in the climate financialization predicament. The concrete quantification of various dimensions of climate performance measured in relation to other countries' performance allows for setting positive incentives to compete over scarce resources. The yearly measurement of changes within countries and the big data age possibility to track changes around the globe in real time as well as innovations in Geographic Information Systems (GIS) all enable to constantly adjust and adapt the country contributions based on performance changes. All these measures combined and the prospect to shift one's position from being a climate bonds contributor to becoming a climate bonds transfer funds recipient holds the opportunity to create a positive race-to-the-top for incentives. Concrete incentives could target at lowering CO_2 emissions or to move production to places that are climate losers in order to help revitalizing economies that have a shrinking prospect under climate change (Puaschunder, 2023).

The idea of differing climate bonds regimes is also extendable to sector-specific bond yield interest rate regimes (Puaschunder, 2023). On a country level, high CO_2 emitting industries should face a climate taxation in order to set market incentives for a transition to renewable energy (Puaschunder, 2023). The revenues generated from the taxation of carbon-intensive industries should be used to offset the losses of climate change and subsidize climate bonds (Puaschunder, 2023).

Within a country, the bonds could be offered by commissioning agents, such as local investment banks, who could install industry-specific premium bonds payments and maturity bond yields based on the environmental sustainability of an industry, e.g., as measured by the European Sustainable Finance Taxonomy (Puaschunder, 2023). The more sustainable an industry performs; the higher bond yield should be granted in sector-specific interest rate regimes within a country (Puaschunder, 2023). This strategy should set positive market incentives via subsidies (Puaschunder, 2023). Funding industries for not polluting could change the traditional race-to-the-bottom price-cutting behavior driving CO_2 emitting energy supply to having industries compete over subsidies for using clean energy (Puaschunder, 2023). In this way, bond yield differences between industries could set positive market incentives for a transitioning to renewable energy productivity solutions (Puaschunder, 2023).

8.3.3 Future Studies

The presented redistribution was based on simplistic models of the variegated climate change economic impacts as a basis for inequality alleviation strategies. Future model sophistications could include different environmental, political, and financial drivers of economic growth, which could also be measured on a per capita basis. Future studies on climate change impacts may address the different tipping point and disaster drivers of temperature rise, weather extremes and sea-level rises as this current model only considered temperature as main determinant of economic growth (Dietz et al., 2021; Puaschunder, 2023). Future model sophistications could entail risk propensities on a country level and include actual climate damage accounting for reducing some of the transfer payment loads. A clear reporting of climate risks and actual damages would not only help map the costs of global warming around the world, but it would also hone an understanding of the long-term implications and economic burden of climate change throughout the world. If this risk and damage deduction is institutionalized within the tax-and-bonds transfer scheme, it would gain international standing and may even convince regimes that are agnostic about climate change to act on climate stabilization. An international account of the variegated risks of global warming would also help derive generalizations and inferences for combating climate change with mitigation and adaptation. Cross-pollination in the efficient response to climate change will only be possible if risks are made transparent and adaptation outcome results are discussed on an international level due to incentivized reporting of climate risks.

In a heterodox opening of the macroeconomic aggregate, future Law and Economics extensions could unravel disparate impacts of climate policies. Attention should be paid to the disparate impact on marginalized and vulnerable societal groups with particular concern for gender and race. Legal advancements of commitment bond strategies should be explored in future applied model versions (Armour et al., 2021; Ayres & Abramowicz, 2011; Bishop, 2019; Omarova, 2020). The incentive schemes could be extended for concrete industry impacts on climate change. Financial investment communities could further explore green innovations to complement and be financed by some of the expected climate change economic gains. The variegated impact of climate, weather, and coastal risk should be integrated as a tax deduction strategy in accounting circles, who could thereby help balance the economic gains and losses around the world in the common pursuit of climate justice.

8.3.4 Conclusion

The proposed climate redistribution strategy offers a quantifiable and measurable ranking of responsibility burdens for all countries of the world. Shedding scientific light at various factors that may have contributed to the climate crisis but also allowing for future planning of propensities to lead and administer positive change around the globe is meant to provide the means to find a viable and sustainable solution to the climate problem. Highlighting potentials of specific nations to lead on climate science diplomacy is the first step in directions of climate finance governance. Future model sophistications may consider integrating risks and may work on flexibly allowing for redistribution breaks if being impacted by climate change in an extraordinary hard way in comparison to other countries.

The implementation of the proposed strategies will depend on the national policies and the local design, adoption, and interaction in pursuing a common goods plan. While the climate taxation-bonds structure is more housed in the global governance and international development arena, the local policies, regulations, and standards but also the prevailing tax authorities' impetus will determine the likelihood of success in the implementation of a common shared climate regime. As previously seen with international environmental treaties and tradable permits, financial incentives are as important as informal agreements and the dissemination structure of the plan in order to accomplish ambitious goals.

Scientific discourse as well as research and development can help with technological advancements and innovative transformations to foster climate stabilization goals in unprecedented ways. Government have played a distinctive role in subsidizing basic research to enhance climate innovations in the past and can lead forward in finding—for instance—respective technologies to mitigate and adapt to climate change efficiently and effectively.

References

Armour, J., Enriques, L., & Wetzer, Th. (2021, July 2). Corporate carbon reduction pledges: Beyond greenwashing. Business Law and the Net Zero Economy, Oxford Business Law Blog. https://www.law.ox.ac.uk/business-law-blog/blog/2021/07/corporate-carbon-reduction-pledges-beyond-greenwashing

Ayres, I., & Abramowicz, M. (2011). Commitment bonds. *Georgetown Law Journal, 100*, 605–656.

Bishop, N. (2019). Green bond governance and the Paris Agreement. *New York University Environmental Law Journal*, 377–411. https://www.nyuelj.org/wp-content/uploads/2019/10/Bishop_Green-Bond-Governance-and-the-Paris-Agreement.pdf

Braga, J. A., Fischermann, T., & Semmler, W. (2020, March 5). Ökonomie und Klimapolitik: So könnte es gehen. Die Zeit. https://www.zeit.de/2020/11/europaeischer-green-deal-emissionshandel-wirtschaft-klimaschutz

Chappe, R. (2014). The legal framework of global environment governance on climate change: A critical survey. In L. Bernard & W. Semmler (Eds.), *The Oxford handbook of the macroeconomics of global warming* (pp. 603–638). Oxford University Press.

Chappe, R. (2021). *Climate policies and global agreements*. Guest lecture notes. The New School for Social Research.

Dietz, S., Rising, J., Stoerk, Th., & Wagner, G. (2021). Economic impacts of tipping points in the climate system. *Proceedings of the National Academy of Sciences, 118*(34), 1–9.

Flavelle, Ch. (2021, December 3). Billions for climate protection fuel new debate: Who deserves it most. *The New York Times*. https://www.nytimes.com/2021/12/03/climate/climate-change-infrastructure-bill.html

Flavelle, Ch. (2021, December 8). The climate bill includes billions in funding: Will it be spent fairly? *The New York Times*. https://www.nytimes.com/2021/12/03/climate/climate-change-infrastructure-bill.html

Hansen, J. E. (2014). Environment and development challenges: The imperative of a carbon fee and dividend. In L. Bernard & W. Semmler (Eds.), *The Oxford handbook of the macroeconomics of global warming* (pp. 639–646). Oxford University Press.

Haurie, A., Babonneau, F., Edwards, N., Holden, Ph., Kanudia, A., Labriet, M., Pizzileo, B., & Vielle, M. (2014). Fairness in climate negotiations: A meta-game analysis based on Community Integrated Assessment. In L. Bernard & W. Semmler (Eds.), *The Oxford handbook of the macroeconomics of global warming* (pp. 170–203). Oxford University Press.

Heal, G. (2000). *Valuing the future: Economic theory and sustainability*. Columbia University Press.

Jonas, H. (1979). *Das Prinzip Verantwortung: Versuch einer Ethik für die technologische Zivilisation* (p. 1979). Insel.

Kaldor, N. (1961). Capital accumulation and economic growth. In F. A. Lutz & D. C. Hague (Eds.), *The theory of capital* (pp. 177–222). St. Martin's Press.

Kant, I. (1783/1993). *Grounding for the metaphysics of morals*. Hackett.

Law, J., & Smullen, J. (2008). *A dictionary of finance and banking*. Oxford University Press.

Moneer, Z. (2022, September 28). COP27: A time to address the "triple inequality". Middle East Institute. https://www.mei.edu/publications/cop27-time-address-triple-inequality

Nikulin, D. (2021). Responsibility and hope. Bollettino della Società Filosofica Italiana, Carocci editore Roma, Nuova Serie n. 232, Gennaio/Aprile 2021.

Nordhaus, W. D. (2008). *A question of balance: Weighing the options on global warming policies*. Yale University Press.

Nordhaus, W. D. (2013). *The climate casino: Risk, uncertainty and economics for a warming world*. Yale University Press.

Omarova, S. T. (2020, August). The climate case for a national investment authority. Data for Progress. https://www.filesforprogress.org/memos/white-paper-nia.pdf

Orlov, S., Rovenskaya, E., Puaschunder, J. M., & Semmler, W. (2024). Green bonds, transition to a low-carbon economy, and intertemporal welfare allocation: Evidence from an extended DICE model. *AIMS Environmental Science, 11*, 4, 628–648. https://doi.org/10.3934/environsci.2024031. International Institute for Applied Systems Analysis (IIASA) Working paper: Green bonds, transition to a low-carbon economy, and intergenerational welfare allocation: Evidence from an extended DICE model.

Puaschunder, J. M. (2016). On eternal equity in the fin-de-millénaire: Rethinking capitalism for intergenerational justice. *Journal of Leadership, Accountability and Ethics, 13*(2), 11–24.

Puaschunder, J. M. (2020a). *Behavioral economics and finance leadership: Nudging and winking to make better choices*. Springer Nature.

Puaschunder, J. M. (2020b). *Governance and climate justice: Global South and developing nations*. Palgrave Macmillan; Springer Nature.

Puaschunder, J. M. (2022). *Ethics of inclusion: The cases of health, economics, education, digitalization and the environment in the post-COVID-19 era*. Ethics International.

Puaschunder, J. M. (2023). *The future of resilient finance: Finance politics in the age of Sustainable Development*. Palgrave Macmillan.

Rawls, J. (1971). *A theory of justice*. Harvard University Press.

Sachs, J. D. (2021, November 17). Fixing climate finance. *Social Europe: Politics, Economy and Employment & Labor*.

Tokyo Climate Center. (2022). *Japan Meteorological Agency*. https://ds.data.jma.go.jp/tcc/tcc/products/gwp/temp/list/year_wld.html

CHAPTER 9

Global Climate Change-Induced Migration and Financial Flows

9.1 Introduction

The Earth's climateis changing at a rate that has exceeded most scientific forecasts (Ferreira, 2017). Climate change accounts for one of the most pressing problems in the age of globalization as for exacerbating more complex risks than ever before. As never before in history since the birth of the earth, there is an environmental sensitivity to economic growth (Rovenskaya, 2008; The World Economic Forum Report, 2015). Today climate control needs have reached unprecedented momentum. Climate change has leveraged intergenerational equity as contemporary challenge of modern democracy and temporal justice an ethical obligation for the sake of sustainable development (Puaschunder, 2016d).

In the current post-COP21 Paris agreementclimate change mitigation and adaptation efforts, the financialization of the ambitious goals has leveraged into an internationally challenging demand. As an innovative approach to raise climate mitigation and adaptation financialization attention is paid to global warming gains. State-of-the-art welfare function measurements and economic productivity parameters thereby become the basis for conclusions about a fair spread of the gains and benefits from a warming earth. Contemporary Gross Domestic Product (GDP) measurements served as basis for estimations about the productivity of the agriculture, industry and service sectors around the world in a changing temperature. Based on the optimal temperatures for the agriculture,

J. Puaschunder, *Governance & Climate Justice*,
https://doi.org/10.1007/978-3-031-87971-5_9

industry and service sectors productivity, the average temperature per country around the world as well as climate projections of the year 2100 under the business-as-usual path, Chapter 6 revealed climate winners and losers around the world. Overall and simply seen from a narrow-minded GDP perspective, the world will macroeconomically benefit more from climate change until 2100 than lose. Winning and losing from a warming earth is significantly positively correlated with the Paris COP 21 emissions country percentage of Greenhouse Gas (GHG) for ratification, leading to the conclusion that the countries that have the longest time horizon regarding a warming earth also lack motivation to mitigate global climate change.

The conclusion of global warming gains and losses being distributed unequally around the world leads to the climatorial imperative—advocating for the need for fairness in the distribution of the global earth benefits among nations (Puaschunder, 2017b). Based on Kant's (1785/1993) imperative to only engage in actions one wants to experience themselves being done to them, passive neglect of action on climate mitigation is an active injustice to others (Puaschunder, 2017b). While the method to measure the gains from climate change can certainly be refined in future studies, this preliminary line of research is meant as very first preliminary step to open a gate to legal codifications and policy work to settle for a right, just and fair distribution of benefits from our common warming mother earth. The introduction of the gains from climate change is a novel approach that should solely be seen in connection to the imperative to distribute the short-term wins in a fair manner among all world inhabitants in order to avert climate inequality. The gains should offset climate change mitigation and adaptation costs in those countries that suffer the most and should help avert a deeper climate crisis in the future.

In order to manifest the climate inequality in the twenty-first century, two additional gravitational factors are considered in this chapter: Migration and financial flows. The chapter argues that by this time already, the effects of global environmental change are altering patterns of human migration and Foreign Direct Investments (FDI). Studying contemporary climate-induced migration streams therefore allows predictions of future human capital flows. While climate change exacerbates some of the most pressing issues of our time, the inequality arising from global warming is hardly captured. After outlining pure economic GDP growth perspectives on a warming earth, this chapter therefore examines the relation between climate change and migration as well as financial inflows in order to provide a nuanced description of the unequal distribution of climate change benefits around the world.

To this day, institutional and legal systems remain inadequately prepared to protect people that are exposed to climate threats. Global governance entities have recently started considering redistributing the gains from a warming earth around the world. Outlining the manifold climate gain and losses distributions may therefore serve as basis for well-tempered and rightly-balanced climate change gains redistributions.

The following chapter will innovatively outline the relation of international migrants and those countries that win from a warming earth. The chapter features climate-induced migration streams in order to consolidate the idea that a warming earth is more beneficial for some countries than to others. The countries, to which migrants flee, are assumed to offer better economic prospects. The territories that gain human capital are therefore argued to have a responsibility to redistribute some of the gains of a warming earth but also legally take care of the human influx appropriately. Monetary redistribution strategies should be coupled with legal automatic asylum options for those who come from areas that will be losing from global warming. Depicting climate-induced human power and capital gains is ultimately used to strengthen the case of climate inequality in the twenty-first century. All these endeavors serve in order to advocate for a fair global warming benefits distribution but also legal codifications of the rights of climate migrants and recognizing climate-induced migration under the Geneva Convention.

The chapter is structured as follows: A theoretical foundation of climate change winners and losers around the world based on GDP and cardinal optimal temperature for agriculture, industry and service sectors is first established in the theoretical Sect. 9.2.1. Global climate inequality demands for a fair climate change benefits distribution around the world, which is described in Sect. 9.2.2. Climate-induced migration (Sect. 9.2.3) but also climate-driven financial flows (Sect. 9.2.4) are theoretically introduced in the following. The research questions (Sect. 9.3) investigate whether there are migration and lasting capital inflows into the countries that will benefit from global warming until 2100. The empirical Sect. 9.4 at first highlights the model used to calculate what countries will economically benefit from a warming earth (Sect. 9.4.1) and outlines the quantification of the imperative to redistribute the unequally-spread gains (Sect. 9.4.2). Sections 9.4.3 and 9.4.4

hold the previous findings of concrete country perspectives of being close to the cardinal optimal temperature around the globe in order to outline concrete country transfer obligations to reach a more right, just and fair gain distribution (Sects. 9.4.4–9.4.5). The current model under scrutiny is presented in Sect. 9.4.6, whereby the world migration flow operationalization is explained in Sect. 9.4.6.1 and world financial flows measurement is featured in Sect. 9.4.6.2. The results (Sect. 9.5) highlight a significant migration inflow into climate change winner countries as well as a positive correlation between climate change gains and foreign direct investment inflows. The chapter discusses (Sect. 9.6) recommendations to transfer the unequally-distributed gains of climate change to those parts of the world that are on the climate change losing edge. Prospects of the findings' impetus regarding climate refugees but also in the financialization of the 2015 Paris accord are given in order to conclude with future research prospects (Sect. 9.7).

9.2 Climate Change

9.2.1 *The Problem of Unequal Climate Change Gains and Losses Distribution*

Climate change has become reality. In the very many contemporary writings about global warming, the discussion has been centered around the negative impacts of climate change. The burden of climate change has been thematized and cost sharing strategies debated on the international level. The need for fairness in sharing the burden of climate change has been argued to considering national economies, the world society but also balance in between generations. In order to ensure interest in ratifying intergovernmental climate agreements, the question arises whether the incentive structure of burden sharing models is the appropriate tool to move the very many we need for climate action. A more engaging prospect is offered in contemplating on the benefits of a warming earth. Opening up the unidimensional climate change discussion, Chapter 6 described benefits arising from a warming earth in order to help build a more whole-rounded participatory incentive structure.

State-of-the-art welfare function measurements and economic productivity parameters became the basis for conclusions about a fair spread of the gains and benefits from a warming earth. Contemporary Gross Domestic Product (GDP) measurements were used to estimate the productivity of the agriculture, industry and service sectors around the world. Based on the cardinal optimal temperatures for the agriculture, industry and service sectors productivity, the average temperature per country around the world as well as climate projections of the year 2100 under the business-as-usual path, Chapter 6 revealed for the very first time climate winners and losers around the world. Overall and simply seen from a narrow-minded GDP perspective, the world will macroeconomically benefit more from climate change until 2100 than lose. Winning and losing from a warming earth is significantly positively correlated with the Paris COP 21 emissions country percentage of Greenhouse Gas (GHG) for ratification, leading to the conclusion that the countries that have the longest time horizon regarding a warming earth also lack motivation to mitigate global climate change based on the short-term benefits.

Puaschunder (2017b) also introduced the climatorial imperative—advocating for fairness in the distribution of the global earth benefits among nations based on Kant's (1785/1993) imperative to only engage in actions one wants to experience themselves being done to oneself. Passive neglect of action on climate mitigation is seen as an active injustice to others. While the method to measure the gains from climate change can certainly be refined in future studies, all these endeavors are a very first preliminary step to open a gate to future research, legal codifications and policy work to settle for a right, just and fair distribution of benefits and gains from a warming earth. The introduction of the gains from climate change is a novel approach that should solely be seen in connection to the imperative to distribute the gains in a fair manner among all world inhabitants and as means to hopefully draw attention to climate change to agnostic market actors or those who shy away from action given the overall negative connotation of burden sharing and loss aversion surrounding global warming.

9.2.2 *Climate Change Benefits Transfers*

Climate justice accounts for the most challenging global governance goal. In the current post-COP21 Paris agreementclimate change mitigation and adaptation efforts, the financialization of the ambitious goals has leveraged into a blatant demand. In the weighting of the burden of global warming, the benefits of a warming earth have been neglected since recently. The implementation of climate stability accounts for the most challenging contemporary global governance predicament that seems to pit world countries but also today's generation against future world inhabitants. In a trade-off of economic growth versus sustainability, a broad-based international coalition could establish climate stability. As a novel angle toward climate justice, the attention to global warming gains and losses being distributed unequally around the globe allows to propose to search for a well-balanced climate mitigation and adaptation public policy mix guided by micro- and macroeconomic analysis results.

Following the introduction of the gains from climate change, Chap. 6 proposes a model to distribute the benefits of a warming earth in a fair way based on which countries are losing and which countries are winning from a warming earth until 2100. A macroeconomic cost-benefit analysis thereby aids to find the optimum solution on how to distribute climate change benefits and burden within society.

When unidimensionally focusing on estimated GDP growth given a warmer temperature, over all calculated models assuming linear, prospect or hyperbolic gains and losses, the world will be gaining more than losing until 2100. Based on the WL index of 188 countries of the world, less countries ($n = 78$) will gain more from global warming until 2100 than more countries ($n = 111$) will lose from a warming earth. Based on the overall WL_{TT} index factored by GDP per inhabitant, global warming benefits are demanded to be redistributed in a fair way to offset climate change loser countries for climate change mitigation and adaptation efforts and to instigate a transition into renewable energy.

Having shed light on the gains of a warming earth demands for the redistribution of climate change benefits to those areas of the world that will be losing from a warming earth. In the implementation, a climate change bonds but also taxation strategies are recommended. Having found that there are gains from a warming earth demands to transfer benefits into areas of the world that will be primarily losing from climate change. In order to avoid governmental expenditure on climate change hindering economic growth (Barro, 1990), the climate transfers should be enacted through bonds and taxes.

Adding onto contemporary climate fund raising strategies ranging from emissions trading schemes (ETS) and carbon tax policies as well as financing climate justice through bonds as viable mitigation and adaptation strategies, climate justice is introduced to comprise fairness within a country but also among different nation states in a unique and unprecedented tax-and-bonds climate change gains and losses distribution (Puaschunder, 2016b).

Thereby, climate change winning countries should be using taxation to raise revenues to offset the losses incurred by climate change. Climate change losers could raise revenues by issuing bonds that have to be paid back by taxing future generations. Regarding taxation, within the winning countries, foremost the gaining GDP sectors should be taxed. Climate justice within a country should also pay tribute to the fact that low- and high-income households share the same burden proportional to their dispensable income, for instance enabled through a progressive carbon taxation. Those who caused climate change could be regulated to bear a higher cost through carbon tax in combination with retroactive billing through inheritance tax. Deriving respective policy recommendations for the wider climate change community in the discussion of the results is aimed at ensuring to share the burden but also the benefits of climate change within society in an economically efficient, legally equitable and practically feasible way. The 'Climate in the 21st Century' idea therefore offers a new way of funding climate change mitigation and adaptation policies but also the transition to renewable energy through broad-based climate stability bonds-and-taxation mix that also involves future generations (World Bank, 2015).

First Jeffrey Sachs (2014) proposed an intergenerational burden sharing idea by presenting a three-model climate change burden sharing through fiscal policy with bond issuing in order to reflect the implementation regarding contemporary finance and growth models with respect for maximizing utility of the model. In an overlapping-generations type model, research should elucidate climate change abatement and mitigation policies to lead to a fairer solution across generations. The current generation mitigates climate change and provides infrastructure against climate risk financed through climate bonds to be paid by future generations. Since for future generations the currently created externalities from economic activities—the effects of CO_2emissions—are removed, this entails that the current generations remain financially as well off as without mitigation while improving environmental well-being of future generations. As Sachs (2014) shows, this intergenerational tax-and-transfer policy turns climate change mitigation and adaptation policy into a Pareto-improving strategy. Shifting the costs for climate abatement to the recipients of the benefits of climate stability appears as novel, feasible and easily implementable solution to nudge many overlapping generations toward future-oriented loss aversion in the sustainability domain (Puaschunder, 2016a).

The respective bonds-and-tax climate stabilityfinancing strategy therefore proposes to bear the burden of climate in a right, just and fair way around the globe. In the climate change winner countries, taxation should become the main driver over financing climate stability strategies. Foremost, the industries winning from a warming climate should be taxed. As the Winner-Loser index is based on the cardinal temperatures for all GDP contributing sectors, agriculture, industry and service, the taxation should be enacted for those sectors having most time ahead. The rational is that these sectors will be gaining the most from a warming earth and will therefore be flourishing. These will also be the sectors that will attract human capital and Foreign Direct Investments (FDI).

The taxation models should aid to share the burden of climate change within society in a fair way. Regarding concrete climate taxation strategies, a carbon tax on top of the existing tax system should be used to reduce the burden of climate change and encourage economic growth through subsidies (Chancel & Piketty, 2015). Within a country, high- and low-income households should face the same burden of climate stabilization adjusted for their disposable income. First, climate justice within a country should pay tribute to the fact that low- and high-income households share the same burden proportional to their dispensable income, for instance enabled through a progressive carbon taxation. Those who caused climate change could be regulated to bear a higher cost through carbon tax in combination with retroactive billing through a corporate inheritance tax whenever there is a board change in the corporation. But also developed and underdeveloped countries as well as various overlapping generations are affected differently. Besides progressive taxation schemes to imbue a sense of fairness in climate change burden sharing, inheritance taxation is also a flexible means to reap past wealth accumulation, which potentially caused environmental damage. The burden of climate change mitigation and adaptation could therefore also be allocated in a fair way within society through contemporary inheritance tax in order to reap benefits of past wealth accumulation.

In addition, finding the optimum balance between consumption tax adjusted for disposable income through a progressive tax scheme will aid to unravel drivers of tax compliance in the sustainability domain. If climate taxation is perceived as fair and just allocation of the climate burden, this could convince tax payers to pay one's share. A novel 'service-and-client' atmosphere could promote taxpayers as cooperative citizens who are willing to comply if they feel their share as fair contribution to the environment. Taxpayers as cooperative citizens would then be willing to comply voluntarily following the greater goal to promote taxpayer collaboration and enhance tax morale in the environmental domain. International comparisons of tax behavior also reveal tax norms being related to different stages of institutional development of the government, which is an essential consideration in sharing the climate change burden in a fair manner between countries.

To shed light on the benefits of a warming earth, underlying all these taxation and bonds benefit redistribution efforts is a fairly novel strategy to derive fair climate gains distribution strategies around the world. In order to more vividly manifest the manifold gains of climate change and explore the impetus of climate change benefits around the globe, innovatively climate-induced migration and financial flows should be integrated.

9.2.3 Climate-Induced Migration

Globalization led to unprecedented mobility and monetary transfer possibilities (Centeno & Tham, 2012). International human and capital transfer reached deeper into more sectors of national economies than ever before as human labor mobility became reality (Held & McGrew, 2007). Mass transportation means and labor market flexibility but also instant monetary banking options hold enormous societal advancement and socio-economic development potentials (Gilpin, 2001; Puaschunder, 2017a).

Human migration is the long-distance movement by people from one country to another with the intentions of settling temporarily or permanently in the new location. According to the World Bank's Migration and Remittances Factbook 2016, more than 250 million people, or 3.4% of the world population, live outside their countries of birth. About 6% of international migrants are refugees, who flee conflict, violence or persecution across an international border (World Bank Group Migration and Development Brief 26, 2016). Asylum seekers are those in process of having their refugee status determined (World Bank Group Migration and Development Brief 26, 2016). Apart from political reasons, migration can also be driven by economic factors, family reunification or other circumstances, such as climate change, which is currently not included in the definition of causing refugees. By the end of 2014, there were 19.5 million refugees with a rising tendency peaking at estimated 25 million by the end of 2015 (World Bank Group Migration and Development Brief 26, 2016; United Nations High Commissioner for Refugees Report, 2015).

In the enormous movement of people, little attention has been paid to climate-induced migration. In light of migration, the developed world has identified the need to better manage the flows of refugees and migrants but concrete drivers for migration stream predictions are missing. Investigating sources of migration allows retrieving migration stream estimates in the future. Better estimating what kind of populations will be on the flow into what territories of the world allows to prepare for coming migration streams and therefore reduce human tragedy, conflict and economic inefficiencies during the relocation processes (Burke, Hsiang, & Miguel, 2015). In the focal point of questions centered around cutting-edge migration support and relocation strategies, this chapter investigates the demographics of contemporary migrants induced by global warming. Providing migration estimates of the future serves as a legal foundation for recognizing climate refugees under the Geneva Convention—which rules immediate asylum for those fleeing from armed conflict.[1]

9.2.4 *Climate-Induced Financial Flows*

Foreign Direct Investments (FDIs)are an indicator of economic productivity. FDIs are investments made by a company or individual in one country to another country based on business interests. As an international factor movement, FDI is a controlling ownership in a business enterprise in one country by an entity based in another country. FDI is thereby distinguished from foreign portfolio investment and other passive financial transfers in the securities of another country such as buying stocks and/or bonds by the active element of business control.

The controlled ownership can either be inorganic by buying a company or organic by expanding operations of an existing business in that country. FDIs can thereby either be establishing business operations or acquiring business assets in a foreign country through ownership or controlling interest in a foreign company.[2] FDIs establish either effective control of

[1] United Nations High Commissioner for Refugees, The 1951 Refugee Convention retrieved 4/20/2016 at http://www.unhcr.org/pages/49da0e466.html

[2] http://www.investopedia.com/terms/f/fdi.asp

or substantial influence over the decision making of a foreign business. Naturally, FDIs are made in countries with flourishing and/or prospected economically advancing areas of the world. Lasting management interests leading to FDIs are commonly made in open economies that offer a skilled workforce and above average economic growth prospects for investors.[3] Not only capital investments but also provision of infrastructure, management, technology, expertise and human capital are involved in the wake of FDIs.[4]

Concrete FDIs comprise the opening of subsidiaries or associate companies in a foreign territory, acquiring a controlling interest in an existing foreign company or a merger, acquisition, conglomerate or joint venture of a foreign company.[5] A minimum of 10% ownership at stake in a foreign-based entity is required by definition of the Organization of Economic Cooperation and Development (OECD).[6] Typically that resembles investors acquiring 10% or more of the ordinary shares or voting shares of a foreign company.[7] Effective controlling interest in a firm can also be established with less than 10% of the company's voting shares.[8] Concrete FDI driving forces are tax advantages, economic protection zones, investment subsidies, free land and land subsidies, relocation opportunities, expatriation in the original home territory, infrastructure subsidies, R&D support, human power, energy, derogation from regulation and marketing efforts.

FDIs are macroeconomic country efficiency factors and trade balance predictors. FDI inflow countries often feature absolute cost advantages, production differentiation advantages and economies of scale. FDIs usually also go along with firm-specific advantages such as extension of market power and competitive market gain and an internationalization strategy to mitigate market risks (Dunning & Pitelis, 2008). FDIs are associated with improved economic growth due to the influx of capital and increased tax revenues for the host country that often channels FDI into lasting infrastructure that boost development.

[3] http://www.investopedia.com/terms/f/fdi.asp

[4] http://www.investopedia.com/terms/f/fdi.asp

[5] http://www.investopedia.com/terms/f/fdi.asp

[6] http://www.investopedia.com/terms/f/fdi.asp

[7] http://www.investopedia.com/terms/f/fdi.asp

[8] http://www.investopedia.com/terms/f/fdi.asp

In neoclassical economics, competition from external companies is believed to lead to productivity gains and knowledge transfer opportunities that will eventually lead to greater efficiency in the host country. Application or contrast of a foreign entity's policies to a domestic subsidiary may improve overall corporate governance standards. Soft skill transfer through training and job creation but also the availability of more advanced technology for the domestic market and access to research and development resources are additional benefits in the wake of FDIs.[9] Besides employment and training that the local population enjoy with new business investments, new technology and management offer development opportunities for underdeveloped regions. FDI into developing and transition countries robustly increases local productivity growth (Havranek & Irsova, 2011). At the same time, FDI can become the reason for risk concern, political tensions and domestic frustration over productivity losses to foreign territories.

Around the world countries are competing over FDIs. In September 2013, the United States House of Representatives voted to pass the Global Investment in American Jobs Act of 2013 (H.R. 2052; 113th Congress), a bill which would direct the United States Department of Commerce to 'conduct a review of the global competitiveness of the United States in attracting foreign direct investment,' in order to aid job creation.[10]

In 2012, FDI in China reached $19.1 billion in the first six months of 2012, making China the largest recipient of FDI and topping the United States which had USD 17.4 billion of FDI.[11] In 2015, India emerged as top FDI destination surpassing China and the USA. India attracted FDI of USD 31 billion compared to USD 28 billion and USD 27 billion of China and the US respectively.[12] In 2016, China reportedly surpassed the USA to become the world's largest assets acquirer, measured by the value of corporate takeovers (Hellström, 2016). UK Prime Minister Theresa May has outlined that Britain's largest infrastructure including energy and skyscrapers have been built on FDI. A common critique is that FDIs are

[9] http://unctad.org/en/Docs/ciiem2d2_en.pdf

[10] https://www.congress.gov/bill/113th-congress/house-bill/2052/text

[11] http://www.reuters.com/article/us-china-us-investment-idUSBRE89N0EZ20121024

[12] http://timesofindia.indiatimes.com/india/India-pips-US-China-as-No-1-foreign-direct-investment-destination/articleshow/49160838.cms

increasing inequality in the distribution of globalization gains and losses (Dunning & Pitelis, 2008).

In addition to FDI, remittances are international monetary transfers. While FDI implies corporate sponsors and control, remittances are more likely monetary gifts sent home by international migrants. Remittances of migrating populations were recently found to be connected to intergenerational transfers (Puaschunder, 2017a). As such remittances of migrating populations could potentially counterbalance losses countries incur that have a human capital outflow. Just imagine a migrant worker, who sends remittances to his family back in his former home country. The international remittance flows from migrants to developing countries are estimated at 432 billion USD in 2015, which could counterweight negative externalities of brain drain in the wake of migration but also balance out competitive advantages of FDIs.

Around the world, India was the largest remittance-receiving country, with 69 billion USD in 2015, followed by China with 64 billion USD and the Philippines ranging at 28 billion USD. The growth rate of remittances to developing countries has fallen from 3.2% in 2014 to 0.4% in 2015 (World Bank Group Migration and Development Brief 26, 2016). For instance, remittances to India decreased by 21% in 2015, plummeted from 8% to 2.5% for Bangladesh, from 16.7% to 12.8% for Pakistan and from 9.6% to −0.5% in Sri Lanka (World Bank Group Migration and Development Brief 26, 2016). The slowdown in remittances is largely attributed to economic weakness in the major remittance-sending countries. Weak oil prices and inflated currencies in many remittance-source countries further depressed remittances flows (Puaschunder, 2017a).

Remittances play an important role in recipient countries as for supporting domestic consumption, economic growth and aiding a current account surplus. Remittances can also boost local businesses and real estate markets, production and banking sectors. Most recent international development efforts are therefore tightening regulations governing remittance transfers featuring the closure of bank accounts of money-transfer operators to prevent money laundering and terrorist financing. In addition, the Valletta Summit on Migration of November 2015 expressed the goal to 'promote cheaper, safer, legally-compliant and faster transfers of remittances and facilitate productive domestic investment.'[13] The

[13] http://www.consilium.europa.eu/en/meetings/international-summit/2015/11/11-12/

United Nations Sustainable Development Goals, which were incepted in September 2015, also put an emphasis on migration including explicit targets to ensure safe, orderly and regular migration, well-managed migration policies, efforts to end human trafficking, promoting decent labor conditions for migrant workers, including women migrants, reducing the costs of remittance transfers and the collection of statistics on migration.[14] The World Bank Report on Remittances (2016) also adds information on how migration and remittances help cope with natural disasters and epidemics.

With the recent introduction of the gains and benefits of global warming, the question arises whether the climate change winners will also be gaining from favorable positive FDI inflows into their territories. Shedding light on an inequality in regards to FDI drain into those parts of the world that will be winning from a warming earth as well as lacking remittances to counterbalance this trend provides multi-faceted insights into climate inequality in the twenty-first century.

9.3 Research Questions and Hypotheses

A warming global earth was recently revealed to hold economic advantages. Based on state-of-the-art welfare function measurements and economic productivity parameters, the economic gains and productivity-efficiency benefits from a warming earth were captured. Contemporary Gross Domestic Product (GDP) measurements served as basis for estimations about the productivity of the agriculture, industry and service sectors around the warming globe. Based on the cardinal, hence optimal, temperatures for the agriculture, industry and service sectors' productivity, the average temperature per country around the world as well as climate projections of the year 2100 under the business-as-usual path, climate winners and losers were outlined around the world from now on until the year 2100 in order to argue for fair climate change benefits distribution around the globe. The climatorial imperative was defined as ethical obligation to avert climate change concertedly (Puaschunder, 2017b). Even if being in a part of the world that is still gaining from a warming climate, people should be striving to avert actions that cause the earth to warm. Based on the categorical imperative, no one should participate in

[14] https://sustainabledevelopment.un.org/?menu=1300

actions with harmful consequences to others. One should only engage in those activities, which consequences one wishes to incur to oneself. The climatorial imperative serves as basis for climate change benefits transfers around the globe (Puaschunder, 2017b). In order to manifest the established facts that some parts of the world will be gaining from a warming earth and some are losing in economic terms, the following chapter investigates climate-induced migration streams and financial flows.

With the recent introduction of the gains and benefits of global warming, the question arises whether the climate change winners will also be gaining from favorable positive FDI inflows into their territories and if remittances may counterweight this tendency. Shedding light on an inequality in regards to human capital but also FDIs into those parts of the world that will be winning from a warming earth will highlight variegated climate inequalities in the twenty-first century.

In order to investigate whether the warming of the globe leads to migration into those parts of the world that are expected to gain economically from climate change, the following part will analyze if climate change winners also benefit from population inflows and monetary drain. Population inflow would underline the argument that some countries benefit from a warming earth and others lose out more nuancedly.

In order to clarify whether an unequal gains and loss distribution on a warming earth leads to unequal human capital movements, hypothesis $H1_{.1}$ investigates:

- Countries that economically benefit from climate change feature a significant migration inflow of human capital.

In addition, financial inflows are a measurement of economic prospect and imply that some countries gain and others lose from investment opportunities in the shadow of a warming globe.

In order to clarify whether an unequal gains and loss distribution on a warming earth leads to unequal financial capital movements, hypothesis $H2_{.1}$ investigates the following:

- Countries that economically benefit from climate change feature a significant foreign direct investment (FDI) inflow that is not counterweighted by a corresponding remittances outflow into the climate change losing parts of the world.

The prospective results offer insights for international development since brain drain and capital outflows hold unfavorable developments in fragile economies if not compensated by respective remittances (Puaschunder, 2017a).

9.4 Method

9.4.1 *Overall Model Assumptions*

Building on Chapters 6 and 7, economic output was measured under projected conditions of a warming earth. For 189 countries the GDP agriculture, industry and service sector composition was retrieved from the Central Intelligence Agency World Factbook.[15] Contemporary climate change projections estimate a mean temperature rise of approximately 4.24 degrees Celsius. The cardinal temperature C per GDP sector was related to the projected temperature in 2100 per country. From the current world temperature mean per country and the estimate of a 4.25 degree Celsius climate change until 2100, the closeness to cardinal temperature per sector was calculated for each country by using the following formula (9.1):

$$T_A = C_A - T_e \tag{9.1}$$

whereby T_A represents the closeness to the optimum cardinal temperature for agriculture calculated by the cardinal temperature for agriculture C_A, which equals to 28.5 degrees Celsius, subtracted by T_e as the temperature estimated for the year 2100 per country. The result for the distance to the optimum temperature for each sector for each country was then multiplied by the GDP contribution percentage of the sector using the following formula (9.2):

For the agricultural sector, formula (9.2) reads

$$WL_A = T_A \times GDP_A, \tag{9.2}$$

whereby WL_A stands for the climate change winner and loser index for agriculture per country comprised of the distance to the optimum cardinal temperature per GDP sector T_A multiplied by the percentage of agriculture contributing to the overall GDP indicated by GDP_A.

[15] https://www.cia.gov/the-world-factbook/

For the industry sector, formula (9.3) reads

$$WL_I = T_I \times GDP_I, \tag{9.3}$$

whereby WL_A stands for the climate change winner and loser index for industry per country comprised of the distance to the optimum cardinal temperature per GDP sector T_I multiplied by the percentage of industry contributing to the overall GDP indicated by GDP_I.

For the service sector, formula (9.4) reads

$$WL_S = T_S \times GDP_S \tag{9.4}$$

whereby WL_A stands for the climate change winner and loser index for service per country comprising the distance to the optimum cardinal temperature per GDP sector T_S multiplied by the percentage of the service sector contributing to the overall GDP indicated by GDP_S.

The overall Winner-Loser WL_T index was calculated per GDP sector leading the sector-specific Winner-Loser indices WL_A for the agriculture sector, WL_I for the industry sector and WL_S for the service sector, which were then added up into the Winner-Loser total index WL_T based on the following formula (9.5):

$$WL_T = WL_A + WL_I + WL_S \tag{9.5}$$

whereby WL_A denotes the Winner-Loser index for the agriculture sector, WL_I the index for the industry sector and WL_S the index for the service sector. All indices were calculated per country for the year 2100 business-as-usual projection. To account for gain-loss-prediction model differences, Chapter 6 outlined three model variants over gain and loss perspectives. A linear model included gains and losses based on formulas (9.6) and (9.7):

$$W = T^1 \times GDP \tag{9.6}$$

For the losers' index L, linear growth is assumed by calculating using formula (9.7) for each respective sector:

$$L = -T^1 \times GDP \tag{9.7}$$

per country and per GDP sector, which were then added up into the Winner-Loser total index WL_T based on formula (9.5).

A prospect model included concave gains and convex losses based on formulas (9.8) and (9.9):

$$W = T^{0.5} \times GDP \tag{9.8}$$

For the losers' index L, linear growth is assumed by calculating using formula (4.9) for each respective sector:

$$L = -T^2 \times GDP \tag{9.9}$$

per country and per GDP sector, which were then added up into the Winner-Loser total index WL_T based on formula (9.5).

A hyperbolic model included gains and losses based on formulas (9.10) and (9.11):

$$W = T^2 \times GDP \tag{9.10}$$

For the losers' index L, linear growth is assumed by calculating using formula (9.11) for each respective sector:

$$L = -T^2 \times GDP \tag{9.11}$$

per country and per GDP sector, which were then added up into the Winner-Loser total index WL_T based on formula (9.5).

The overall Winner-Loser WL index was calculated per GDP sector leading the sector-specific Winner-Loser indices WL_A for the agriculture sector, WL_I for the industry sector and WL_S for the service sector, which were then added up into the Winner-Loser total index WL_T based on formula (9.5). The mean overall Winner-Loser WL index was retrieved from summing the WL_T indices for the linear, prospect and hyperbolic models per country divided by $n = 3$.

Overall, a positive WL index result indicates a distance to the optimum ahead, whereas a negative index would be considered as a negative time prospect or overshooting regarding optimum temperature for GDP production. In sum, the more positive the index, the longer the period of time the country could expect to have a favorable climate for agriculture, industries or service production until peak condition; and the more negative the index, the sooner the country would (have) run out of efficiency time. In sum, the Winner-Loser index is an indicator of how much cool or moderate climate time a country still has ahead regarding the prospect of an optimum GDP productivity temperature and the assumption that the earth is warming. For a detailed explanation of the creation of the WL_T index, see Chapter 6.

9.4.2 *Modeling Distribution of Climate Change Gains and Losses*

In order to estimate a fair distribution of climate change gains and losses, for each model variant, the gains and losses were first calculated.

9.4.2.1 *Linear Model*

When unidimensionally focusing on estimated GDP growth given a warmer temperature and estimating a linear growth of losses and wins, overall the world will be gaining more than losing until 2100. Based on the *WL* index of 198 countries of the world and under the assumption of linear gains and losses in light of climate change, fewer countries ($n = 79$) will win more from global warming until 2100 than more countries ($n = 119$) will lose from a warming earth. In particular, seventy-nine countries of the world will gain $WL_L = 78{,}139.08$ in GDP output, whereas 119 countries of the world will lose an estimated $WL_L = -52{,}061$.

9.4.2.2 *Prospect Convex Losses and Concave Gains Model*

When unidimensionally focusing on estimated GDP growth given a warmer temperature and estimating an exponential growth of losses and concave wins, overall the world will be losing more than gaining until 2100. Based on the *WL* index of 188 countries of the world, more countries ($n = 113$) will lose more from global warming until 2100 than fewer countries ($n = 75$) will win from a warming earth. In particular, seventy-five countries of the world will gain $WL_P = 22{,}717.161$ in GDP output, whereas 113 countries of the world will lose an estimated $WL_P = -353{,}175.32$.

9.4.2.3 *Hyperbolic Model*

When unidimensionally focusing on estimated GDP growth given a warmer temperature and estimating hyperbolic growth, overall the world will be gaining more than losing until 2100. Based on the *WL* index of 188 countries of the world, fewer countries ($n = 79$) will gain more from global warming until 2100 than more countries ($n = 109$) will lose from a warming earth. In particular, seventy-nine countries of the world will gain $WL_H = 1{,}037{,}192$ in GDP output, whereas 109 countries of the world will lose an estimated $WL_H = -3{,}52{,}088$.

9.4.2.4 *Total Estimate*

The total Winner-Loser *WL* index was calculated per country based on the mean of the overall Winner-Loser index WL_L for the linear model, the overall Winner-Loser index WL_P for the prospect model, and the overall Winner-Loser index WL_H for the hyperbolic model, which were then added up into the Winner-Loser total index WL_{TT} based on the following formula (9.12):

$$WL_{TT} = (WL_L + WL_P + WL_H)/n \tag{9.12}$$

whereby WL_L denotes the Winner-Loser index for the linear model, WL_P the index for the prospect model and WL_H the index for the hyperbolic model and $n = 3$. All indices were calculated per country for the year 2100 business-as-usual projection.

9.4.3 *Global Warming Winners and Losers Around the World*

When unidimensionally focusing on estimated GDP growth given a warmer temperature, over all calculated models assuming linear, prospect or hyperbolic gains and losses, the world will be gaining more than losing until 2100. Based on the *WL* index of 188 countries of the world, fewer countries ($n = 78$) will gain more from global warming until 2100 than more countries ($n = 111$) will lose from a warming earth. In particular, seventy-eight countries of the world will gain $WL_G = 354{,}039{,}6345$ in GDP output, whereas 111 countries of the world will lose an estimated $WL_L = -232{,}613{,}188$.

Chapter 6 investigated the relation between GDP growth prospects in light of climate change. The relation between GDP growth prospects in light of climate change and percentage of GHG for ratification was studied based on the total and percentage of GHG emissions communicated by the Paris COP 21 Parties to the Convention retrieved in their national communications and GHG inventory reports as of December 2015.[16] Over a sample of 181 countries of the world, a highly significant correlation of $r_{\text{Pearson}(181)} = 0.215$, $p < 0.004$ between the *WL* index over all models and the self-reported percentage of GHG emissions for ratification was found. As a cross-validation check, the percentage of GHG emissions for ratification was significantly positively correlated

[16] http://unfccc.int/resource/docs/2015/cop21/eng/10.pdf#page=30

($r_{\text{Pearson}(181)} = 0.178$, $p < 0.016$) with self-reported GHG emissions per country.

This result leads to the conclusion that those countries that emit more GHG are the ones with a positive GDP prospect on the warming earth until 2100. The more time countries seem to have in a favorable climate for production, the likelier they are to emit GHG and hence contribute to global warming. All these results outline the need for attention to climate justice.

9.4.4 *Fair Climate Change Gains Distribution*

In order to offset the losses from climate change based on the overall WL_{TT} index, global warming benefits could be redistributed in a fair way to compensate climate change loser countries for climate change mitigation and adaptation efforts and to instigate a transition to renewable energy.

To ensure a fair benefit transfer strategy, the difference of $\Delta WL =$ 121,426,447 should be distributed based on the following criteria. To ensure a fair benefit transfer strategy between countries, the per-country contributions to alleviate the losses caused by global warming should only come from those countries that win from a warming earth, that is, seventy-eight nation-states. The winning countries' contribution should be made in relation to the other winning countries' contributions. Each country's gain in relation to the overall gain from a warming earth should be calculated based on formulas (9.13) and (9.14).

In order to account for country differences in offsetting global warming through GDP growth (especially on the winners' side) and the country differences in their ability to transfer to renewable energy, the overall GDP per inhabitant was factored into the transfer equation as outlined in formula (9.13):

$$R^W = \left(\frac{P_W}{WL_G}\right) \times \left(\frac{GDP_c}{I_c}\right) \tag{9.13}$$

whereby R^W denotes the total transfer of climate change wins obligation per winning country, and P_W is the country-specific GDP win divided by all wins $WL_G = 354{,}039{,}6345$ from a warming earth for the year 2100 under business-as-usual projection in order to derive the percentage fraction of wins per country. GDP_C equals the per-country GDP estimate per inhabitant I_c per country for the year 2016.

Losses were calculated based on formula (9.14):

$$R^L = \left(\frac{P_L}{P_{LT}}\right) \times \left(\frac{GDP_c}{I_c}\right) \tag{9.14}$$

whereby R^L denotes the total transfer of climate change compensation, and P_L is the country-specific GDP loss divided by all losses WL_L = −232,613,188 from climate change for the year 2100 business-as-usual projection in order to derive the relative fraction of losses per country. GDP_C equals the per-country GDP estimate per inhabitant I_c per country for the year 2016.

The overall transfers R^T equate to the sum of contributions from the countries benefiting from global warming and the losses incurred by the countries that lose from global warming until 2100. Overall, the transfers R^T derive from R^W and R^L, which must be equal based on the following formula (9.15):

$$R^T = R^W + R^L \tag{9.15}$$

whereby R^T denotes the total transfer of wins for compensation of losses.

9.4.5 Country Contributions

9.4.5.1 Climate Change Transfers

Based on the WL_{TT} index, which measures the distance to cardinal temperature per GDP sector in 188 countries of the world, there are 77 climate change transfer grantor countries and 109 climate change transfer beneficiary countries. The top 19 transfer grantor countries are Canada (WL_{TTB} = −7299.04), Russia (WL_{TTB} = −6999.3), Mongolia (WL_{TTB} = −5032.75), Kyrgyzstan (WL_{TTB} = −4154.31), Tajikistan (WL_{TTB} = −4089.42), Iceland WL_{TTB} = −3887.91), Finland (WL_{TTB} = −3798.36), Norway (WL_{TTB} = −3708.43), Sweden (WL_{TTB} = −3644.39), Latvia (WL_{TTB} = −2465.58), Georgia (WL_{TTB} = −2452.19), North Korea (WL_{TTB} = −2428.53), Switzerland (WL_{TTB} = −2394.05), Estonia (WL_{TTB} = −2342.77), Liechtenstein (WL_{TTB} = −2318.88), Lithuania (WL_{TTB} = −2188.99), Nepal (WL_{TTB} = −2131.26), Austria (WL_{TTB} = −2103.02), Belarus (WL_{TTB} = −2077.12) and Kazakhstan (WL_{TTB} = −2039.07).

The major transfer beneficiaries are Qatar (WL_{TT} = 2911.69), Bahrain (WL_{TT} = 2793.87), Brunei (WL_{TT} = 2591.98), United Arab Emirates (WL_{TT} = 2446.67), Kiribati (WL_{TT} = 2370.55), Mauritania (WL_{TT} = 2311.14), Tuvalu (WL_{TT} = 2228.85), Djibouti (WL_{TT} = 2140.24), Senegal (WL_{TT} = 2091.74), Burkina Faso (WL_{TT} = 2009.04), Maldives (WL_{TT} = 2008.46), Trinidad and Tobago (WL_{TT} = 1997.28), Mali (WL_{TT} = 1922.99), Equatorial Guinea (WL_{TT} = 1875.88), Sri Lanka (WL_{TT} = 1855.24), Palau (WL_{TT} = 1838.85), Seychelles (WL_{TT} = 1775.79), Saint Vincent and the Grenadines (WL_{TT} = 1772.60), Samoa (WL_{TT} = 1760.75) and Guinea (WL_{TT} = 1753.07).

Based on the WL_{TT} index weighted per GDP and per inhabitant, which measures the distance to cardinal temperature per GDP sector in 186 countries of the world, there are 76 climate change transfer grantor countries and 110 climate change transfer beneficiary countries.

The top twenty transfer grantor countries per capita are Liechtenstein (WL_{TTB} = −10.34), Canada (WL_{TTB} = −8.88), Norway (WL_{TTB} = −7.65), Iceland WL_{TTB} = −6.35), Switzerland (WL_{TTB} = −5.67), Luxembourg (WL_{TTB} = −5.52), Finland (WL_{TTB} = −3.77), Monaco (WL_{TTB} = −3.41), Denmark (WL_{TTB} = −3.19), Austria (WL_{TTB} = −3.03), the United States (WL_{TTB} = −2.87), Ireland (WL_{TTB} = −2.18), Germany (WL_{TTB} = −2.15), the United Kingdom (WL_{TTB} = −1.80), Belgium (WL_{TTB} = −1.44), New Zealand (WL_{TTB} = −1.41), Estonia (WL_{TTB} = −1.38), France (WL_{TTB} = −1.18), Latvia (WL_{TTB} = −1.15) and Slovenia (WL_{TTB} = −1.15).

The major transfer beneficiaries are Qatar (WL_{TT} = 19.06), United Arab Emirates (WL_{TT} = 9.33), Singapore (WL_{TT} = 7.90), Brunei (WL_{TT} = 6.06), Bahrain (WL_{TT} = 5.90), Kuwait (WL_{TT} = 3.94), Trinidad and Tobago (WL_{TT} = 2.98), Saudi Arabia (WL_{TT} = 2.92), Mauritania (WL_{TT} = 2.51), Seychelles (WL_{TT} = 2.39), Equatorial Guinea (WL_{TT} = 2.29), Palau (WL_{TT} = 2.27), Oman (WL_{TT} = 2.00), Antigua and Barbuda (WL_{TT} = 1.99), Barbados (WL_{TT} = 1.70), Venezuela (WL_{TT} = 1.29), Saint Kitts and Nevis (WL_{TT} = 1.28), Panama (WL_{TT} = 1.27), Grenada (WL_{TT} = 1.26) and Malaysia (WL_{TT} = 1.24).

For a more detailed description of the methodology of climate change winners and losers around the world, see Chapter 6, and for the distribution of global warming gains and losses, see Chapter 7.

9.4.6 *Sophisticated Climate Change Winners and Losers*

9.4.6.1 Climate Change-Induced Migration Patterns

Based on a world countries dataset, the relationship between being a climate winner or loser around the globe and migration and financial flows was investigated.

In order to operationalize hypothesis H1.1, whether countries that economically benefit from climate change feature a significant migration inflow of human capital, worldwide migration streams for the year 2013 were retrieved from the World Bank database. The bi-lateral migration inflow and outflow matrix for 2013 served as an indicator of whether a country is a positive net-migration inflow or net-migration outflow country based on formula (9.16):

$$M = M_I - M_O \tag{9.16}$$

whereby M denotes the net migration movement of a country, M_I the migration inflow into a country and M_O the migration outflow of a country.

9.4.6.2 Climate Change-Induced Financial Flow Patterns

In order to operationalize hypothesis H2.1, whether countries that economically benefit from climate change feature a significant financial inflow, worldwide FDI[17] and remittances[18] streams for the year 2015 were retrieved from the World Bank database.

Hypothesis H2.1 investigated whether countries that economically benefit from climate change feature a significant FDI inflow that does not have a corresponding remittances outflow from the FDI receiving territories into the FDI sponsoring countries.

9.5 Results

Human migration: When investigating the relation of human migration and climate change winners and losers in 187 countries of the world, a highly significant correlation ($r_{\text{Pearson}}(187) = 0.184$, $p < 0.006$) was found between being climate change winner and human migration

[17] http://data.worldbank.org/indicator/BX.KLT.DINV.CD.WD

[18] http://data.worldbank.org/indicator/BX.TRF.PWKR.CD.DT

inflows. A highly significant independent *t*-test with $t_{(185)} = -2.401$ and $p = 0.017$ of 110 climate change loser and 77 winner countries consolidates the finding.

Financial inflows: When investigating the relation of financial inflows and climate change winners and losers including 181 countries of the world, a significant correlation ($r_{Pearson}(181) = 0.180$, $p < 0.016$) was found between being climate change winner and FDI inflows. A highly significant independent *t*-test with $t_{(179)} = 2.680$ and $p = 0.008$ of 110 climate change loser and 77 winner countries consolidates the finding. As a cross validation of lasting financial inflows into climate change winner countries, a non-significant correlation and non-significant independent *t*-tests are reported.

9.6 Discussion

Climate change has become reality. Climate stability accounts for the most challenging global governance goal. In the very many contemporary writings about global warming, the discussion has been centered around the negative impacts of climate change. The burden of climate change has been thematized and cost-sharing strategies debated. The need for fairness in sharing the burden of climate change has been argued to consider national economies, the world society but also balance in between generations.

Given a potential neglect of interest in ratifying intergovernmental climate agreements, the question arises whether the incentive structure of burden sharing models is appropriate to move the very many we need for climate stabilization. Considering climate change only as problematic appears unidimensional. Researchers are therefore advised to pursue a more creative approach to overcome the obstacle of climate change.

While the contemporary climate negotiations may be stuck in the one-sided grilling to share burden, which lowers participation motivation, unraveling the unprecedentedly described benefits arising from a warming earth helps build a more whole-rounded participatory incentive structure. Shedding novel light on global warming from an almost unprecedentedly covered angle of outlining the ample and hidden benefits of a warming earth is targeted at finding solutions for a fair global warming benefits distribution.

This chapter innovatively integrated two contemporary societal worldwide crises in order to outline the climate-induced hegemony of certain countries. The empirical investigation covered climate change and migration in order to prepare for future migrant and financial flow prospect predictions but also global governance recommendations. A significant positive inflow of migrants was found into the climate change winner countries. A statistically significant correlation holds that there is a positive Foreign Direct Investment (FDI) financial inflow into the winning countries, which is not counterweighted by remittances. These migration and monetary flows are argued to vividly outline the competitive advantages of the climate change winner countries. The results target at underlining the importance of climate change benefits transfers to offset the losses incurred in the global warming-burdened areas of the world.

The introduction of the gains from climate change is a novel approach that should solely be seen in connection to the imperative to distribute the gains in a fair manner among all world inhabitants. Focusing on gains may serve as means to hopefully draw attention to climate change of agnostic market actors or those who shy away from action given the overall negative connotation of climate change burden sharing and economic productivity loss aversion. The novel argumentation point is targeted at helping to build a more whole-rounded participatory incentive structure as the basis for contemporary climate change negotiations. The provocative results therefore have to be seen with the caution of fairness in between society and the generations to come. In its entirety, the chapter therefore offered the first introduction of climate-induced financial flows. Highlighting climate-induced migration streams serves as unique justification for climate benefits transfers around the globe. The findings will also help recognize climate refugees under the Geneva Convention and protecting the rights of violently forced migration refugees in order to ensure climate equality in the twenty first century.[19]

The climatorial imperative advocates for the need for fairness in the distribution of the global earth benefits among nations based on Kant's (1785/1993) imperative to only engage in actions one wants to experience being done to oneself (Puaschunder, 2017b). Passive neglect of action on climate mitigation is argued as an active injustice to others. Countries passive or agnostic about global warming mitigation that reap benefits from a warming earth should therefore be obliged to

[19] United Nations High Commissioner for Refugees, The 1951 Refugee Convention retrieved 4/20/2016 at http://www.unhcr.org/pages/49da0e466.html

finance international aid for those countries that are directly and negatively impacted by climate change. In addition, building on common and international law, those countries that have better means of protection or conservation of the common climate should also face a greater responsibility to protect the earth (Puaschunder, 2016a, 2016b).

Innovatively differing from the contemporary climate change literature and discussions centered around the problem of a warming earth as prospected in losses from climate change demanding for a burden sharing strategy, investigating underlying benefits of climate change allows to derive fair benefits transfer distribution mechanisms. This innovative way to frame the prisoner's dilemma of a warming common climate as a benefits distribution conundrum proposes a completely new incentive structure, which will hopefully embrace a larger audience to be nudged through intriguing gains perspectives. As such, this chapter advocates for opening up unidimensional climate negotiations that are one-sidedly focused on climate stabilization problem solving through burden sharing.

The results also underline the importance of attention to climate refugees and to advocate for international protection of climate-induced asylum seekers. The findings also stress the need to consider climate refugees under the Geneva Convention asylum protectorate. Based on the United Nations High Commissioner for Refugees 1951 Geneva Refugee Convention[20] and amended by the 1967 United Nations Office of the High Commissioner for Human Rights New York Protocol relating to the Status of Refugees[21] as well as the European Union Qualification Directive 2011/95/EU[22] and the Dublin Regulation (2013), asylum is 'granted to a person who is unable to seek protection in his/her country of citizenship and/or residence, in particular for fear of being persecuted for reasons of race, religion, nationality, membership of a particular social group, or political opinion.' Until today climate refugees are not considered as refugees under the Geneva Convention and thus lack instant asylum protection. In the followingly prospected consideration of climate

[20] United Nations High Commissioner for Refugees, The 1951 Refugee Convention retrieved 4/20/2016 at http://www.unhcr.org/pages/49da0e466.html

[21] United Nations High Commissioner for Human Rights, Protocol relating to the Status of Refugees retrieved 4/20/2016 at http://www.ohchr.org/EN/ProfessionalInterest/Pages/ProtocolStatusOfRefugees.aspx

[22] European Union Eur-Lex retrieved 4/20/2016 at http://eur-lex.europa.eu/legal-content/EN/TXT/?uri=celex%3A32011L0095

refugees under the Geneva Convention accord, a first guide how to distributing climate refugees in a fair manner around the globe is given through the global warming-induced migration patterns outlined. The demand for those countries that will be winning from global warming, should be obliged to accept climate refugees, is firstly quantified in the presented work.

While the method to measure the gains from climate change can certainly be refined in future studies, this chapter is meant as very first preliminary step to open a gate to legal codifications and policy work to settle for a right, just and fair distribution of benefits but also legally-sound and economically-beneficial relocation strategies on our common warming mother earth.

Future research studies could draw a more refined map of where migrants are moving within a country, with particular focus on the rural versus metropolitan areas. This will hold invaluable opportunities for city developers as well as governance officials to concert a harmonious integration of climate migrants into the productive workforce. Security concerns of a warming temperature could be added to the future research endeavors in order to cover positive as well as negative impacts of a warming globe with a focus on climate refugees (Burke et al., 2015).

Additional research studies could investigate remittances as innovative way to implement climate justice that may counterweight the financial and human capital losses in climate change loser areas. Especially intergenerational responsibility and interfamilial ties that solicit remittances could be introduced as natural international development enhancement. For instance, remittances are found to occur if moved population parts are conscientious of their family members abroad and attentive to their former home countries in monetary gifts (Puaschunder, 2017b). Since their strong intergenerational ties are likely to elicit empathy with their family members, climate refugees may send back parts of their earnings to their former home territories. Intergenerational care and family ties may thus lead to more remittances and also help migrants advocate for international help for their former homelands in their new territories. In this light, intergenerational ties would connect and therefore be considered as pure—as for being free of technocratic, bureaucratic and institutional downfalls, such as corruption—direct and long-term oriented international development aid through remittances (Puaschunder, 2017b). If so, intergenerational ties could be portrayed as to psychologically last stronger than nation states, on which refugees often have turned their back. The intergenerational glue could therefore be unraveled as an unprecedentedly described implicit international development means and global climate change losses aversion (Puaschunder, 2017b).

Future research should also clarify real-life influence factors on intergenerational responsibility and how intergenerational ties can be strengthened in the family compound but also over distance and time. Future studies could therefore also aim at deriving public policy recommendations on how to harmoniously implement intergenerational equity and foster socio-economic mobility in a changing world climate. Practitioners could gain from information on climate-induced migration streams in order to be better equipped to integrate migrants in society. Meaningfully relocation of migrants into the local labor markets will enhance their opportunity to send remittances to those family members who are left in climate losing territories that are physically shunned from the benefits of a warming earth.

For international institutions, the results offer a legitimization of the importance of considering climate refugees under international law protection in order to argue for certain countries to let migrants in. In addition, climatorial transfer implementations and potential downfalls should be investigated. Methodological extensions could embrace a larger dataset to run a Bayesian analysis, with the hope to get closer to the true parameter value (Gelman et al., 2013). In addition to cross-sectional analyses, time series could add more information on an eternal effect of climate on migration and financial flows.

In addition to methodological advancements of the measurement of climate gains and losses, future research could more sophistically unravel the unequal gravitation of climate change gains and losses around the globe and over time. The role of international trade regimes and implicit trade benefits and burdens should be integrated in a more real-world adapted investigation of global warming.

All these efforts should alleviate the contemporary global governance predicament that seems to pit today's generation against future world inhabitants in a trade-off of economic growth versus sustainability (Puaschunder, 2016c). Deriving respective policy recommendations and incentives to spread the prospective global warming gains for the wider climate change community around the globe and over time is aimed at ensuring that the shared benefits of climate change reach all contemporary and future world inhabitants around the globe in an economically efficient, legally sound and ethically equitable but also practically feasible way (Puaschunder, 2016f).

9.7 Conclusion

Overall, capturing the current Zeitgeist on climate change and migration provides a real-world relevant contemporary snapshot of the importance of responsibility for finding an equal spread of climate-induced welfare around the globe (Puaschunder, 2016e). Theoretically supporting the idea of gains of a warming earth fortifies the climatorial imperative's application in the fair distribution of climate-induced gains (Puaschunder, 2017b). Empirically, the results offer an internationally validated climate-induced migration and financial inflow model that supports previous findings of the gains of a warming earth holistically alongside elucidating the importance of fair climate gains transfer demand for international development. From a practitioners' standpoint, the chapter gives a descriptive case of contemporary migration patterns in relation to a potential driver of migration in the twenty-first century. All these invaluable findings will help craft validated models for the transfers of a warming earth's benefits as an international law and international development means. Remittances and the intergenerational glue may become especially important when to compensate for brain drain and financial losses. Capturing intergenerational responsibility as a form of international development provides avenues for harmoniously integrating intergenerational equity in contemporary global governance paradigms. Overall, the research aims at fostering hope in intergenerational care spreading prosperity over distance and time following the greater goal of international advancement in times of global warming-induced migration and financial flows (Graph 9.1).

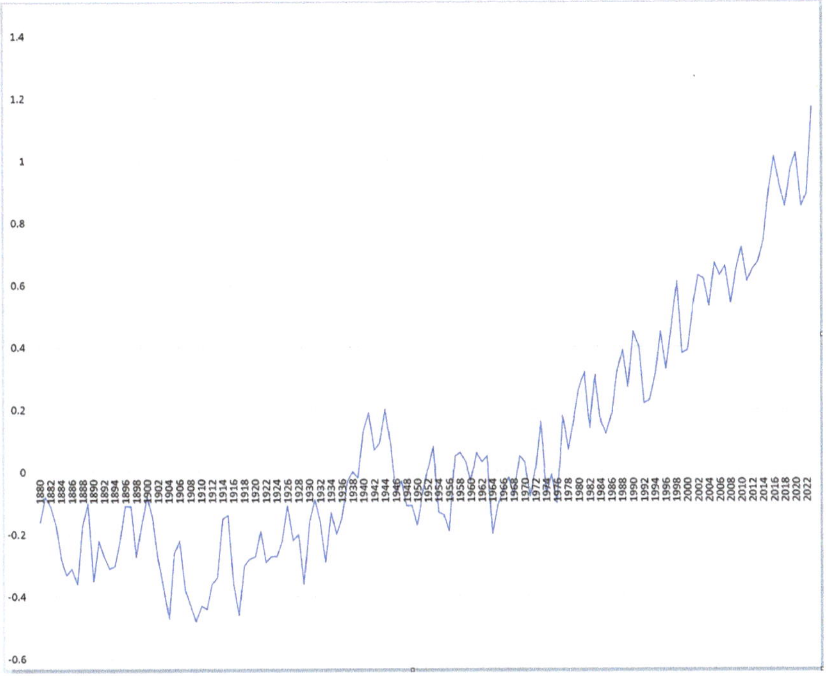

Graph 9.1 Global land-ocean temperature index from 1880–2023

References

Barro, R. (1990). Government spending in a simple model of endogenous growth. *Journal of Political Economy, 98*, 103–125. https://doi.org/10.1086/261726

Burke, M., Hsiang, S. M., & Miguel, E. (2015). Climate and conflict. *Annual Review of the Economics, 7*(8), 577–617. https://doi.org/10.1146/annurev-economics-080614-115430

Centeno, M. A., & Tham, A. (2012). *The emergence of risk in the global system*. Working paper. Princeton University.

Chancel, L., & Piketty, T. (2015). *Carbon and inequality: From Kyoto to Paris*. Paris School of Economics.

Dunning, J., & Pitelis, C. N. (2008). Stephen Hymer's contribution to international business scholarship and assessment and extension. *Journal of International Business Studies, 39*(1), 167–173. https://doi.org/10.1057/palgrave.jibs.8400328

Ferreira, V. (2017). Climate induced migration: Legal challenges. In J. M. Puaschunder (Ed.), *Intergenerational responsibility in the 21st century* (pp. 107–121). Vernon.

Gelman, A., Carlin, J. B., Stern, H. S., Dunson, D. B., Vehtari, A., & Rubin, D. B. (2013). *Bayesian data analysis*. Chapman & Hall. https://doi.org/10.1201/b16018

Gilpin, R. (2001). *Global political economy: Understanding the economic international order*. Princeton University Press.

Havranek, T., & Irsova, Z. (2011). *Which foreigners are worth wooing? A meta-analysis of vertical spillovers from FDI*. IES Working Paper, No. 16/2010.

Held, D., & McGrew, A. G. (2007). *A new world economic order? Global markets and state power: Beyond globalization/anti-globalization: Beyond the great divide*. Polity.

Hellström, J. (2016). China's acquisitions in Europe: European perceptions of Chinese investments and their strategic implications. https://www.foi.se/download/18.6f5585fa15a36ec7a961b9b/1487592009526/foir_4384.pdf

Kant, I. (1785/1993). *Grounding for the metaphysics of morals*. Hackett.

Puaschunder, J. M. (2016a). *Intergenerational transfer model*. Unpublished manuscript. The New School.

Puaschunder, J. M. (2016b). Intergenerational climate change burden sharing: An economics of climate stability research agenda proposal. *Global Journal of Management and Business Research: Economics and Commerce, 16*(3), 31–38.

Puaschunder, J. M. (2016c). Mapping climate justice. In *Proceedings of the 2016 Young Scientists Summer Program Conference*, International Institute for Applied Systems Analysis (IIASA), Laxenburg, Austria.

Puaschunder, J. M. (2016d). On eternal equity in the fin-de-millénaire: Rethinking capitalism for intergenerational justice. *Journal of Leadership, Accountability and Ethics, 13*(2), 11–24.

Puaschunder, J. M. (2016e). Putty capital and clay labor: Differing European Union capital and labor freedom speeds in times of European migration. *The New School Economic Review: A Journal of Critical Economics at The New School, 8*(3), 147–168.

Puaschunder, J. M. (2016f). The call for global responsible intergenerational leadership in the corporate world: The quest for an integration of intergenerational equity in contemporary Corporate Social Responsibility (CSR) models. In D. Jamali (Ed.), *Comparative perspectives in global corporate social responsibility* (pp. 275–288). IGI Global Advances in Business Strategy and Competitive Advantage Book Series.

Puaschunder, J. M. (2017a). Gifts without borders: Intergenerational glue connecting over distance and time as pure international development in the age of migration. In J. M. Puaschunder (Ed.), *Intergenerational responsibility in the 21st century*. Vernon.

Puaschunder, J. M. (2017b). The climatorial imperative. *Agriculture Research & Technology, 7*(4), 1–2.
Rovenskaya, E. (2008, January). Optimal economic growth under stochastic environmental impact: Sensitivity analysis. *International Institute for Applied Systems Analysis Interim Report*. Laxenburg, Austria.
Sachs, J. D. (2014). Climate change and intergenerational well-being. In L. Bernard & W. Semmler (Eds.), *The Oxford handbook of the macroeconomics of global warming* (pp. 248–259). Oxford University Press.
The World Bank. (2015). Green bonds attract private sector climate finance. World Bank Brief.
The World Economic Forum Report. (2015). Davos, Switzerland: World Economic Forum.
United Nations High Commissioner for Refugees Report. (2015). Retrieved October 5, 2016, from http://www.unhcr.org/56701b969.html
World Bank Group Migration and Development Brief 26. (2016, April). *Migration and remittances: Recent development and outlook*. International Bank for Reconstruction and Development, World Bank Group.

CHAPTER 10

Looking Forward to World Peak: Climate Change-Induced Market Prospects

10.1 Introduction

Climate change has become a reality (Nature Climate Change, 2018). Scientists around the world are documenting the impacts of global warming and collecting evidence of this climatorial shift to be human-caused (Savva, 2018). In the manifold writings about environmental changes, a main discourse has embarked on the different facets a hotter climate will impose on the world population and its natural habitat (Environmental Defense Fund, 2018). Climate change is already and further predicted to change weather patterns, farmland, plant and animal habitats (Environmental Defense Fund, 2018). Human lives are said to be affected daily due to warming temperature (Environmental Defense Fund, 2018). Extreme novel risks impose dangers for humans, habitat, plants, infrastructure and transportation (Centeno et al., 2013; Centeno & Tham, 2012; Environmental Defense Fund, 2018; Strobl, Elliott, & Cole, 2018). Higher heat alters ice, creates extreme weather conditions and disturbs oceans (Environmental Defense Fund, 2018). Natural habitats are prospected to become more hostile in the future as the ecosystem faces more risk of collapsing (Environmental Defense Fund, 2018). Human lives and prosperity are assumed to suffer due to declining health conditions and livelihoods. What follows is a naturally logic consensus to avert the warming of our common earth.

J. Puaschunder, *Governance & Climate Justice*,
https://doi.org/10.1007/978-3-031-87971-5_10

The economic discourse on climate change is historically and traditionally focused on discounting future anticipated changes and losses (Allais, 1947; Bauer, 1957; Harrod, 1948; Klaassen & Opschoor, 1991; Kruse & Wegener, 2017; Nordhaus, 2008; Orlov et al., 2024; Ramsey, 1928; Rovenskaya, 2008). Contemporary energy efficiency in a transitioning economy under climate uncertainty is widely debated (Adao et al., 2018; Antonietti & Fontini, 2018; Batten, 2018; Benchekroun et al., 2018; Buso et al., 2018; Court & Bovari, 2018; Egli et al., 2018; Forbes & Zampelli, 2018; Ielpo & Sevi, 2013; Kilian & Zhou, 2017; Massol et al., 2018; Orttung, 2009, 2011a, 2011b, 2011c, 2011d, 2013a, 2017; Shahbaz et al., 2018; Talevi, 2018). Restructuring industrial systems needs a diverse, cross-sector approach so that green industrial policy frameworks can become a valuable tool for all economies (Altenburg & Assmann, 2018). The policy recommendations range from taxation, transfer payments and bond solution proposals to ensure climate stabilization and intergenerational fairness (Bernarda & Kichian, 2017; Bürgenmeier, 1994; Cambini & Soroush, 2018; Chancel & Piketty, 2015; Douenne, 2018; Kirat, 2018; Marron & Morris, 2016; Najm, 2018; Puaschunder, 2016a, 2017a; Sachs, 2014; The World Bank, 2015). Intergenerational responsibility has been defined as imperative in the twenty-first century (Puaschunder, 2016c, 2018).

Most recently, economists have turned to measuring the actual gains and losses of a warming globe for the global economy. Dell, Jones, and Olken (2012) use annual variations in temperature and rainfall to identify causal impacts of weather shocks on per capita Gross Domestic Product (GDP) growth. Burke, Hsiang, and Miguel (2015) estimate weather and precipitation effects on economic output for 180 countries. Contemporary International Monetary Fund (IMF) research concludes that in low-temperature countries, an increase in temperature boosts economic activity, while in higher temperature countries, an increase in temperature will dampen economic activity (Acevedo et al., 2018). The findings of uneven economic effects are robust over the different models and backed by regional cluster analysis and geographic case studies (Somanathan et al., 2014). The IMF model now sophisticates these findings with insights about the effect of temperature on productivity, capital and labor.

In a focused macroeconomic model, Puaschunder (2016b) maps climate justice based on the conclusion that low-temperature countries closer to the poles of the earth have more time until the optimal cardinal temperature for economic output peak condition. State-of-the-art welfare function measurements and economic productivity parameters serve as basis for conclusions about climate change benefits and losses. Contemporary GDP measurements help infer future prospects about the productivity of the agriculture, industry and service sectors around the world in a changing temperature. The cardinal temperatures for the agriculture, industry and service sectors productivity, the average temperature per country around the world as well as climate projections of the year 2100 under the business-as-usual path determine climate winners and losers around the world from now on until the year 2100.[1] When unidimensionally focusing on estimated GDP growth given a warmer temperature, over all calculated models assuming linear, prospect or hyperbolic gains and losses, the world will be gaining more than losing from a warming earth until 2100. Based on the *WL* index of 188 countries of the world, less countries ($n = 78$) will gain more from global warming until 2100 than more countries ($n = 111$) will lose from a warming earth. Winning and losing from a warming earth is significantly positively correlated with the Paris COP 21 emissions country percentage of Greenhouse Gas (GHG) for ratification, leading to the conclusion that the countries that have the longest time horizon regarding a warming earth also lack motivation to mitigate global climate change.

The conclusions from all these novel insights are that the contemporary discussion of climate change is biased toward reporting negative aspect of global warming, while the immediate future holds short-term gains of a warming globe for at least some countries. Ignorance about the change of climate not only takes away the opportunity to reap benefits of an unprecedented market condition that will bestow invaluable assets for those with foresights, who move first. But also not reporting and investigating how climate change gains and losses are distributed around the globe takes away the opportunity to derive clear strategies on how to redistribute gains from the winning nation-states to areas that will be losing economically from a hotter temperature. From an inequality alleviation perspective, shedding light at the gains of a warming globe is therefore an ethical imperative (Puaschunder, 2017a, 2017b, 2017c). For the global governance community, outlining gains will ensure to secure the necessary funding for stabilizing the climate and grant means

[1] https://juliampuaschunder.com/mapping-climate-justice/.

to financially support a transition to renewable energy (Giraud et al, 2018; Oppenheimer et al., 2011).

Drawing on the conclusion of the climatorial imperative—advocating for the need for fairness in the distribution of the global earth benefits among nations based on Kant's (1785/1993) imperative to only engage in actions one wants to experience being done to oneself—the redistribution of climate gains and losses is argued philosophically and ethically to alleviate climate inequality (Puaschunder, 2017a, 2017b, 2017c). In a macroeconomic cost-benefit analysis, Puaschunder (2017a) sketched a solution on how to distribute climate change benefits and burden within society. Based on the overall WL_{TT} index factored by GDP per inhabitant, global warming benefits are demanded to be redistributed in a fair way to offset the costs of climate change loser countries for climate change mitigation and adaptation efforts and to instigate a transition into renewable energy. Climate justice is thereby proposed in an unprecedented tax-and-bonds climate change gains and losses distribution strategy. Climate change winner countries are advised to use taxation to raise revenues to offset the losses incurred by climate change. Climate change losers could raise revenues by issuing bonds that have to be paid back by taxing future generations.[2]

However, in the economic distribution of climate change transfer grantors and beneficiaries around the world, political considerations may limit the actual feasibility to carry through with plans. Throughout the history of modern humankind, politics have influenced transfer payments and remittances (Orttung, 2017). The concept of nation-states has directly determined access to resources and energy control of the populace (Orttung, 2014a, 2014b; Orttung & Wenger, 2012). Cooperation and conflict around the preservation and distribution of common environmental conditions appear to obey more the laws of politics than to follow from rational economic calculus (Orttung, 2013a, 2013b; Orttung et al., 2012; Puaschunder & Schwarz, 2012).

The following chapter therefore builds on previous findings of global climate change winners and losers around the world. In advocating for stepping away from country denominations when transferring some of the expected climate change gain also winning marketindustries could be taxed to raise funds for offsetting costs arising in climate change losing

[2] http://blogs.cuit.columbia.edu/jmp2265/.

markets. But who will be the climate change winning market actors, agents and industries? This book aims to unravel this in the following chapter.

The chapter starts with outlining the mapping climate justice approach to divide all nation-states into climate change (Sect. 10.2.1) winning (Sect. 10.2.2) and losing GDP prospects areas arising from global warming (Sect. 10.2.3). In the following, a transfer payments scheme will be introduced to offset the economic losses with economic gains from a warming earth (Sect. 10.2.5) based on moral and ethical imperatives to redistribute gains to offset for losses of global warming in a right, just and fair way (Sect. 10.2.4). The book then builds a politically more neutral redistribution dynamism based on climate change market winners and losers (Sect. 10.3). A novel model featuring market actors, agencies and industries that may gain and lose from a temperature rise will be envisioned (Sect. 10.4) based on actual GDP components (Sect. 10.5). The end of this chapter then discusses the results (Sect. 10.6) and grants prospective future research avenues with a particular focus on financial market outcome prospects in the light of a warming globe (Sect. 10.7).

10.2 Theoretical Background

10.2.1 Climate Change

Climate change is the change in average weather conditions, which is currently observed in climate record data and predicted to be long-term. The term 'climate change' was proposed by the World Meteorological Organization in 1966 to encompass all forms of climatic variability on time-scales longer than 10 years but regardless of cause (Hulme, 2016). Internal and external forcing mechanisms are attributed to cause global warming. Internal forcing mechanisms are natural processes within the climate, such as thermohaline circulation. External forcing mechanisms can be either human-caused—such as human-made greenhouse gas emissions—or natural based on the solar output, the earth's orbit and volcano eruptions. Today, climate change monitoring and resilience is foremost incorporated in the title of the Intergovernmental Panel on Climate Change (IPCC) and the UN Framework Convention on Climate Change (UNFCCC).

The contemporary climate change discourse is widely focused on the measurement of climate change, the contributing forcing factors to global warming and the ample effects a heating up earth will have on humans, animals and the natural habitat now and for centuries to come. Currently, record breaking temperatures are recorded since temperature measurement started in the late 19th century (Potter et al., 2017). Global warming is also found in Asian monsoon records, Northern Hemisphere temperature recordings, Alpine glacier extent and Arctic temperature anomalies over a 100-year period as estimated by NASA.[3] The impacts of climate change are reported to be unpredictable. The scientific literature mentions rising sea level, tide changes, coral reef destruction, soil, vegetation and precipitation changes, rainforest collapse, habitat isolation and the extinction of plants and animal species, but also drought, pests, diseases and forest fire imposing additional threats to human, animals, vegetation, forests, oceans, sediments and their populations around the world (Konnert et al., 2015).

The contemporary climate consequence literature is coupled with insights on how to avert global warming in climate change mitigation and adaptation efforts in the global governance, legal, research, public administration and policy domains (Nordhaus, 1994; Puaschunder, 2018). Case studies and teaching insights follow (Engel et al., 2017; Helm, 2009; Nordhaus, 1994). Ethical imperatives and human rights consequences are currently derived from the insights on global warming (Ferreira, 2017; Pope Francis, 2015; Puaschunder, 2016d, 2017b, 2017c; Rawls, 1971). Adding onto contemporary climate fund-raising strategies ranging from emissions trading schemes (ETS) and carbon tax policies to financing climate justice through bonds as viable mitigation and adaptation strategies, climate justice is introduced to comprise fairness within a society, between countries and also over generations (Puaschunder, 2016d, 2017b, 2017c). In all these writings, information about the gains of global warming appears to be missing. A primary focus on the downsides of global warming, however, appears critical in light of human tendencies of loss aversion and avoidance of negative enforcement. Highlighting the positive gains from a warming globe offers unprecedented opportunities to engage the many stakeholders needed for solving an as complex challenge as climate change appears to be imposing (Puaschunder, 2010).

[3] https://science.nasa.gov/climate-change/evidence/.

10.2.2 Climate Change Gains

Building on the previous chapters, economic output was measured under projected conditions of a warming earth. For 189 world countries, the GDP agriculture, industry and service sector composition was retrieved from the Central Intelligence Agency World Factbook.[4] Contemporary climate change projections estimate a mean temperature rise of approximately 4.24 degrees Celsius (°C) or 39.63 degree Fahrenheit (°F). The cardinal temperature C per GDP sector was related to the projected temperature in 2100 per country. From the current world temperature mean per country and the estimate of a 4.24 °C (39.63 °F) climate change until 2100, the closeness to cardinal temperature per sector was calculated for each country by using the following formula (10.1):

$$T_A = C_A - T_e \tag{10.1}$$

whereby T_A represents the closeness to the optimum cardinal temperature for agriculture calculated by the cardinal temperature for agriculture C_A, which equals 28.5 °C or 83.3 °F based on wheat and rice production capacities, subtracted by T_e as the temperature estimated for the year 2100 per country. The result for the distance to the optimum temperature for each sector for each country was then multiplied by the GDP contribution percentage of the sector using the following formulas (10.2–10.4):

For the agricultural sector, formula (10.2) reads

$$WL_A = T_A \times GDP_A, \tag{10.2}$$

whereby WL_A stands for the climate change Winner and Loser index for agriculture per country comprising the distance to the optimum cardinal temperature per GDP sector T_A multiplied by the percentage of agriculture contributing to the overall GDP indicated by GDP_A.

For the industry sector, formula (10.3) reads

$$WL_I = T_I \times GDP_I, \tag{10.3}$$

whereby WL_A stands for the climate change Winner and Loser index for industry per country comprising the distance to the optimum cardinal

[4] https://www.cia.gov/the-world-factbook/.

temperature per GDP sector T_I multiplied by the percentage of industry contributing to the overall GDP indicated by GDP_I.

For the service sector, formula (10.4) reads

$$WL_S = T_S \times GDP_S, \tag{10.4}$$

whereby WL_A stands for the climate change Winner and Loser index for service per country comprising the distance to the optimum cardinal temperature per GDP sector T_S multiplied by the percentage of the service sector contributing to the overall GDP indicated by GDP_S.

The overall Winner-Loser WL_T index was calculated per GDP sector leading to the sector-specific Winner-Loser indices WL_A for the agriculture sector, WL_I for the industry sector and WL_S for the service sector, which were then added up into the Winner-Loser total index WL_T based on the following formula (10.5):

$$WL_T = WL_A + WL_I + WL_S, \tag{10.5}$$

whereby WL_A denotes the Winner-Loser index for the agriculture sector, WL_I the index for the industry sector and WL_S the index for the service sector. All indices were calculated per country for the year 2100 business-as-usual projection. The following three specific model variants account for gain-loss-prediction model differences. A linear model included gains and losses based on formulas (10..6) and (10.7):

$$W = T^1 \times GDP \tag{10.6}$$

For the losers' index L, linear growth is assumed by calculating using formula (10.7) for each respective sector:

$$L = -T^1 \times GDP, \tag{10.7}$$

per country and per GDP sector, which were then added up into the Winner-Loser total index WL_T based on formula (10.5).

A prospect model included concave gains and convex losses based on formulas (10.8) and (10.9):

$$W = T^{0.5} \times GDP \tag{10.8}$$

For the losers' index L, linear growth is assumed by calculating using formula (10.9) for each respective sector:

$$L = -T^2 \times GDP, \tag{10.9}$$

per country and per GDP sector, which were then added up into the Winner-Loser total index WL_T based on formula (10.5).

A hyperbolic model included gains and losses based on formulas (10.10) and (10.11):

$$W = T^2 \times GDP \tag{10.10}$$

For the losers' index L, linear growth is assumed by calculating using formula (10.11) for each respective sector:

$$L = -T^2 \times GDP, \tag{10.11}$$

per country and per GDP sector, which were then added up into the Winner-Loser total index WL_T based on formula (10.5).

The overall Winner-Loser *WL* index was calculated per GDP sector leading the sector-specific Winner-Loser indices WL_A for the agriculture sector, WL_I for the industry sector and WL_S for the service sector, which were then added up into the Winner-Loser total index WL_T based on formula (10.5). The mean overall Winner-Loser *WL* index was retrieved from summing the WL_T indices for the linear, prospect and hyperbolic models per country divided by $n = 3$.

Overall, a positive *WL* index result would indicate a long distance to the optimum, whereas a negative index would be considered as negative prospect. The more positive the index, the longer time the country could expect to have a favorable climate for agriculture, industries or service production until peak condition and the more negative the index, the sooner the country would (have) run out of efficiency time. In sum, the Winner-Loser index is an indicator of how much cool time a country still has ahead regarding an optimum GDP productivity temperature and the assumption that the earth is warming. For a detailed explanation of the WL_T index creation, see the previous chapters.

In order to estimate a fair distribution of climate change gains and losses, for each model variant, the economic gains and losses were first calculated.

10.2.2.1 *Linear Model*

When unidimensionally focusing on estimated GDP growth given a warmer temperature and assuming a linear growth of losses and wins, overall the world will be gaining more than losing until 2100. Based on the *WL* index of 198 countries of the world and under the assumption of linear gains and losses in light of climate change, fewer countries (n = 79) will win more from global warming until 2100 than more countries (n = 119) will lose from a warming earth. In particular, 79 countries of the world will gain WL_L = 78,139.08 in GDP output, whereas 119 countries of the world will lose an estimated WL_L = −52,061.

10.2.2.2 *Prospect Convex Losses and Concave Gains Model*

When unidimensionally focusing on estimated GDP growth given a warmer temperature and assuming an exponential growth of losses and concave wins, overall the world will be losing more than gaining until 2100. Based on the *WL* index of 188 countries of the world, more countries (n = 113) will lose more from global warming until 2100 than fewer countries (n = 75) will win from a warming earth. In particular, 75 countries of the world will gain WL_P = 22,717.161 in GDP output, whereas 113 countries of the world will lose an estimated WL_P = −353,175.32.

10.2.2.3 *Hyperbolic Model*

When unidimensionally focusing on estimated GDP growth given a warmer temperature and assuming a hyperbolic growth, overall the world will be gaining more than losing until 2100. Based on the *WL* index of 188 countries of the world, fewer countries (n = 79) will gain more from global warming until 2100 than more countries (n = 109) will lose from a warming earth. In particular, 79 countries of the world will gain WL_H = 1,037,192 in GDP output, whereas 109 countries of the world will lose an estimated WL_H = −352,088.

10.2.2.4 *Total Estimate*

The total Winner-Loser *WL* index was calculated per country based on the mean of the overall Winner-Loser index WL_L for the linear model, the overall Winner-Loser index WL_P for the prospect model, and the overall Winner-Loser index WL_H for the hyperbolic model, which were then added up into the Winner-Loser total index WL_{TT} based on the following formula (10.12):

$$WL_{TT} = (WL_L + WL_P + WL_H)/n, \tag{10.12}$$

whereby WL_L denotes the Winner-Loser index for the linear model, WL_P the index for the prospect model and WL_H the index for the hyperbolic model and $n = 3$ as the overall number of models. All indices were calculated per country for the year 2100 business-as-usual projection.

10.2.3 Global Warming Winners and Losers Around the World

When unidimensionally focusing on estimated GDP growth given a warmer temperature, over all calculated models assuming linear, prospect or hyperbolic gains and losses, the world will be gaining more than losing until 2100. Based on the *WL* index of 188 countries of the world, fewer countries ($n = 78$) will gain more from global warming until 2100 than more countries ($n = 111$) will lose from a warming earth. In particular, 78 countries of the world will gain $WL_G = 3{,}540{,}396{,}345$ in GDP output, whereas 111 countries of the world will lose an estimated $WL_L = -232{,}613{,}188$. Climate change winners and losers are displayed in Graph 10.1: red countries are those that have run out of time, yellow countries are the ones around peak condition and green countries have the most time ahead.

The relation between GDP growth prospects in light of climate change and percentage of GHG for ratification was studied based on the total and percentage of GHG emissions communicated by the Paris COP 21 Parties to the Convention retrieved in their national communications and GHG inventory reports as of December 2015.[5] Over a sample of 181 countries of the world, a highly significant correlation of $r_{\text{Pearson (181)}} = 0.215$, $p < 0.004$ between the *WL* index over all models and the self-reported percentage of GHG emissions for ratification was found. As a cross-validation check, the percentage of GHG emissions for ratification was significantly positively correlated ($r_{\text{Pearson (181)}} = 0.178$, $p < 0.016$) with self-reported GHG emissions per country.

This result leads to the conclusion that those countries that emit more GHG are the ones with a positive GDP prospect on the warming earth until 2100. The more time countries seem to have in a favorable climate

[5] http://unfccc.int/resource/docs/2015/cop21/eng/10.pdf#page=30.

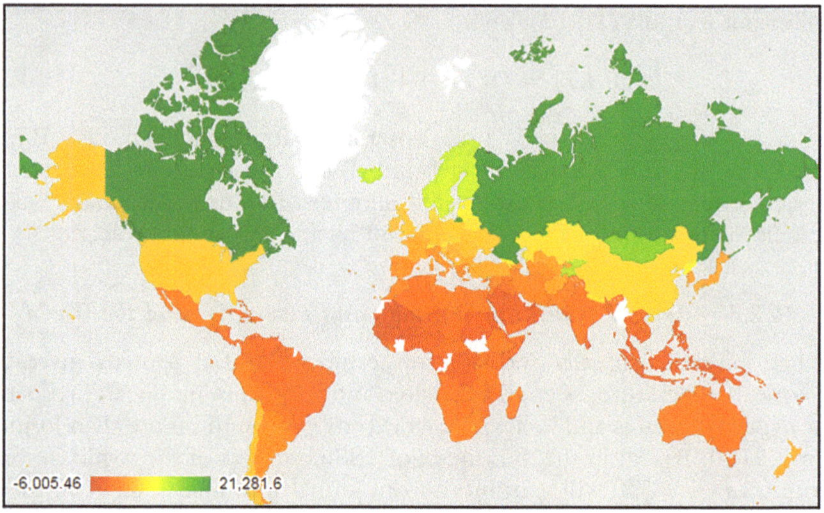

Graph 10.1 Climate change winners (green) and losers (red) around the world

for production, the likelier they are to emit GHG and hence contribute to global warming. All these results outline the need for attention to climate justice.

10.2.4 Fair Climate Change Gains Distribution

In order to offset the losses from climate change based on the overall WL_{TT} index, global warming benefits need to be redistributed in a fair way to support climate change loser countries for climate change mitigation and adaptation efforts and to instigate a transition to renewable energy.

To ensure a fair benefit transfer strategy, the difference of ΔWL = 121,426,447 between losses and gains from a warming earth until 2100 should be distributed based on the following criteria. To ensure a fair benefit transfer strategy between countries, the per-country contributions to alleviate the losses caused by global warming should only come from those countries that win from a warming earth, that is, 78 nation-states. The winning countries' contribution in relation to the other winning countries' contribution, as share of all wins, is calculated based on the percentage of wins and losses based on formulas (10.13) and (10.14).

In order to account for country differences in offsetting global warming through GDP growth (especially on the winners' side) and the country differences in their ability to transfer to renewable energy, the overall GDP per inhabitant was factored into the transfer equation as outlined in formula (10.13):

$$R^W = \left(\frac{P_W}{WL_G}\right) \times \left(\frac{GDP_c}{I_c}\right), \tag{10.13}$$

whereby R^W denotes the total transfer of climate change wins obligation per winning country and P_W is the country-specific GDP win divided by all wins $WL_G = 3{,}540{,}396{,}345$ from a warming earth for the year 2100 business-as-usual projection in order to derive the percentage fraction of wins per country. GDP_C equals the per-country GDP estimate per inhabitant I_c per country for the year 2016.

Losses were calculated based on formula (10.14):

$$R^L = \left(\frac{P_L}{P_{LT}}\right) \times \left(\frac{GDP_c}{I_c}\right), \tag{10.14}$$

whereby R^L denotes the total transfer of climate change compensation and P_L is the country-specific GDP loss divided by all losses $WL_L = -232{,}613{,}188$ from climate change for the year 2100 business-as-usual projection in order to derive the relative fraction of losses per country. GDP_C equals the per-country GDP estimate per inhabitant I_c per country for the year 2016.

The overall transfers R^T equate to the sum of contributions from the countries benefiting from global warming and the losses incurred by the countries that lose from global warming until 2100. Overall, transfers R^T derive from R^W and R^L, which must be equal based on the following formula (10.15):

$$R^T = R^W + R^L, \tag{10.15}$$

whereby R^T denotes the total transfer of wins for compensation of losses.

10.2.5 Climate Change Transfers

Based on the WL_{TT} index, which measures the distance to cardinal temperature per GDP sector in 188 countries of the world, all 77 climate change transfer grantor countries and 109 climate change transfer beneficiary countries are outlined in Graph 10.2. Graph 10.2 provides an overview of all climate change winning transfer grantors in green and yellow and all climate change loser country transfer beneficiaries in orange and red.

Graph 10.2 features the countries with the most time to a favorable climate weighted by overall GDP and their overall transfer burden in green. The top twenty transfer grantor countries are proposed to be Canada ($WL_{TTB} = -7299.04$), Russia ($WL_{TTB} = -6999.3$), Mongolia ($WL_{TTB} = -5032.75$), Kyrgyzstan ($WL_{TTB} = -4154.31$), Tajikistan ($WL_{TTB} = -4089.42$), Iceland ($WL_{TTB} = -3887.91$), Finland ($WL_{TTB} = -3798.36$), Norway ($WL_{TTB} = -3708.43$), Sweden ($WL_{TTB} = -3644.39$), Latvia ($WL_{TTB} = -2465.58$), Georgia ($WL_{TTB} = -2452.19$), North Korea ($WL_{TTB} = -2428.53$), Switzerland ($WL_{TTB} = -2394.05$), Estonia ($WL_{TTB} = -2342.77$), Liechtenstein ($WL_{TTB} = -2318.88$),

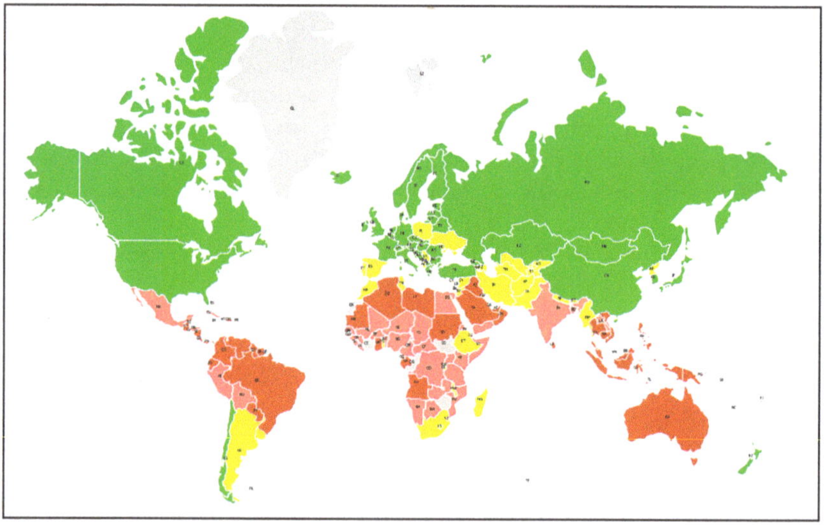

Graph 10.2 Climate change grantor (green) and beneficiary (red) countries

Lithuania (WL_{TTB} = −2188.99), Nepal (WL_{TTB} = −2131.26), Austria (WL_{TTB} = −2103.02), Belarus (WL_{TTB} = −2077.12) and Kazakhstan (WL_{TTB} = −2039.07).

The major transfer beneficiaries are proposed to be Qatar (WL_{TT} = 2911.69), Bahrain (WL_{TT} = 2793.87), Brunei (WL_{TT} = 2591.98), United Arab Emirates (WL_{TT} = 2446.67), Kiribati (WL_{TT} = 2370.55), Mauritania (WL_{TT} = 2311.14), Tuvalu (WL_{TT} = 2228.85), Djibouti (WL_{TT} = 2140.24), Senegal (WL_{TT} = 2091.74), Burkina Faso (WL_{TT} = 2009.04), Maldives (WL_{TT} = 2008.46), Trinidad and Tobago (WL_{TT} = 1997.28), Mali (WL_{TT} = 1922.99), Equatorial Guinea (WL_{TT} = 1875.88), Sri Lanka (WL_{TT} = 1855.24), Palau (WL_{TT} = 1838.85), Seychelles (WL_{TT} = 1775.79), Saint Vincent and the Grenadines (WL_{TT} = 1772.60), Samoa (WL_{TT} = 1760.75), and Guinea (WL_{TT} = 1753.07). Graph 10.2 features the countries with the most time to a favorable climate weighted by overall GDP and their overall transfer burden.

All global warming transfer beneficiary countries weighted per inhabitant are outlined in Graphs 10.3 and 10.4. Based on the WL_{TT} index weighted per GDP and per inhabitant, which measures the distance to cardinal temperature per GDP sector in 186 countries of the world, there are 76 climate change transfer grantor countries and 110 climate change transfer beneficiary countries.

Graph 10.3 features the highest grantor countries weighted by GDP and per inhabitant. The top twenty transfer grantor countries are proposed as Liechtenstein (WL_{TTB} = −10.34), Canada (WL_{TTB} = −8.88), Norway (WL_{TTB} = −7.65), Iceland (WL_{TTB} = −6.35), Switzerland (WL_{TTB} = −5.67), Luxembourg (WL_{TTB} = −5.52), Finland (WL_{TTB} = −3.77), Monaco (WL_{TTB} = −3.41), Denmark (WL_{TTB} = −3.19), Austria (WL_{TTB} = −3.03), the United States (WL_{TTB} = −2.87), Ireland (WL_{TTB} = −2.18), Germany (WL_{TTB} = −2.15), the United Kingdom (WL_{TTB} = −1.80), Belgium (WL_{TTB} = −1.44), New Zealand (WL_{TTB} = −1.41), Estonia (WL_{TTB} = −1.38), France (WL_{TTB} = −1.18), Latvia (WL_{TTB} = −1.15) and Slovenia (WL_{TTB} = −1.15).

All global warming transfer beneficiary countries are outlined in Graph 10.4. The major transfer beneficiaries are proposed as Qatar (WL_{TT} = 19.06), United Arab Emirates (WL_{TT} = 9.33), Singapore (WL_{TT} = 7.90), Brunei (WL_{TT} = 6.06), Bahrain (WL_{TT} = 5.90), Kuwait (WL_{TT} = 3.94), Trinidad and Tobago (WL_{TT} = 2.98), Saudi Arabia (WL_{TT} =

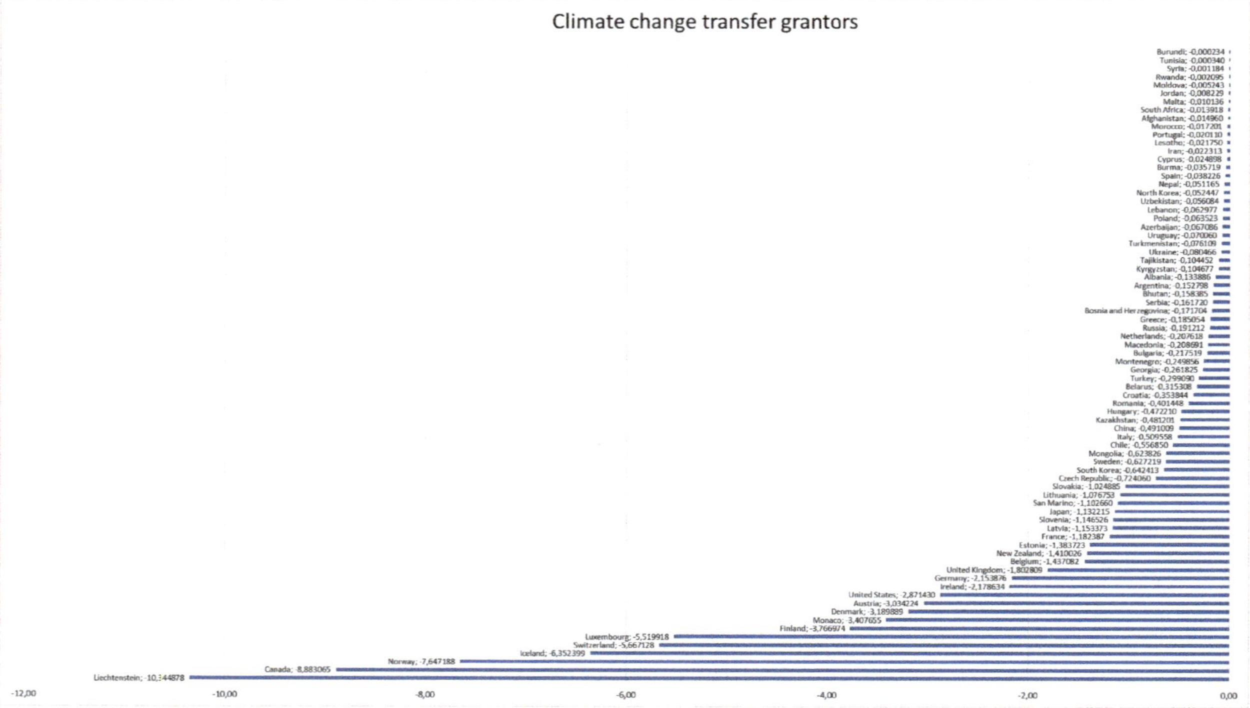

Graph 10.3 Climate change transfer grantors weighted by GDP and per inhabitant

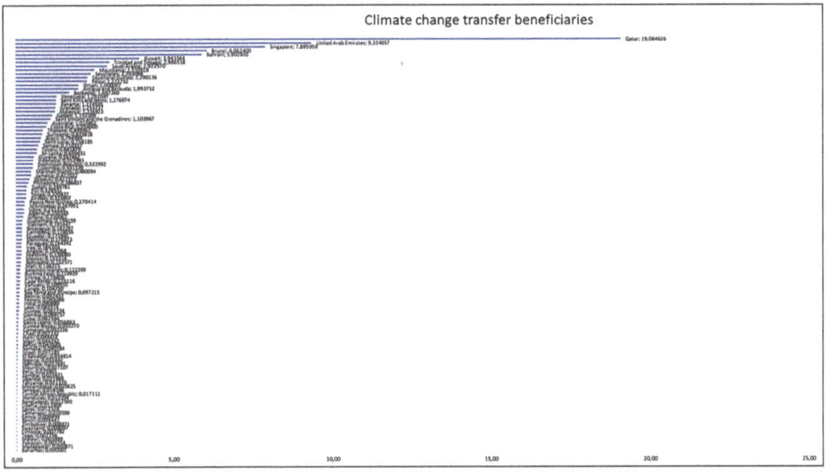

Graph 10.4 Climate change transfer payment beneficiaries weighted by GDP and per inhabitant

2.92), Mauritania (WL_{TT} = 2.51), Seychelles (WL_{TT} = 2.39), Equatorial Guinea (WL_{TT} = 2.29), Palau (WL_{TT} = 2.27), Oman (WL_{TT} = 2.00), Antigua and Barbuda (WL_{TT} = 1.99), Barbados (WL_{TT} = 1.70), Venezuela (WL_{TT} = 1.29), Saint Kitts and Nevis (WL_{TT} = 1.28), Panama (WL_{TT} = 1.27), Grenada (WL_{TT} = 1.26) and Malaysia (WL_{TT} = 1.24).

The history of humankind outlines that politics and the idea of nation-states often trumps economic calculus and rationale. In the economic distribution of climate change transfer grantors and beneficiaries around the world, political considerations may limit the actual feasibility of carrying out plans. Throughout the history of modern humankind, politics have influenced transfer payments and remittances (Orttung, 2017). The concept of nation-states has directly determined access to resources and energy control of the populace (Orttung, 2014a, 2014b; Orttung & Wenger, 2012). Cooperation and conflict around the preservation and distribution of common environmental conditions appear to obey the laws of politics rather than following from rational economic calculus (Orttung, 2013b; Orttung et al., 2012; Puaschunder & Schwarz, 2012).

10.3 Research Question

The following chapter builds on the outlined previous findings of global climate change winners and losers around the world in order to advocate for stepping away from country denominations. The measured economic gains from climate change are then advocated to be transferred into losing areas. Economically winning market industries should be taxed to raise funds for offsetting costs arising in climate change losing markets. Coupling the findings on economic redistribution pledges with a purely market-based climate change winners and losers model will aid in advocating for nation-state blind climate change gains and losses redistribution in the hope to make climate justice more approachable (Puaschunder, 2017b). But who will be the climate change winning market actors, agents and industries? The impact of weather on market decisions can serve as basis for global governance determining how to raise the necessary funds for climate stabilization. Quantifying the prospective gains from global warming will aid policy makers to understand how to foster a transition into renewable energy and a sustainable economy. The monetary gains from global warming will also inform financial market experts to guide stock management decisions and marketers to use temperature forecasts and weather conditions to streamline procurement and contextualize campaigns.

10.4 Method

10.4.1 Overall Model Assumptions

The presented model is built on GDP optimal cardinal temperature for production. The GDP pillars of agriculture, industry and service are the foundation of the model. While the previous model divided the world production by nation-states, the following model extension will focus on agriculture, industry and service sector representing economic growth factors.

Methodologically, the optimum cardinal temperatures for specific agriculture products, industry representations and service sector providers will be retrieved from literature sources. A future market prospect in light of global temperature rising will then be given based on the mean temperature of the world now and as estimated in the year 2100. Prospective market shifts in the GDP composition but also in agricultural products, industry sector representatives and service providers will be provided. Underlying to all these future prospects will be the idea that GDP production and its manifold components are enabled at best at a certain temperature. Therefore, specific markets will gravitate toward these temperatures and fade away if the temperature starts deviating from the optimum environmental condition. As for key industries, energy prospects will also be taken into consideration.

10.5 Results

The following qualitative meta-analysis results presentation features insights about temperature and the GDP production as comprising the three pillars of agriculture, service and industry sectors.

10.5.1 Agriculture

The optimal cardinal temperature for agriculture production was derived for the agriculture sector. Regardless of how favorable light and moisture conditions may be, plant growth ceases when the air and leaf temperature drops below a certain minimum or exceeds a certain maximum value.[6] Between these limits, there is an optimum temperature at which growth proceeds with greatest rapidity. Cardinal temperatures are—at least approximately—known for most plant species and crops. Temperatures between the optimum minimum and maximum define the range of temperature where the crop performs the best. Controversy exists in the literature if it is the mean point in between minimum and maximum feasibility temperatures. Reproductive stage of summer crops has relatively lower optimal temperatures compared with vegetative stage (Luo, 2011). The cardinal temperatures may also vary with the stage of development.

In general, summer cereal crops—such as maize, rice and sorghum—have relatively higher optimal temperatures for crop growth and development compared with that of winter cereal crops, such as wheat. Cool-season crops—such as oats, rye, wheat and barley—have low cardinal temperatures: minimum 0° to 5 °C (32° to 41 °F), optimum 25 °C to 31 °C (77 °F to 88 °F) and maximum 31° to 37 °C (88 °F to 99 °F). Peas start to germinate at temperatures as low as −2 °C to 10 °C (28 °F to 50 °F); the optimum temperature ranges from 20 °C to 27 °C (68 °F

[6] https://www.britannica.com/technology/agricultural-technology/Regional-variations-in-technique#ref558352.

to 81 °F) and maximum from 29 °C to 40 °C (84 °F to 104 °F) (Raveneau et al., 2011). Milk thistle starts from 0 and 3 °C (32° and 37 °F), with optimal range between 30 °C and 33 °C (86 °F and 91 °F) and an upper temperature cut-off at 40 °C (104 °F) (Heidari et al., 2014). Parsnip has a base temperature of 0 °C (32 °F), optimum temperature at 15 °C (59 °F) and maximum temperature at 37 °C (99 °F) (McCormick et al., 2014). Carrots have a base temperature of 0 °C (32 °F), optimum temperature at 31 °C (88 °F) and maximum temperature at 41 °C (106 °F) (McCormick et al., 2014). Ryegrass has a base temperature of 1 °C (34 °F), optimum temperature at 29 °C (84 °F) and maximum temperature at 40 °C (104 °F) (McCormick et al., 2014). Onions have a base temperature of 1 °C (34 °F), optimum temperature at 22 °C (72 °F) and maximum temperature at 40 °C (104 °F) (McCormick et al., 2014). Cabbage has a base temperature of 1 °C (34 °F), optimum temperature at 31 °C (88 °F) and maximum temperature at 42 °C (108 °F) (McCormick et al., 2014). Red radish has a base temperature of 2 °C (36 °F), optimum temperature at 29 °C (84 °F) and maximum temperature at 42 °C (108 °F) (McCormick et al., 2014). Pak choi has a base temperature of 3 °C (37 °F), optimum temperature at 29 °C (84 °F) and maximum temperature at 46 °C (115 °F) (McCormick et al., 2014). Asian radish has a base temperature of 3 °C (37 °F), optimum temperature at 30 °C (86 °F) and maximum temperature at 41 °C (106 °F) (McCormick et al., 2014). Red beet has a base temperature of 4 °C (39 °F), optimum temperature at 36 °C (97 °F) and maximum temperature at 44 °C (111 °F) (McCormick et al., 2014). Chinese cabbage has a base temperature of 5 °C (41 °F), optimum temperature at 35 °C (95 °F) and maximum temperature at 43 °C (109 °F) (McCormick et al., 2014). Beans start at 5 °C (41 °F) to 9 °C (48 °F), the optimum temperature ranges from 30 °C to 36 °C (86–97 °F) and maximum from 46 °C to 50 °C (115–122 °F) (Raveneau et al., 2011). Mustard has a base temperature of 6 °C (43 °F), optimum temperature at 31 °C (88 °F) and maximum temperature at 41 °C (106 °F) (McCormick et al., 2014). Wheat like wild barley, barley grass and hoary cress have an optimum temperature ranging from 10 °C to 30 °C (50–86 °F) (Hosseini et al., 2017).

Corn and related crops have an absolute minimum temperature of 10 °C (50 °F), an optimum minimum of 18 °C (64 °F), an optimum maximum of 33 °C (91 °F) and an absolute maximum of 47 °C (117 °F) (Hollinger & Angel, 2009). Soybean has absolute minimum temperatures ranging from 2 °C to 5 °C (36 °F to 41 °F), optimum minimums from 15 °C to 20 °C (59 °F to 68 °F), optimum maximums from 23 °C to 33 °C (73 °F to 91 °F) and absolute maximum at 38 °C (100 °F) (Hollinger & Angel, 2009).

Cool-season crops include asparagus, broccoli, Brussels sprouts, cabbage, chives, collards, corn salad, garlic, horseradish, kale, kohlrabi, leeks, mustard, onions, parsley, peas, radishes, rhubarb, rutabaga, spinach and turnips, which grow from 5 °C (40 °F) and grow best when temperature highs range between 21 °C and 24 °C (70 and 75 °F), with maximum temperature for growth at 25 °C (77 °F) (Albert, 2008; Hatfield & Prueger, 2015). Half-hardy and semi-hardy vegetables—such as beets, carrots, cauliflower, celery, celeriac, chard, Chinese cabbage, chicory, globe artichokes, endive, lettuce, parsnips, potatoes, salsify, sorrel, and hardy herbs—grow best at minimum temperature between 5 °C and 10 °C (40 and 50 °F) (Albert, 2008). Cool-season crops usually stop producing when daytime temperatures reach 26 °C (80 °F) or higher (Albert, 2008).

Cotton growth gradually increases with temperatures up to 29 °C to 32 °C (84 to 90 °F) and then sharply decreases when temperature increases beyond this threshold (Schlenker & Roberts, 2009). *Arabica coffee* has an optimum temperature range between 15 °C and 24 °C (59 °F and 75 °F), while *Robusta coffee* has an optimum temperature range from 22 °C to 26 °C (72 °F to 79 °F) and the species is less tolerant of very high as well as very low temperatures than is *Arabica* (Baker & Haggar, 2007). The optimum temperature range in citrus fruits is 22 °C to 27 °C (72 °F to 81 °F) (Cole & McCloud, 1985). Apples appear to have a cardinal temperature at 25 °C (77 °F) (Rodrigues et al., 2016).

For hot-season crops, such as melons and sorghum, the span of cardinal temperatures is much higher. Warm-season crops range from 10 °C (50 °F) and have an optimum temperature at 24 °C (75 °F). Tender vegetables prefer temperatures between 21 °C and 35 °C (70° and 95 °F) and require daytime temperatures of 15 °C (60 °F) or greater to thrive and include beans, celery, sweet corn, cucumbers, New Zealand spinach, summer squash and tomatoes. Very-tender vegetables require daytime temperatures consistently warmer than 13 °C (55 °F) and include lima beans, cantaloupe, eggplant, muskmelon, okra, peppers, pumpkins, winter squash, sweet potatoes and watermelon (Albert, 2008). Pumpkins start at 5–18 °C (41–64 °F), have cardinal temperature at 28 °C (82 °F) and stop growing at 36 °C (97 °F) (Zehtab-Salmasi, 2006). Sorghum has a base temperature of 8 °C (46 °F) and a top temperature of 34 °C (93

°F), with an optimal temperature of 31 °C (88 °F) and optimal range of 26–34 °C (79 °F to 93 °F) (Alagarswamy & Ritchie, 1991; Luo, 2011). Rice is reported to have a minimum range between 10 °C and 12 °C (50 °F to 54 °F), optimum range between 30 °C and 32 °C (86 °F to 90 °F) and maximum range between 36 °C and 38 °C (97 °F to 100 °F).[7] Maize has an optimal temperature range from 33 °C to 38 °C (91 °F to 100 °F) (Crafts-Brandner & Salvucci, 2002). Peanut and groundnut range between 10 °C (50 °F) and 36 °C (97 °F) (Bolhuis & deGroot, 1959), with an optimal temperature of 33 °C (91 °F) (Luo, 2011). Coconut starts at 16 °C (60 °F), has cardinal temperature between 26 °C and 30 °C (79–86 °F) and ends to grow at 42 °C (108 °F) (Ranasinghe, Kumarathunge, & Kiriwandeniya, 2017). Sugarcane grows from 20 °C (68 °F), and the optimum temperature for the germination of stem cuttings is 32 °C to 38 °C (90 to 100 °F).[8] Table 10.1 summarizes the meta-analysis literature review findings for the agricultural sector.

The contemporary world mean temperature over 191 world countries was derived to be 18.6 °C (65.48 °F) based on the information retrieved from the *List of Countries byAverage Yearly Mean Temperature*.[9] The average yearly temperature is calculated by averaging the minimum and maximum daily temperatures in the country, averaged for the years 1961–1990, based on gridded climatology from the Climatic Research Unit elaborated in 2011. Contemporary climate change projections estimate a mean temperature rise of approximately 4.24 °C (39.63 °F) in 2100, leading to the prediction of the world mean temperature by 2100. The cardinal temperature for agriculture is 26.12 °C (79 °F), which is derived from the mean over 59 crops of all agriculture data sources listed.[10]

Graph 10.5 outlines the different crops' base temperature for germination (green), optimum temperature range (yellow), the optimum temperature peak (blue), beyond peak condition (pink) and maximum temperature (red) for crop growth. The contemporary mean temperature of 18.6 °C (65.48 °F) as well as the predictions of estimated 22.92 °C (73.26 °F) mean temperature around the globe in 2100 is outlined in

[7] http://agritech.tnau.ac.in/agriculture/agri_agrometeorology_temp.html.

[8] http://www.daff.gov.za/Daffweb3/Portals/0/BrochuresandProductionguidelines/sugarcaneproductionguideline.pdf.

[9] https://en.wikipedia.org/wiki/List_of_countries_by_average_yearly_temperature.

[10] http://agritech.tnau.ac.in/agriculture/agri_agrometeorology_temp.html.

Table 10.1 Minimum, maximum and cardinal temperature for agriculture production products

Minimum, maximum and cardinal temperatures for agriculture production			
Agriculture	*Temperature*		
Product	*Minimum*	*Optimum*	*Maximum*
Peas	−2–10 °C (28–50 °F)	20–27 °C (68–81 °F)	29–40 °C (84–104 °F)
Parsnip	0 °C (32 °F)	15 °C (59 °F)	37 °C (99 °F)
Milk thistle	0–3 °C (32–37 °F)	30–33 °C (86–91 °F)	40 °C (104 °F)
Carrots	0 °C (32 °F)	31 °C (88 °F)	41 °C (106 °F)
Wild barley		10–30 °C (50–86 °F)	
Barley grass		10–30 °C (50–86 °F)	
Hoary cress		10–30 °C (50–86 °F)	
Arabica coffee		15–24 °C (59–75 °F)	
Robusta coffee		22–26 °C (72–79 °F)	
Oats	0–5 °C (32–41 °F)	25–31 °C (77–88 °F)	31–37 °C (88–99 °F)
Rye	0–5 °C (32–41 °F)	25–31 °C (77–88 °F)	31–37 °C (88–99 °F)
Wheat	0–5 °C (32–41 °F)	25–31 °C (77–88 °F)	31–37 °C (88–99 °F)
Barley	0–5 °C (32–41 °F)	25–31 °C (77–88 °F)	31–37 °C (88–99 °F)
Ryegrass	1 °C (34 °F)	29 °C (84 °F)	40 °C (104 °F)
Onions	1 °C (34 °F)	22 °C (72 °F)	40 °C (104 °F)
Cabbage	1 °C (34 °F)	31 °C (88 °F)	42 °C (108 °F)
Red radish	2 °C (36 °F)	29 °C (84 °F)	42 °C (108 °F)
Soybean	2–5 °C (36–41 °F)	15–33 °C (59–91 °F)	34–38 °C (93–100 °F)
Pak choi	3 °C (40 °F)	29 °C (84 °F)	46 °C (115 °F)
Asian radish	3 °C (40 °F)	30 °C (86 °F)	41 °C (106 °F)
Red beet	4 °C (40 °F)	36 °C (97 °F)	44 °C (111 °F)
Asparagus	5 °C (40 °F)	21–24 °C (70–75 °F)	25 °C (77 °F)
Chinese cabbage	5 °C (40 °F)	35 °C (70–75 °F)	43 °C (109 °F)
Broccoli	5 °C (40 °F)	21–24 °C (70–75 °F)	25 °C (77 °F)
Brussels sprouts	5 °C (40 °F)	21–24 °C (70–75 °F)	25 °C (77 °F)
Cabbage	5 °C (40 °F)	21–24 °C (70–75 °F)	25 °C (77 °F)
Chives	5 °C (40 °F)	21–24 °C (70–75 °F)	25 °C (77 °F)
Collards	5 °C (40 °F)	21–24 °C (70–75 °F)	25 °C (77 °F)
Corn salad	5 °C (40 °F)	21–24 °C (70–75 °F)	25 °C (77 °F)
Garlic	5 °C (40 °F)	21–24 °C (70–75 °F)	25 °C (77 °F)
Horseradish	5 °C (40 °F)	21–24 °C (70–75 °F)	25 °C (77 °F)

(continued)

Table 10.1 (continued)

Minimum, maximum and cardinal temperatures for agriculture production			
Agriculture	*Temperature*		
Product	*Minimum*	*Optimum*	*Maximum*
Kale	5 °C (40 °F)	21–24 °C (70–75 °F)	25 °C (77 °F)
Kohlrabi	5 °C (40 °F)	21–24 °C (70–75 °F)	25 °C (77 °F)
Leeks	5 °C (40 °F)	21–24 °C (70–75 °F)	25 °C (77 °F)
Onions	5 °C (40 °F)	21–24 °C (70–75 °F)	25 °C (77 °F)
Parsley	5 °C (40 °F)	21–24 °C (70–75 °F)	25 °C (77 °F)
Radishes	5 °C (40 °F)	21–24 °C (70–75 °F)	25 °C (77 °F)
Rhubarb	5 °C (40 °F)	21–24 °C (70–75 °F)	25 °C (77 °F)
Rutabaga	5 °C (40 °F)	21–24 °C (70–75 °F)	25 °C (77 °F)
Spinach	5 °C (40 °F)	21–24 °C (70–75 °F)	25 °C (77 °F)
Turnips	5 °C (40 °F)	21–24 °C (70–75 °F)	25 °C (77 °F)
Beans	5–9 °C (40–48 °F)	30–36 °C (86–97 °F)	46–50 °C (115–122 °F)
Beets	5–10 °C (40–50 °F)		26 °C (80 °F)
Carrots	5–10 °C (40–50 °F)		26 °C (80 °F)
Cauliflower	5–10 °C (40–50 °F)		26 °C (80 °F)
Celery	5–10 °C (40–50 °F)		26 °C (80 °F)
Celeriac	5–10 °C (40–50 °F)		26 °C (80 °F)
Chard	5–10 °C (40–50 °F)		26 °C (80 °F)
Chinese cabbage	5–10 °C (40–50 °F)		26 °C (80 °F)
Chicory	5–10 °C (40–50 °F)		26 °C (80 °F)
Globe artichokes	5–10 °C (40–50 °F)		26 °C (80 °F)
Endive	5–10 °C (40–50 °F)		26 °C (80 °F)
Lettuce	5–10 °C (40–50 °F)		26 °C (80 °F)
Parsnips	5–10 °C (40–50 °F)		26 °C (80 °F)
Potatoes	5–10 °C (40–50 °F)		26 °C (80 °F)
Salsify	5–10 °C (40–50 °F)		26 °C (80 °F)
Sorrel	5–10 °C (40–50 °F)		26 °C (80 °F)
Hardy herbs	5–10 °C (40–50 °F)		26 °C (80 °F)
Mustard	6 °C (40 °F)	31 °C (70–75 °F)	41 °C (106 °F)
Citrus fruits		22–27 °C (72–81 °F)	
Sorghum	8 °C (46 °F)	31 °C (88 °F)	34 °C (93 °F)
Peanut	10 °C (50 °F)	33 °C (91 °F)	36 °C (97 °F)
Groundnut	10 °C (50 °F)	33 °C (91 °F)	36 °C (97 °F)
Warm-season crops	10 °C (50 °F)	24 °C (75 °F)	
Rice	10–12 °C (50–54 °F)	30–32 °C (86–90 °F)	36–38 °C (97–100 °F)
Maize		33–38 °C (91–100 °F)	
Apples		25 °C (77 °F)	
Lima beans	13 °C (55 °F)		
Cantaloupe	13 °C (55 °F)		
Eggplant	13 °C (55 °F)		
Muskmelon	13 °C (55 °F)		

(continued)

Table 10.1 (continued)

Minimum, maximum and cardinal temperatures for agriculture production			
Agriculture	*Temperature*		
Product	*Minimum*	*Optimum*	*Maximum*
Okra	13 °C (55 °F)		
Peppers	13 °C (55 °F)		
Pumpkins	5–18 °C (41–64 °F)	28 °C (82 °F)	36 °C (97 °F)
Winter squash	13 °C (55 °F)		
Sweet potatoes	13 °C (55 °F)		
Watermelon	13 °C (55 °F)		
Beans	15 °C (60 °F)	21–35 °C (70–95 °F)	
Celery	15 °C (60 °F)	21–35 °C (70–95 °F)	
Sweet corn	15 °C (60 °F)	21–35 °C (70–95 °F)	
Cucumbers	15 °C (60 °F)	21–35 °C (70–95 °F)	
New Zealand spinach	15 °C (60 °F)	21–35 °C (70–95 °F)	
Summer squash	15 °C (60 °F)	21–35 °C (70–95 °F)	
Tomatoes	15 °C (60 °F)	21–35 °C (70–95 °F)	
Coconut	16 °C (60 °F)	26–30 °C (79–86 °F)	42 °C (108 °F)
Corn	10 °C (50 °F)	18–33 °C (64–91 °F)	47 °C (117 °F)
Sugarcane	20 °C (68 °F)	32–38 °C (90–100 °F)	
Cotton	29 °C (91 °F)	31 °C (88 °F)	32 °C (91 °F)

purple to put the crop growth in perspective with overall current and future temperature conditions. The cardinal temperature for agriculture is 26.12 °C (79 °F), which is derived from the mean over 59 crops of all agriculture data sources listed and outlined in light blue color.[11]

10.5.2 Industry

The optimal cardinal temperature for industry production was derived for the industry sector. Deviations from the optimal temperature for the industry sector are primarily based on energy intensity, for example, engines that run too hot and break or machines that need to be warmed up before running, or are assumed to become riskier to operate due to heat or cold and thus are prone to have accidents and fail. The costs for insurances, monitoring and control as well as depreciation and replacement are thus believed to rise below and beyond peak performance temperature. Energy consumption corresponds to

[11] http://agritech.tnau.ac.in/agriculture/agri_agrometeorology_temp.html.

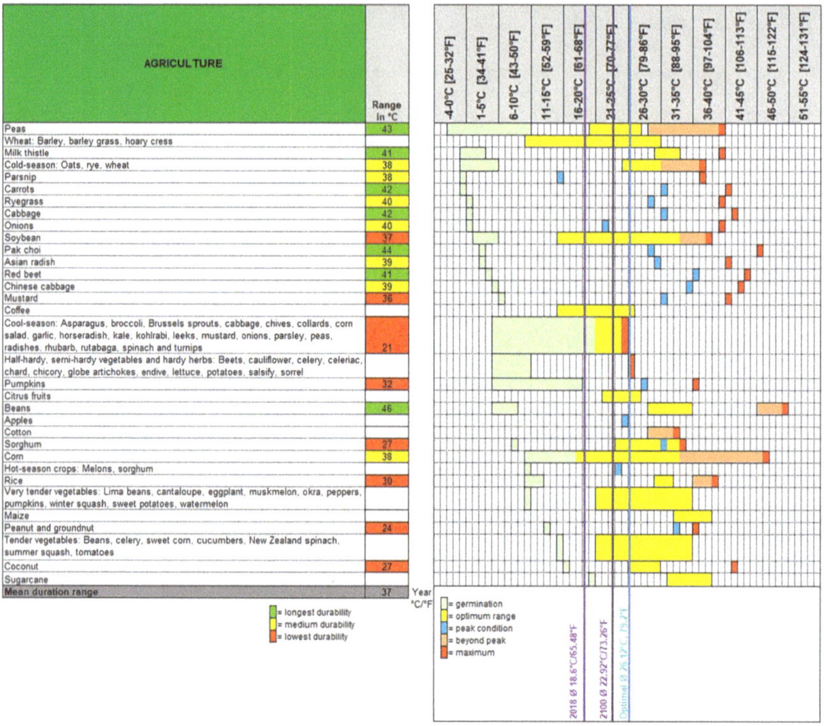

Graph 10.5 Agriculture sector temperature-based condition in relation to world mean temperature (*Source* Author's own)

space heating and cooling and electricity for pumps, motors, cooling/refrigeration, office equipment and lighting (Naegle et al. 2015). For example, air-conditioned buildings use about twice as much energy as naturally-ventilated buildings.[12]

In addition, deviations beyond the optimal temperature are expected to increase industrial waste heat (Papapetrou, Kosmadaki, Cipollina, La Commare, & Micale, 2018). The waste heat of most industry sectors is within the temperature level of 100–200 °C (212–392 °F), which matches the temperature of the exhaust gases of boilers and standard heating

[12] www.carbontrust.co.uk.

equipment. The temperature of the waste heat is increased to more than 500 °C (932 °F) in the industrial sectors: chemical, iron and steel and non-metallic mineral (Papapetrou et al., 2018). The literature on the top temperature for different industry sector components assumes the peak temperature to be constant in the short-run, but only when there is no transition to other industries.[13]

The cardinal temperature for the industry sector comprising of construction, automotive, airframe, locomotive, petroleum, coal and chemicals was found to be 21 °C (70 °F) (Russell, 1957). Most industrial sectors require all ranges of heat, with food, tobacco, textiles and leatherindustries having heat demand mostly up to 100 °C (212 °F). The majority of industries lie in the temperature range of 100–200 °C (212–392 °F), spread over most industrial sectors, while below 100 °C (212 °F), the potential is rather limited, concentrated in the food, drink and clothing sectors, mostly from drying and preheating processes (Papapetrou et al., 2018). At temperature level of 100–200 °C (212–392 °F) are chemicals and non-metallic minerals industries. Higher temperature levels over 500 °C (932 °F) are required for the steel industry (Papapetrou et al., 2018). Within the 200–500 °C (392–932 °F) range lie the pulp, paper, iron and steel industries. Pulp and paper industries require temperatures up to 500 °C (932 °F), while industries for manufacturing metals, iron and steel require mostly heat with temperature being over 500 °C (932 °F) (Brückner et al., 2014; Lauterbach et al., 2012). The 500–1000 °C (932–1832 °F) temperature range features industries such as cement, iron and steel sectors. High-temperature industries include iron and steel production, foundries and glass manufacturers as well as motor-run engines especially if equipped with pumps and fans.[14] Non-metallic minerals and iron and steel sectors require high temperatures due to larger heat losses (Papapetrou et al., 2018). Table 10.2 summarizes the meta-analysis literature review findings for the industry sector.

Regarding maximum temperatures for industry production, the so-called maximum service temperature is a measure indicating the maximum continuous temperature at which an insulation material can operate without any loss in thermal insulation performance. Cold or fresh-air ducts run out of temperature after 20 °C (68 °F). Technical buildings

[13] www.carbontrust.co.uk.

[14] www.carbontrust.co.uk.

Table 10.2 Minimum, maximum and cardinal temperature for industry production products

Temperature requirements and maximum temperatures for industry production		
Industry	*Temperature*	
Product	*Temperature requirements*	*Maximum*
Drink	below 100 °C (212 °F)	
Food	below and up to 100 °C (212 °F)	
Tobacco	up to 100 °C (212 °F)	
Textiles	below and up to 100 °C (212 °F)	
Leather	up to 100 °C (212 °F)	
Chemicals	100–200 °C (212–392 °F)	
Non-metallic minerals	100–200 °C (212–392 °F)	
Pulp	200–500 °C (212–932 °F)	up to 500 °C (932 °F)
Paper	200–500 °C (212–932 °F)	up to 500 °C (932 °F)
Iron	over to 500 °C (932 °F)	
Steel	over to 500 °C (932 °F)	
Cement	500–1000 °C (932–1832 °F)	
Iron	500–1000 °C (932–1832 °F)	
Steel	500–1000 °C (932–1832 °F)	
Foundries	500–1000 °C (932–1832 °F)	
Glass	500–1000 °C (932–1832 °F)	
Motor-run engines equipped with pumps and fans	500–1000 °C (932–1832 °F)	

Source Author's own

equipment such as pipes and ducts in rooms of power plant components can run up to 100 °C (212 °F). Boiler walls and roof areas have a maximum operating temperature of up to 400–520 °C (752–968 °F). Containers, pipes and gas-turbine tiles range from 400–520 °C (752–968 °F). Dead rooms, flue-gas ducts, hot-air ducts and steel chimneys work properly at 400–640 °C (752–1184 °F).

Table 10.3 summarizes the literature review findings on maximum temperatures for industry production.

Table 10.3 Maximum temperature for industry production products

Temperature requirements and maximum temperatures for industry production	
Industry Product	*Temperature Maximum*
Cold or fresh-air ducts	up to 20 °C (68 °F)
Technical buildings equipment, such as pipes and ducts	up to 100 °C (212 °F)
Power plant equipment	up to 100 °C (212 °F)
Boiler walls	400–520 °C (752–968 °F)
Roof areas	400–520 °C (752–968 °F)
Containers	400–520 °C (752–968 °F)
Pipes	400–520 °C (752–968 °F)
Gas-turbine tiles	400–520 °C (752–968 °F)
Dead rooms	400–640 °C (752–1184 °F)
Flue-gas ducts	400–640 °C (752–1184 °F)
Hot-air ducts	400–640 °C (752–1184 °F)
Steel chimneys	400–640 °C (752–1184 °F)

Source Author's own

When the temperature is high, high-temperature industries are the most energy efficient and thus considered to be saving energy, cutting costs and thereby increasing profit margins. Industrial buildings in the high-temperature sector typically require large areas for storage and production and often have high ceilings and considerable ventilation rates.[15]

Regarding commercial heating and cooling, climate determines heating and cooling days, which appears to have the switching temperature at 18 °C or 64 °F. During heating days, fuel is required as primary space heating resource; during cooling days, electricity, oil and natural gas consumption tends to increase (Pardo et al., 2012).

Temperatures are assumed to influence economic human output through the impact on worker productivity in a non-linear relation (Sudarshan & Tewari, 2014). In the Indian manufacturing sector, plant-level manufacturing output was found to respond negatively in high temperatures in the order of 2% of daily output per degree Celsius (Sudarshan & Tewari, 2014). Cross-country evidence from 111 countries for a time period from 1997 to 2006 undermines that higher temperature

[15] www.carbontrust.co.uk.

conversely affects labor productivity (Yildirim et al., 2009). The temperature level of a country is assumed as the second most contributing factor to labor productivity. Focusing on non-agricultural output reveals that hot temperature exerts a significant impact on local labor productivity, with substantially larger effects in highly exposed industries such as construction, manufacturing and transportation (Park, 2017). For US data, Park (2016) found that an additional day above 32 °C (90 °F) results in a −0.048% decline in payroll per capita that year. These effects are explained by hot temperatures making workers face binding physiological constraints that prevent them from exerting additional effort in response to higher wages (Stevens, 2016). Temperature effects on productivity were also evidenced in a dataset of production of half million Chinese manufacturing plants during 1998–2007 (Zhang et al., 2017). An inverted U-shaped relationship exists between temperature and total factor productivity as found in China, Germany and a cross-country dataset (Heal et al., 2017). This temperature-output effect is found in labor- and capital-intensive firms, which exhibit sensitivity to high temperatures (Zhang et al., 2017). The negative effects of temperatures above 32 °C (90 °F) suggest that a day with temperature larger than 32 °C (90 °F) reduces economic output by 0.45% relative to an extra day with temperatures between 10–16 °C (50–60 °F) across various industries (Zhang et al., 2017). Using country-level panel data on average temperature and income from 1950 to 2005 illustrated that hotter-than-average years are associated with lower output per capita for already hot countries and higher output per capita for cold countries, with approximate 3–4% in each direction (Heal & Park, 2014). Household data on air-conditioning and heating expenditures underline this trend and effect as well (Heal & Park, 2014).

Graph 10.6 outlines the different industry sector temperature requirements with the room heating/cooling switch point in green color, service range in yellow, the optimum cardinal temperature peak in blue and beyond service range temperature in red for industry production. The contemporary mean world temperature of 18.6 °C (65 °F) and the prediction of estimated 22.92 °C (73.26 °F) mean temperature around the globe in 2100 are outlined in purple to put the industry production in perspective with overall current and future temperature conditions.

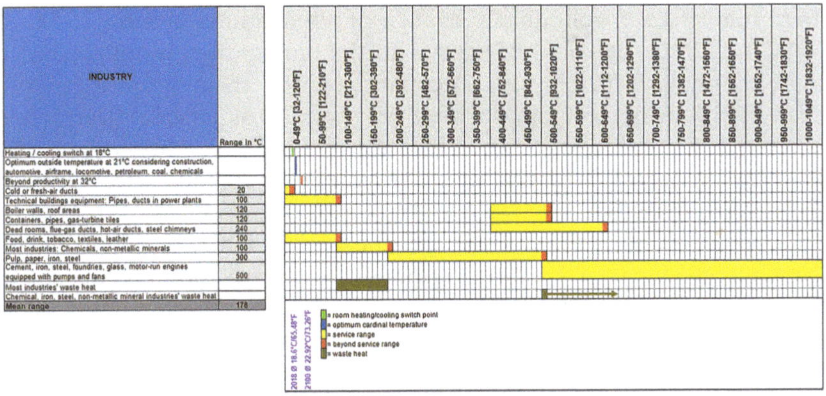

Graph 10.6 Industry sector, temperature-based condition in relation to world mean temperature (*Source* Author's own)

10.5.3 Service

The optimal cardinal temperature for service production depends on worker productivity, retail, food and leisure activities, such as vacation. Overall estimations based on storage, food production and cooling define the cardinal temperature for the service sector at 25 °C (77 °F) (Somanathan et al., 2014).

10.5.3.1 Labor Productivity

Thermal comfort is a condition that expresses satisfaction with the thermal environment. This subjective evaluation is assessed with the ANSI/ASHRAE Standard 55: Thermal Environmental Conditions for Human Occupancy, which is an American National Standard for occupants of buildings, as well as the European EN 15251[16] and ISO 7730,[17] respectively, for European and international measurement standards. Maintaining this standard of thermal comfort for occupants of buildings or other enclosures and heating, ventilation and air-conditioning is enabled through design engineers. Most people will feel comfortable at

[16] https://www.rehva.eu/fileadmin/hvac-dictio/04-2012/revision-of-en-15251_rj1204.pdf.

[17] https://www.iso.org/standard/39155.html.

room temperature, colloquially a range of temperatures around 18 °C or 20 °C to 22 °C (68 to 72 °F) and with the biological optimum assumed to be around 37 °C (99 °F). Occupants control their thermal environment by means of clothing, operable windows, fans, personal heaters and sunshades. Overall influencing factors are ventilation conditions, personal and environmental factors, clothing insulation, metabolic rate, air speed and relative humidity.

Thermal well-being influences productivity and health. According to the United Kingdom Health and Safety Executive, optimal workplace conditions are between 13 °C and 16 °C (55–61 °F) for physical labor and if much of the work involves rigorous physical effort. If workers are able to adapt to the change of varying temperature, they will be more productive. If a significant number of employees are complaining about thermal discomfort, employers are advised to carry out a risk assessment. Office workers who are satisfied with their thermal environment are more productive (Huizenga et al., 2006). However, this estimate may vary greatly between individuals depending on factors such as activity level, clothing and humidity (Heal & Park, 2014).

In economic terms, humans do not perform well under thermal stress. People's performance under thermal stress is about 11% lower than their performance at normal thermal wet conditions. Human performance in relation to thermal stress varies greatly by the type of task which the individual is completing. Some of the physiological effects of thermal heat stress include increased blood flow to the skin, sweating and increased ventilation (Hancock et al., 2007; Leon, 2008). Thermal discomfort has also been known to lead to sick building syndrome (Myhren & Holmberg, 2008). The combination of high temperature and high relative humidity reduces thermal comfort and indoor air quality (Fang et al., 2004). Large differences in the thermal radiation of the surfaces surrounding a person may also cause local discomfort or reduce acceptance of the thermal conditions. ASHRAE Standard 55 sets limits on the allowable temperature differences between various surfaces. Because people are more sensitive to some asymmetries than others, for example, that of a warm ceiling versus that of hot and cold vertical surfaces, the limits depend on which surfaces are involved. The ceiling is not allowed to be more than +5 °C (9.0 °F) warmer, whereas a wall may be up to +

23 °C (41 °F) warmer than the other surfaces in the UK.[18] Thermal stratification that results in the air temperature at the head level being higher than at the ankle level may cause thermal discomfort. ASHRAE Standard 55 recommends that the difference should not be greater than 3 °C (5.4 °F) for seated occupants or 4 °C for standing occupants (7.2 °F).[19] Floors that are too warm or too cool may cause discomfort, depending on footwear. ASHRAE 55 therefore recommends that floor temperatures stay in the range of 19–29 °C (66–84 °F) in spaces where occupants will be wearing lightweight shoes.[20] The ASHRAE-55 2010 Standard introduced the preferred mean temperature greater than 10 °C (50.0 °F) and less than 33.5 °C (92.3 °F).[21] Occupants in naturally ventilated buildings appear to be more tolerant of a wider range of temperatures (de Dear & Brager, 1998). ASHRAE-55 2010 Standard states that differences in recent thermal experiences, changes in clothing, availability of control options and shifts in occupant expectations can change people's thermal responses.[22]

The United Kingdom National Health Service Estimates Health Guidance Note refers to water and surface temperatures for safe use. Under the Registered Home Act 1984 (Ref 3), 44 °C (111 °F) is recommended for an unassisted bath fill, 46 °C (115 °F) for an assisted bath fill, 41 °C (106 °F) for shower and washbasin applications and 38 °C (100 °F) for bidet applications as well as 60 °C (140 °F) for cleaning (Pardo et al., 2012).[23]

Residential buildings in particular can vary much more in thermal comfort than public and commercial buildings due to their smaller size, the variations in clothing worn and the different uses of each room. The main rooms of concern are bathrooms and bedrooms. Bathrooms need to be at a temperature comfortable for a human with or without clothing.

[18] https://www.ashrae.org/technical-resources/bookstore/standard-55-thermal-environmental-conditions-for-human-occupancy.

[19] https://www.ashrae.org/technical-resources/bookstore/standard-55-thermal-environmental-conditions-for-human-occupancy.

[20] https://www.ashrae.org/technical-resources/bookstore/standard-55-thermal-environmental-conditions-for-human-occupancy.

[21] https://www.ashrae.org/technical-resources/bookstore/standard-55-thermal-environmental-conditions-for-human-occupancy.

[22] https://www.ashrae.org/technical-resources/bookstore/standard-55-thermal-environmental-conditions-for-human-occupancy.

[23] http://www.legislation.gov.uk/ukpga/1984/23/part/I/enacted.

Bedrooms need to accommodate different levels of clothing and also different metabolic rates of people asleep or awake (Peeters et al., 2009).

Regarding water risks, *Legionella* is a bacteria that naturally occurs in many water systems and can cause Legionnaires' disease. The UK Health and Safety Executive states that *Legionella* bacteria will not multiply at temperatures below 20 °C (68 °F), but will remain as a potential threat in the system. *Legionella* bacteria will proliferate between 20 °C (68 °F) and 45 °C (113 °F) and die above 60 °C (140 °F).[24] Therefore, hot water should be stored above 60 °C (140 °F) to eliminate *Legionella* and stored at a minimum of 55 °C (131 °F). Cold water storage should never exceed 20 °C (68 °F).

Regarding psychological adaptation, an individual's comfort level in a given environment may change and adapt over time due to psychological factors. Subjective perception of thermal comfort may be influenced by the memory of previous experiences. Habituation takes place when repeated exposure moderates future expectations and responses to sensory input. Thermal comfort is a psychological term. Among the factors that affect the condition of mind (in the laboratory) are a sense of control over the temperature, knowledge of the temperature and the appearance of the (test) environment. A thermal test chamber that appeared residential 'felt' warmer than the one which looked like the inside of a refrigerator (Rohles, 2007). Weather is related to individual happiness insofar as 14 °C (57 °F) appears to be the happiness peak temperature (Tsutsui, 2012).

The human body has physiological adaptation mechanisms to survive in drastic temperature environments. In a cold environment, the body utilizes vasoconstriction, which reduces blood flow to the skin, skin temperature and heat dissipation. In a warm environment, vasodilation will increase blood flow to the skin, heat transport and skin temperature and heat dissipation (Szokolay, 2010). If there is an imbalance despite the vasomotor adjustments listed above, in a warm environment, sweat production will start and provide evaporative cooling. If this is insufficient, hyperthermia will set in, body temperature may reach 40 °C (104 °F) and heat stroke may occur. In a cold environment, shivering will start, involuntarily forcing the muscles to work and increasing the heat production by up to a factor of 10. If equilibrium is not restored, hypothermia can set in, which can be fatal (Szokolay, 2010).

[24] http://www.hse.gov.uk/pubns/books/l8.htm.

Long-term exposure to extreme temperatures, of a few days to six months, may result in cardiovascular and endocrine adjustments. A hot climate may create increased blood volume, improving the effectiveness of vasodilation, enhanced performance of the sweat mechanism and the readjustment of thermal preferences. In cold or underheated conditions, vasoconstriction can become permanent, resulting in decreased blood volume and increased body metabolic rate (Szokolay, 2010). Behavioral adjustments are naturally ventilated buildings, operating windows and fans, adjusting blinds/shades, changing clothing and consuming cold or hot food and drinks.

As for individual differences, thermal sensitivity has been noticed in pregnant and menopausal women, disabled as well as individuals whose age is below fourteen or above sixty, which is considered not to fall into the adult range. Existing literature provides consistent evidence that sensitivity to hot and cold surfaces usually declines with age. There is also some evidence of a gradual reduction in the effectiveness of the body in thermo-regulation after the age of sixty (Lenzuni et al., 2009). This is mainly due to a more sluggish response of the counteraction mechanisms in lower parts of the body that are used to maintain the core temperature of the body at ideal values (Lenzuni et al., 2009). Seniors prefer warmer temperatures than young adults: 24 °C vs. 22 °C (76 vs. 72 °F) (Rohles, 2007). Males report discomfort due to rises in temperature much earlier than females. Males also estimate higher levels of their sensation of discomfort than females (Wyon et al., 2009). Males also tend to be more sensitive to relative humidity levels (Karjalainen, 2007; Lan et al., 2007).

10.5.3.2 Retail

Weather has the biggest influence on consumer behavior after the economy, according to the British Retail Consortium. It affects consumers' emotional state, drives their purchase decisions and dictates how much they are willing to spend. In the USA, it was estimated that one-third of all business activities have some form of weather sensitivity, with weather accounting for a swing factor of 3.4% of the total GDP.[25]

[25] http://www.weatherads.io/blog/2015/august/the-impact-of-weather-on-retail-sector-in-the-uk.

Recent research into the US retail sector's susceptibility to weather variability showed that 2.3% of sales are weather-sensitive.[26] A seasonal temperature, just 1 °C (34 °F) higher or lower than average, typically causes a 1% fluctuation in sales.[27]

For retailers, having an informed understanding of how weather affects sales helps them deploy weather-responsive advertising and capitalize on weather-driven trends based on procurement and logistical foresight. Weather analytics is an equally useful tool for identifying optimum windows of opportunity for marketing specific goods and services in weather-sensitive marketing and procurement strategies. Through real-time weather-responsive marketing, retail websites can leverage forecast information to anticipate periods of hot weather, reaching out to their audience to advise them to stock up on key products online before the weather arrives. By targeting consumers with relevant deals and promotions beforehand, online retailers can mitigating the negative impact good weather can have on their online sales, before it even hits them.[28] By deploying weather-adjusted online and mobile advertising strategies in weather-responsive email campaigns and weather-adaptive website content, retailers can drive sales of their weather-sensitive products.

Different sectors within the retail industry have specific periods of the year that are pivotal to their annual success.[29] Overall important appears seasonality. Unpredicted changes to seasonal weather conditions lead to problems in procurement and retail trade.[30] As for concrete examples, garden centers experience a boom in spring—but when the weather turns temperate again following a frosty winter, people begin to contemplate gardening once more. For the clothing and apparel sectors, spring and autumn are the two key seasons heralding the arrival of new collections for the upcoming summer or winter months.

26 http://www.weatherads.io/blog/2015/august/the-impact-of-weather-on-retail-sector-in-the-uk.

27 http://www.weatherads.io/blog/2015/august/the-impact-of-weather-on-retail-sector-in-the-uk.

28 http://www.weatherads.io/blog/2014/august/weather-and-ecommerce-how-weather-impacts-retail-website-traffic-and-online-sales.

29 http://www.weatherads.io/blog/2015/august/the-impact-of-weather-on-retail-sector-in-the-uk.

30 http://www.weatherads.io/blog/2015/august/the-impact-of-weather-on-retail-sector-in-the-uk.

For the pharmaceutical industry, summer and winter are the biggest grossing seasons, with cold and fluremedies being prominent in winter times, and antihistamines, suncream and hair removal products selling well during summers.[31] Unseasonable weather conditions tend to lower demand expectations.[32]

Certain sectors within the retail industry are more weather-sensitive than others. For instance, perishable items like food, which have limited stock durations, have far higher weather risk attached to them than, for instance, decorative and do-it-yourself (DIY) goods or consumer electronics—where holding large volumes of stock is a common practice.[33] Cost-intensive climate sensitivities estimated for the UK retail market are food, apparel, home improvement, accessories/jewelry/luggage, consumer electronics, garden and outdoor furniture and pharmaceuticals.[34]

On the demand side, weather not only shapes consumers' desire for particular products and services, but equally determines the consumers' purchase channels and methods. Inclement weather conditions drop sales at high street shops and increase purchases at shopping centers and malls.[35]

Online retailers tend to profit during inclement weather, and particularly in extreme weather events.[36] Conversely, during the four warmest days of April 2012, online retail searches plummeted.[37] About 60% of consumers change their shopping habits when it is hot or when it rains. When it is raining, a third of consumers shop less or in covered shopping areas, such as large shopping centers, rather than in high street shops.[38]

[31] http://www.weatherads.io/blog/2015/august/the-impact-of-weather-on-retail-sector-in-the-uk.

[32] http://www.weatherads.io/blog/2015/august/the-impact-of-weather-on-retail-sector-in-the-uk.

[33] http://www.weatherads.io/blog/2015/august/the-impact-of-weather-on-retail-sector-in-the-uk.

[34] http://www.weatherads.io/blog/2015/august/the-impact-of-weather-on-retail-sector-in-the-uk.

[35] http://www.weatherads.io/blog/2015/august/the-impact-of-weather-on-retail-sector-in-the-uk.

[36] https://www.tradedoubler.com/en/.

[37] https://store.mintel.com/seasonal-shopping-uk-august-2014.

[38] https://store.mintel.com/seasonal-shopping-uk-august-2014.

In a revenue by temperature study, online sales through e-commerce channels drop increasingly with temperature rises when considering the 5–30 °C (41–86 °F) range.[39] So consumers like to shop online when the temperature is colder and prefer to shop outside when it is warm and sunny. This relationship between weather and online retail sales was backed by the e-retail industry.[40] Weather-driven demand varying by industry warrants for further market research on weather-impacts on consumer behavior and sales.

Weather also influences consumer behavior through its effect on mood. Studies show that temperature, humidity, air pressure, snowfall and especially sunlight can have a huge impact on a consumer's mind frame and by extension their spending. Seasonality and sector are the two key considerations. If the weather is consistently wet and cold, barbecues, swimming devices and gardening utensils sales drop.[41] Shops being too stuffy and warm, particularly in clothing stores, are one of the biggest turn-offs for consumers. If they find a shop uncomfortable, people spend less time looking around and will likely not use the change rooms. Exposure to sunlight dramatically increases levels of consumption as well as the amount spent per item.[42] Many retailers are savvy to this phenomenon and use bright halogen lighting which mimics the effect of sunlight in their stores.[43] Consumers' mind frame can also be determined by seasonal weather events—so if there is snowfall in late October, this gets consumers into the Christmas spirit early and therefore boost pre-Christmas sales.[44]

Weather is a critical driver of product demand. If temperatures reach over 18 °C (64 °F) in the UK, supermarkets experience a 22% increase

[39] http://www.weatherads.io/blog/2014/august/weather-and-ecommerce-how-weather-impacts-retail-website-traffic-and-online-sales.

[40] http://www.weatherads.io/blog/2014/august/weather-and-ecommerce-how-weather-impacts-retail-website-traffic-and-online-sales.

[41] https://store.mintel.com/seasonal-shopping-uk-august-2014.

[42] http://www.weatherads.io/blog/2015/august/the-impact-of-weather-on-retail-sector-in-the-uk.

[43] http://www.weatherads.io/blog/2015/august/the-impact-of-weather-on-retail-sector-in-the-uk.

[44] http://www.weatherads.io/blog/2015/august/the-impact-of-weather-on-retail-sector-in-the-uk.

in cool drinks, 20% increase in juices and 90% increase in garden furniture.[45] A 1 °F increase in temperature is associated with 2% soft drink, 24% air conditioner, 11% suncare, 4% infant apparel and 13% hedge trimmers.[46] Likewise, a 1 °F drop in temperature in the USA leads to increase in sales of soup by 2%, porridge, portable heaters by 15%, mousetraps by 25%, softline goods—such as sneakers—by 2.5% and lip-care products.[47] Snowy days are related to exceptionally high click through rates on Internet retailers and holiday bookings, which have also been proven to be inextricably linked to periods of inclement domestic weather.[48] If the temperature stayed low, there was an average order plus of 10–12 percentage points for home and furniture, wholesales and clothing e-retailers and a 3–5 percentage point plus for big box retailers.[49]

As for overall effects of weather on consumer purchase decisions, consumers either prolong or fast-track their shopping experience depending on the clemency of weather conditions.[50] During mild autumn, consumers will postpone their purchase of winter clothing. During an unseasonably warm spring, they will fast-track their purchase of gardeningequipment. Consumers directly respond to weather—for instance, buying sunscreen and cold drinks when it is hot and sunny or rain jackets when it has been consistently raining. Shoppers may align food with weather and substitute barbecued meats for roasted meats, or salad for soup, when the weather turns cold.[51] After exposure to sunlight, consumers are more willing to self-reward (i.e. make purchases), and are

[45] http://www.weatherads.io/blog/2015/august/the-impact-of-weather-on-retail-sector-in-the-uk.

[46] http://www.weathertrends360.com/?ref=www.wxtrends.com.

[47] http://www.weatherads.io/blog/2015/august/the-impact-of-weather-on-retail-sector-in-the-uk.

[48] http://www.businessinsider.com/the-effect-of-weather-on-purchasing-decisions-2012-11.

[49] http://www.weatherads.io/blog/2014/august/weather-and-ecommerce-how-weather-impacts-retail-website-traffic-and-online-sales.

[50] http://www.weatherads.io/blog/2014/august/weather-and-ecommerce-how-weather-impacts-retail-website-traffic-and-online-sales.

[51] http://www.weatherads.io/blog/2014/august/weather-and-ecommerce-how-weather-impacts-retail-website-traffic-and-online-sales.

also willing to pay more for goods and services.[52] When consumers are in a warm environment, their purchase intent increases and they tend to be more susceptible to marketing messages.[53]

10.5.3.3 Vacation

Temperature and weather conditions impact outdoor and sport opportunities. For instance, low temperatures, ice and snow determine winter sports possibilities. Climate change has continuously led to missing snow over the last decades, and subsequently winter sports have experienced a decline in popularity due to a price spike to maintain artificial snow conditions (Emmert, 2018). Millennials are losing interest in paying high prices for snow activities. Winter sport resorts are forced to make use of high energy snowmaking technology to sustain the industry, which also account for an enormously ecologically-damaging process (Emmert, 2018).

It takes about 75,000 gallons (285,000 liters) of water to create a 6-inch blanket of snow covering a 200 × 200-foot area (Emmert, 2018). Snow guns run on diesel engines, which expel extremely high levels of pollution. Furthermore, due to warmer temperatures, snowmakers have concocted chemical compounds that allow the snow to stay frozen above 0 °C (32 °F) (Emmert, 2018). One of the chemicals used is a salt, which has a similar ecological impact as road salt. When this snow melts, it can contaminate human drinking water and wildlife. The clear-cutting of vegetation to put in water pipes for snowmaking apparatuses disturbs wildlife and causes soil erosion. Yet, snowmaking provides economic predictability and therefore stabilizes local economies by workforce job security for those whose income depends on snow tourism (Emmert, 2018).

Diversification strategies for alternative sport opportunities but also diversified vacation opportunities are currently used to combat weather risk in the sports industry. Price mechanisms may be an alternative natural solution. The price for winter vacation has gone up in recent years and is expected to continuously rise until consumers hit a price level that

[52] http://www.weatherads.io/blog/2014/july/weather-and-marketing-how-temperature-can-trigger-more-sales.

[53] http://www.weatherads.io/blog/2014/july/weather-and-marketing-how-temperature-can-trigger-more-sales.

makes it more efficient to substitute this service by other leisure activities (Emmert, 2018).

In the winter sports industry, we can now detect an oligopolization of the market, with only a few consolidated conglomerates owning the entire snow sports market (Emmert, 2018). These conglomerates are believed to diversify their mountain portfolio by geographically spreading their risk of bad weather across territories (Emmert, 2018). In addition, the insurance industry is not more involved in compensating the snow sports industry for losses (Emmert, 2018). By now, 50% of revenue is believed to be made even without snow from food, accommodation and rentals around the sports (Emmert, 2018).

Graph 10.7 outlines some of the quantitative measures for service sector temperature requirements, with the agreeable outside temperature in light yellow shade, the optimum outside temperature range in yellow shade, the optimum cardinal temperature peak in blue, healthy water conditions in green, health risks in orange and excessive health risks in red for service production. The contemporary mean world temperature of 18.6 °C (65 °F) as well as the predictions of estimated 22.92 °C (73.26 °F) mean temperature around the globe in 2100 are outlined in purple to put the service production in perspective with overall current and future temperatureconditions.

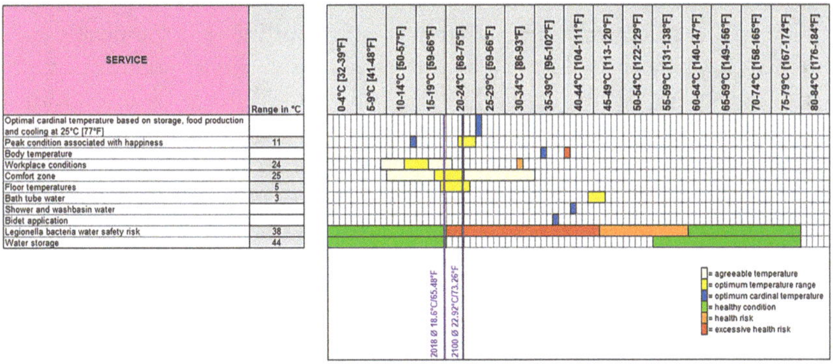

Graph 10.7 Service sector temperature-based condition in relation to world mean temperature (*Source* Author's own)

10.6 Discussion

10.6.1 Climate Change-Induced Market Changes

Overall, the data presented over all three GDP pillars of the agriculture, service and industry sectors suggests that climate change will affect the GDP sectors' feasibility and performance. Contemporary climate change projections estimate a mean temperature rise of approximately 4.24 °C (39.63 °F) by the year 2100. The cardinal temperature per sector was related to the prospected temperature in 2100 per country. The cardinal temperature for agriculture is 26.12 °C (79 °F) derived from the mean over 59 crops of all agriculture data sources listed.[54] The cardinal temperature for the industry sector comprising construction, automotive, airframe, locomotive, petroleum, coal and chemicals was found to be 21 °C (70 °F) (Russell, 1957). The cardinal temperature for the service sector lies at 25 °C (77 °F) (Somanathan et al., 2014). Overall, it can be projected that the world will run out of cardinal time, hence optimum production temperatures, for industry production first, followed by service sector and lastly the agricultural sector. As the temperature will be rising, we may therefore see a shift from industry production to service sector activities and then to agriculture sector productivity. As general market trend indicator, first the earth appears to become too hot for the industry sector, then for the service sector activities and lastly for agriculture production.

Overall, the literature meta-analysis results outline that the peak condition for happiness at 14 °C (57 °F) has passed when considering the world mean temperature of 18.6 °C (65 °F). Also the peak condition for workplace temperature has passed when considering the world mean temperature of 18.6 °C (65 °F). We are currently in the general comfort zone range for agreeable temperature when considering the world mean temperature. So, while the peak temperature time for plant seed growth is still ahead, subjective well-being based on temperature is either currently prevalent or has passed. The optimum temperature for productivity at the workplace has already passed. When it comes to safety, *Legionella* bacteria in the water will soon become a problem, and medical and hygiene markets that combat this risk and disease will likely prosper.

[54] http://agritech.tnau.ac.in/agriculture/agri_agrometeorology_temp.html.

Based on the lowest optimal temperature for the industry sector of 21 °C (70 °F), we can assume that industry will move first to the poles of the world (Russell, 1957). Then, based on the optimal temperature for the service sector of 25 °C (77 °F), we can assume that service sector activities will relocate toward cooler climates (Somanathan et al., 2014). Lastly, agriculture with the relatively high optimal temperature of 26.12 °C (79 °F) will move to cooler areas. Overall, a trend of migration and corporate as well as industry relocation is predicted. Globalization may have exacerbated climate change when considering human-made global warming contributions. But globalization will also be exacerbated by the change in temperature if we consider industries, service sector activities and agricultural patterns being forced to shift in a changing climate.

As for general market trends, there is an overall predicted shift into green and clean energy technologies that may also increase the capacity of non-green sectors for improved competitiveness through higher resource productivity and newly created incomes. On the other hand, environmental policies may negatively affect labor absorption in polluting industries (Esposito et al., 2018). The shift of jobs across enterprises, industries and sectors may entail adjustment costs for enterprises and workers (Esposito et al., 2018).

Overall, geographical mobility is expected to increase and thus should be supported, especially when massive shifts of the industry structure result from green industrial policies. This may be the case, for example, when shifting from fossil to renewable energy. Fossil energy plants are usually located next to energy-intensive industries and cities, whereas renewable power plants are located where the respective power resources are abundant. Shifting energy systems will therefore affect spatial distribution, and workers are expected to move physically accordingly (Esposito et al., 2018).

Since the world mean temperature is somewhat close to the optimum performance temperature, we can assume climate will affect production. Since the continental temperature averages are differing, a drive toward northern and southern poles of the earth with economic productivity is projected. When the earth poles are getting warmer, economic productivity is projected to settle in those newly won arable land, and economic productivity may start in previously too cold territories. Human and Foreign Direct Investment incoming streams were already detected for climate change winning territories.

10.6.1.1 *Agriculture*

Regarding agriculture products, the presented novel results add to the existing literature on agriculture cardinal temperature and refine the estimates in a more comprehensive meta-analysis of the existing literature. The current world mean temperature of 18.6 °C (65 °F) indicates that the world has still time ahead until peak condition for agriculture, which lies at 22.92 °C (73.26 °F) derived from literature data of 59 crops. As the range of time for agriculture production varies by crop, we can divide the different crops in durability classes based on the available literature sources.

The most flexible crops with the most time to be cultivated for production in a changing climate appear to be beans (range r = 46 °C, crops grow from [5 °C up to 50 °C, 40–122 °F]), followed by pak choi (r = 44 °C, [3 °C–46 °C, 40–115 °F]), peas (r = 43 °C, [−2 °C–40 °C, 28–50 °F]), carrots (r = 42 °C, [0 °C–41 °C, 32–106 °F]), cabbage (r = 42 °C, [1 °C–42 °C, 34–108 °F]), milk thistle (r = 41 °C, [0 °C–40 °C, 32–104 °F]) and red beet (r = 41 °C, [4 °C–44 °C, 40–111 °F]).

Medium flexibility exhibit ryegrass (r = 40 °C, [1 °C–40 °C, 34–104 °F]), onions (r = 40 °C, [1 °C–40 °C, 34–104 °F]), Chinese cabbage (r = 39 °C, [5 °C–43 °C, 40–109 °F]), Asian radish (r = 39 °C, [3 °C–41 °C, 40–106 °F]), cold-season crops such as oats, rye and wheat (r = 38 °C, [0 °C–37 °C, 32–99 °F]), parsnip (r = 38 °C, [0 °C–37 °C, 32–99 °F]) and corn (r = 38 °C, [10 °C–46 °C, 50–115 °F]).

Lowest flexibility is a feature of soybean (r = 37 °C, [2 °C–38 °C, 36–100 °F]), pumpkins (r = 32 °C, [5 °C–36 °C, 41–97 °F]), rice (r = 30 °C, [10 °C–40 °C, 50–104 °F]), coconut (r = 27 °C, [16 °C–42 °C, 60–108 °F]), sorghum (r = 27 °C, [8 °C–34 °C, 46–93 °F]) and peanut and groundnut (r = 27 °C, [13 °C–36 °C, 55–97 °F]), and ends with cool-season crops such as asparagus, broccoli, Brussels sprouts, cabbage, chives, collards, corn salad, garlic, horseradish, kale, kohlrabi, leeks, mustard, onions, parsley, peas, radishes, rhubarb, rutabaga, spinach and turnips (r = 21 °C, [5 °C–25 °C, 41–77 °F]). The temperature ranges are outlined in Graph 10.8, which holds the crop durability.

Longest temperature range crops are indicated by green color, medium temperature range in yellow color and shortest temperature range in red color.

The first crops to be produced in glasshouses due to running out of natural agreeable conditions are cool-season crops such as asparagus, broccoli, Brussels sprouts, cabbage, chives, collards, corn salad, garlic,

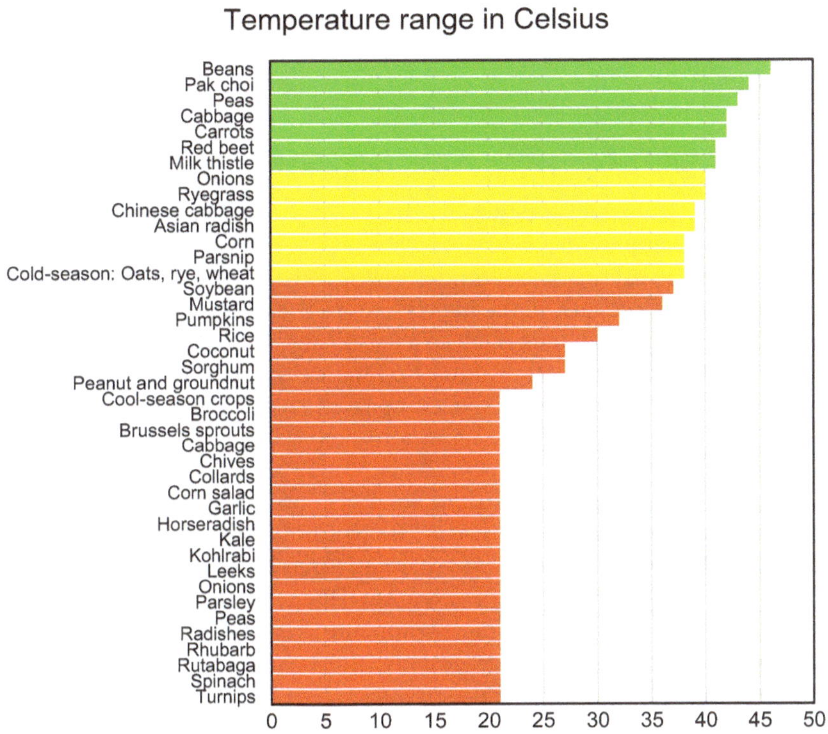

Graph 10.8 Temperature ranges for outlining agriculture durability (*Source* Author's own)

horseradish, kale, kohlrabi, leeks, mustard, onions, parsley, peas, radishes, rhubarb, rutabaga, spinach and turnips (maximum temperature t_{max}= 25 °C, 77 °F). The next crops that run out of natural growing possibilities are half-hardy and semi-hardy vegetables and hardy herbs such as beets, cauliflower, celery, celeriac, chard, chicory, globe artichokes, endive, lettuce, potatoes, salsify and sorrel (t_{max}= 26 °C, 79 °F). Next to follow are sorghum (t_{max}= 34 °C, 93 °F), pumpkins, peanut and groundnut (t_{max}= 36 °C, 97 °F), cold-season crops such as oats, rye, wheat and parsnip (t_{max}= 37 °C, 99 °F), soybean (t_{max}= 38 °C, 100 °F), rice (t_{max}= 39 °C, 102 °F), peas, milk thistle, ryegrass and onions (t_{max}= 40 °C, 104 °F), carrots, Asian radish and mustard (t_{max}= 41 °C, 106 °F), cabbage and coconut (t_{max}= 42 °C, 108 °F), Chinese cabbage

(t_{max}= 43 °C, 109 °F), red beet (t_{max}= 44 °C, 111 °F), pak choi (t_{max}= 46 °C, 115 °F), corn (t_{max}= 47 °C, 117 °F) and beans (t_{max}= 50 °C, 122 °F).

10.6.1.2 Industry

As for industry-sector-related changes, with the switch temperature from heating to cooling being determined at 18 °C (64 °F) and the current world mean temperature being at 18.6 °C (65 °F), we can currently witness a potential market demand change from heating into cooling devices. During heating days, fuel is required as primary space heating resource. During cooling days, electricity, oil and natural gas consumption tends to increase (Pardo et al., 2012). Due to the nature of heating and cooling devices, we may see an accompanied switch from space heating, which is often run on gas, to air-conditioned ventilation more prone to be driven by electricity.

The world mean temperature at 18.6 °C (65 °F) is still ahead of the peak outside temperature at 21 °C (70 °F) for construction, automotive, airframe, locomotive, petroleum, coal and chemicals. In this regard, the world is not at the efficiency frontier yet (Foley, 2007). Cold and fresh air ducts will run out of favorable temperature conditions first, and will soon be followed by human productivity drops, when the mean world temperature is above 32 °C (90 °F).

With the warming of the globe, industrial waste heat is prospected to become a more dangerous negative externality of industry production. It is expected that industries with heightened levels of industry waste heat will become more charged, for example, via taxation, in order to offset for the losses incurred. Most industries' waste heat is within the temperature range of 100–200 °C (212–392 °F) but increased to more than 500 °C (932 °F) in the chemical, iron and steel and non-metallic mineral industrial sectors (Papapetrou et al., 2018).

The global warming accelerating impact of the heavy heat waste producing industries—such as chemical, iron and steel production—is visible in the two temperature spikes around the two world wars, when these industries' production peaked. Transitions into less polluting industries are predicted as heat waste industries are prospected to be considered as polluting industries for their level of heat emitted and eventually may face higher costs and lose competitive advantage (Altenburg & Rodrik, 2018). Already now there is a trend toward using carbon-offsetting materials such as plants rather than cement, for example, at Frankfurt airport.

The textile industry has also picked up on the trend and now creates clothing out of fungi and bamboo, which are carbon-negative and absorb greenhouse gas emissions. The COVID-19 pandemic era has also seen a biophilia trend emerging in interior design, which featured elements close to nature, such as wood, cork, plants and healthy surroundings.

When analyzed purely from a heat-producing perspective, the food, drink, tobacco, textiles and leather production is the least polluting, followed by chemicals, non-metallic minerals, then pulp, paper, iron and steel, followed by cement, iron, steel, foundries, glass and motor-run engines equipped with pumps and fans. The latter industries are expected to become more expensive with time on a warming globe as for contributing the climate change more than other industry sectors. As economic complements, dead rooms, flue-gas ducts, hot-air ducts, steel chimneys, containers, pipes and gas-turbine tiles are expected to rise in price the most, followed by boiler walls and roof areas and technical building equipment, such as pipes and ducts for power plants. For cold and fresh-air production and complements, such as cold or fresh-air ducts, the price is expected to rise only when the world mean temperature is beyond 20 °C (68 °F) or in places where the mean temperature during production will be beyond 20 °C (68 °F).

Unfavorable work conditions due to unpleasant temperatures are assumed to be drivers of labor migration. Chapter 9 already reports a human migration stream into climate change winning countries. Worker transfer into occupations that are performable in agreeable temperatures is expected. If work requires human labor, workers that work and can sustain a good performance under thermal stress are expected to become better remunerated in the future. Robotics are emergingly taking over human work tasks and with this cooling of AI processors and data storage are rising energy consumption in many sectors.

10.6.1.3 Service

The optimal cardinal temperature for service production depends on worker productivity, retail, food and leisure activities—such as vacation. Overall estimations based on storage, food production and cooling define the cardinal temperature for the service sector at 25 °C (77 °F) (Somanathan et al., 2014). The discussed temperature-related labor migration effects may also occur in the service sector. We also see this already in seasonal-labor movement in areas with extreme seasonal changes. When temperature for physical labor will be above 16 °C (61

°F), workers will either seek other cooler workspaces or the remuneration for workers may rise in the long run and make service provision in hotter areas more expensive. Women and elder workers appear to be less distracted by heat and may have better natural conditions to sustain a proper work performance in heat. Pregnant women, disabled as well as those below 14 and above 60 years old will become subject to extended exemptions and workplace protection for exhibiting thermal sensitivity. While seniors prefer warmer temperatures than young adults, 24 °C vs. 22 °C (76 vs. 72 °F), and appear to be less susceptible to thermal stress, their health may be more prone to show signs of weakness (Rohles, 2007).

Thermal heat stress reducing problems of skin, sweating and increased ventilation needs is expected to become subject to more research and market opportunities (Hancock et al., 2007; Leon, 2008). The insurance industry will likely pick up on the trend and offer employers and employees protection for thermal stress-related symptoms and diseases. Architecture will likely see a rising focus on thermal comfort in buildings through construction, heating, ventilation and air-conditioning. Thermal environmental care by means of clothing, operable windows, fans, personal heaters and sunshades are prospected future market opportunities. Behavioral adjustments to high temperatures complementing goods are expected to boom in the future, such as ventilated buildings, operating windows and fans, adjusting blinds/shades, changing clothing, sunscreen and body-temperature-balancing cooling food and drinks. But also health and wellness solutions for well-tempered body temperature conditions are expected to rise in the decades to come. Regarding health risks related to water pollution, bottled water will become more prominent soon as the world mean temperature is approaching conditions for bacteria to naturally germinate and intoxicate water.

Weather-sensitive marketing and advertisement campaigns will become more accentuated as climate extremes kick in.[55]Risk foresight and resilience aid will become more profitable industries. For retailers, having an informed understanding of how weather affects sales helps them deploy weather-responsive advertising and capitalize on weather-driven trends by procurement and logistical foresight. Relocation and procurement flexibility, but also respective insurances that cover climate-related risks, will likely become more prominent as climate changes and weather

[55] http://www.weatherads.io/blog/2015/august/the-impact-of-weather-on-retail-sector-in-the-uk.

becomes more unpredictable. Future groundbreaking research on corporate climate-induced risk management and crises resilience relief action as corporate social responsibility (CSR) initiative appears as a novel and important scientific frontier extension (Ballesteros & Kunreuther, 2018).

Seasonal changes should be considered for the retail sector. Longer summers will bring more demand for certain clothes, apparel, skincare and gardening goods. While hotter water will fortify health risks from water consumption, milder winters will decrease consumption of all flu-related medication. The outdoor sports industry and tourism are projected to experience a shift from high-end winter toward extended summer sports and vacation opportunities. Temperature and weather-sensitive goods—like perishable food—will have a harder time on markets than temperature and weather flexible activities, for example, the Internet. Climate control will become a more profitable industry since we are just discovering the very many behavioral effects temperature has on individuals, ranging from well-being and mood, to productivity, purchasing behavior and consumption, to name a few. When conducting research on climate-sensitive tastes and preferences, complements and substitutes, as well as positive and negative externalities, should be taken into consideration—for example, hand in hand with the decline in winter sport activities comes a rise in exotic beach vacation demand, which has exacerbated beach pollution through sunscreen and plastic in the oceans.

As for retail implications, climate and weather should be integrated in strategic management plans. Weather-based marketing strategies would recommend online retailers and e-commerce websites to maximize by online-time-increasing display during sunny and warm weather that distracts consumers away from the Internet.[56]

10.7 Conclusion and Future Outlook

Transformation and change always create winners and losers (Altenburg & Rodrik, 2018). Overall, the present research sheds novel light on the primarily negative effects focused documentation on climate change but also refines a previous climate change winners and losers model for GDP prospects. Given the detected gain perspectives of a warming globe, contemporary standard integrated assessment models may mis-estimate

[56] http://www.weatherads.io/blog/2015/august/the-impact-of-weather-on-retail-sector-in-the-uk.

the global economic costs of climate change by neglecting to account for the widespread productivity benefit aspects of a warming globe.

Overall, it is predicted that there will be a natural shift on a warming globe from industry to service to agriculture given the varying cardinal temperatures. This novel angle on climate-induced market prospects may be of direct value for industrial policy-making. Governments around the globe currently aim to accelerate structural transformation toward a low-carbon, resource-efficient economy in ways that also enable productivity enhancements in the economy (Altenburg & Rodrik, 2018).

A caveat to all of this is the fact that crops and industries as well as tastes and preferences are not static and have shown to change over time. The model assumes static agriculture, industry and service sectors cardinal temperatures and GDP compositions, which is problematic. Future research may refine the model for backtesting these predictions validated by historic data of agriculture, industry and service sector changes as well as GDP compositional changes over time. Future studies may also try filling data gaps and harmonizing the variant, and sometimes contradictory, literature on optimal cardinal temperatures. In the chapter presented, temperature data is sometimes incomplete or reports controversial or instringent results assumed to stem from local, regional and national differences as well as measurement differences. In order to pay tribute to all sources consulted, the conflicting data was kept in the visual display in the graphs of this chapter.

The average world temperature is calculated by averaging the minimum and maximum daily temperatures in the country for the years 1961–1990, based on gridded climatology from the Climatic Research Unit elaborated in 2011, which is critical. Especially since the turn of the millennium, climate change has exacerbated high-temperature records, which weakens the credibility of the results of world mean temperature in relation to cardinal temperature. Future research should consider the current climate change fluctuations especially visible since the turn of the millennium. A future research gap detected is the missing literature and research on the impact of climate change on animal products due to temperature stress.

Since temperature well-being is also a psychological concept, climate effects on consumer psychology and purchase behavior should become subject to future studies. The prospective results may aid the corporate sector to understand how to respond to natural climate changes and also how climate affects consumer purchase decisions. Factors other than air temperature—for example, radiant temperature, humidity and air velocity—should become subject to scrutiny in the interaction with temperature and subjective well-being in future studies. These insights will aid governance experts to derive optimal sustainable growth predictions during the transition to renewable energy sources (Ha-Huy et al., 2018). Energy security experts may derive information on the impact of climate-related economic peak prospects and the transition to renewable energy in a changing climate (Orttung & Perovic, 2010; The World Economic Forum, 2015). Marketers can derive insights on how to implement weather-responsive marketing campaigns and turn climate change into a strategic advantage or corporate asset. Teaching experts may gain novel insights on how to make climate change insights more tangible for well-rounded innovative online learning experience, for instance enhanced by simulation-based experience learning (Gonglewski & Helm, 2010). In the overall discourse, climate and weather should also sometimes be seen as a competitive advantage and basic foundation for the wealth of nations in the age of global warming and growing advantage in the centuries to come.

References

Acevedo, S., Mrkaic, M, Novta, N., Poplawski-Ribeiro, M., Pugacheva, E., & Topalova, P. (2018, April 12). *How can low income countries cope? The effects of weather shocks on economic activity*. Presentation delivered at The New School Economics of Climate Change Project, The New School, New York.

Adao, B., Narajabad, B., & Temzelides, T. (2018). *Renewable technology adoption and the macroeconomy*. CESifo Working Paper 6372. https://papers.ssrn.com/sol3/papers.cfm?abstract_id=2941401##

Alagarswamy, G., & Ritchie, J. T. (1991). Phasic development in CERES-sorghum model. In T. Hodges (Ed.), *Predicting crop phenology* (pp. 143–152). CRC.

Albert, St. (2008). *Cool-season and warm-season crops: Harvest to table*. https://harvesttotable.com/cool-season_and_warm-seasoncro/

Allais, M. (1947). *Economie et intérêt*. Imprimerie Nationale.

Altenburg, T., & Assmann, C. (2018). *United Nations Environment Programme Report on Green Industrial Policy: Concept, policies, country experiences*. Geneva and Bonn: United Nations: United Nations Environment Programme & German Development Institute/Deutsches Institut für Entwicklungspolitik.

Altenburg, T., & Rodrik, D. (2018). Green industrial policy: Accelerating structural change towards wealthy green economies. In T. Altenburg & C.

Assmann (Eds.), *United Nations Environment Programme Report on Green Industrial Policy: Concept, policies, country experiences* (pp. 2–20). United Nations: United Nations Environment Programme & German Development Institute/Deutsches Institut für Entwicklungspolitik.

Antonietti, R., & Fontini, F. (2018, May 24–25). Does energy price affect energy efficiency? Cross-country panel evidence. In *6th International Symposium on Environment Energy & Finance Issues, ISEFI 2018*, Paris, France.

Baker, P., & Haggar, J. (2007). *Climate change and the coffee industry: The coffee sector and climate change*. International Trade Centre. http://www.intracen.org/coffee-guide/climate-change-and-the-coffee-industry/the-coffee-sector-and-climate-change/

Ballesteros, L., & Kunreuther, H. (2018). *Organizational decision making under uncertainty shocks*. National Bureau of Economic Research Working Paper No. 24924.

Batten, S. (2018). *Climate change and the macroeconomy: A critical review*. Working paper 706, Bank of England, London. https://papers.ssrn.com/sol3/papers.cfm?abstract_id=3104554

Bauer, P. T. (1957). *Economic analysis and policy in underdeveloped countries*. Duke University Press.

Benchekroun, H., van der Meijden, G., & Withagen, C. (2018). *OPEC, shale oil, and global warming on the importance of the order of extraction*. https://ideas.repec.org/p/ces/ceswps/_6746.html

Bernarda, J. T., & Kichian, M. (2017). *Carbon tax saliency. The case of B.C. diesel demand*. USAEE Working paper 17-327. https://papers.ssrn.com/sol3/papers.cfm?abstract_id=3100567

Bolhuis, C. G., & deGroot, W. (1959). Observations on the effect of varying temperature on the flowering and fruit set in three varieties of groundnut. *Netherland Journal of Agricultural Science, 7*, 317–326. https://doi.org/10.18174/njas.v7i4.17669

Brückner, S., Liu, S., Miro, L., Radspieler, M., Cabeza, L. F., & Lävemann, E. (2014). Industrial waste heat recovery technologies: An economic analysis of heat transformation technologies. *Entwurf, 151*, 1–37.

Bürgenmeier, B. (1994). Environmental policy: Beyond the economic dimension. In B. Bürgenmeier (Ed.), *Economy, environment, and technology: A Socio-economic approach* (pp. 166–175). Armonk Sharpe.

Burke, M., Hsiang, S. M., & Miguel, E. (2015). Global non-linear effect of temperature on economic production. *Nature, 527*, 235–239. https://doi.org/10.1038/nature15725

Buso, M., Greco, L., & Bertolini, M. (2018). *Competition and regulation with smart grids*. http://www.funseam.com/images/pdf/Presentaciones_VI_Simposio_Acad%C3%A9mico/Marco_Buso.pdf

Cambini, C., & Soroush, G. (2018, May 24–25). Designing grid tariffs in presence of distributed generation: The economic impact of energy losses and connection costs. In *6th International Symposium on Environment Energy & Finance Issues, ISEFI 2018*, Paris, France.

Centeno, M. A., Creager, A. N., Elga, A., Felton, E., Katz, St. N., Massey, W. A., & Shapiro, J. N. (2013). *Global systemic risk: Proposal for a research community*. Working paper. Princeton University: Princeton Institute for International and Regional Studies, Princeton, NJ.

Centeno, M. A., & Tham, A. (2012). *The emergence of risk in the global system*. Working paper. Princeton University, Princeton, NJ.

Chancel, L., & Piketty, T. (2015). *Carbon and inequality: From Kyoto to Paris*. Paris School of Economics.

Cole, P., & McCloud, P. (1985). Salinity and climatic effects on the yields of citrus. *Australian Journal of Experimental Agriculture, 25*, 711–717. https://doi.org/10.1071/EA9850711

Court, V., & Bovari, E. (2018). *Energy, knowledge, and demo-economic development in the long run: A unified growth model*. https://hal.archives-ouvertes.fr/hal-01698755/document

Crafts-Brandner, S. J., & Salvucci, M. E. (2002). Sensitivity of photosynthesis in a C-4 plant, maize, to heat stress. *Plant Physiology, 129*, 1773–1780. https://doi.org/10.1104/pp.002170

de Dear, R., & Brager, G. (1998). Developing an adaptive model of thermal comfort and preference. *ASHRAE Transactions, 104*(1), 145–167.

Dell, M., Jones, B. F., & Olken, B. A. (2012). Temperature shocks and economic growth: Evidence from the last half century. *American Economic Review: Macroeconomics, 4*(3), 66–95.

Douenne, T. (2018). *The vertical and horizontal distributive effects of energy taxes: A micro-simulation study of a French policy*. https://www.parisschoolofeconomics.eu/docs/douenne-thomas/working-paper%2D%2Dthe-vertical-and-horizontal-distributive-effects-of-energy-taxes.pdf

Egli, F., Polzin, F., Schmidt, T., & Steffen, B. (2018, May 24–25). Which policies trigger RE investment? A structured literature review and risk/return assessment. In *6th International Symposium on Environment Energy & Finance Issues, ISEFI 2018*, Paris, France.

Emmert, H. (2018). *The melting point: The economics and environmental impact of the winter sports industry*. Unpublished working paper. Economics of the Environment, The New School, New York.

Engel, L. C., Short, M. E., Jennings, S. E., Orttung, R. W., & Suter, L. J. (2017). *Digital environmental storytelling connecting to the Arctic: #60Above60 Pilot, 2017 Arctic Yearbook*. https://www.arcticyearbook.com/

Environmental Defense Fund. (2018). *How climate change plunders the planet: A warming earth disturbs weather, people, animals and much more.* https://www.edf.org/climate/how-climate-change-plunders-planet

Esposito, M., Haider, A., Semmler, W., & Samaan, D. (2018). Enhancing job creation through the green transformation. In T. Altenburg & C. Assmann (Eds.), *United Nations Environment Programme Report on Green Industrial Policy: Concept, policies, country experiences* (pp. 50–68). United Nations: United Nations Environment Programme & German Development Institute/Deutsches Institut für Entwicklungspolitik.

Fang, L., Wyon, D. P., Clausen, G., & Fanger, P. O. (2004). Impact of indoor air temperature and humidity in an office on perceived air quality, SBS symptoms and performance. *Indoor Air, 14*(7), 74–81. https://doi.org/10.1111/j.1600-0668.2004.00276.x

Ferreira, V. (2017). Climate induced migration: Legal challenges. In J. M. Puaschunder (Ed.), *Intergenerational responsibility in the 21st century* (pp. 125–141). Vernon.

Foley, D. (2007). *The economic fundamentals of global warming*. Working paper 07-12-044. Santa Fe Institute, Santa Fe, NM. https://www.santafe.edu/research/results/working-papers/the-economic-fundamentals-of-global-warming

Forbes, K., & Zampelli, E. M. (2018, May 24–25). Wind energy, the price of carbon allowances, and CO_2 emissions: Evidence from Ireland. In *6th International Symposium on Environment Energy & Finance Issues, ISEFI 2018*, Paris, France.

Giraud, G., Lantremange, H., Nicolas, E., & Rech, O. (2018). *National carbon reduction commitments: Identifying the most consensual burden sharing.* https://halshs.archives-ouvertes.fr/halshs-01673358

Gonglewski, M., & Helm, A. (2010). An examination of business case methodology: Pedagogical synergies from two disciplines. *Global Business Languages, 15*(3), 17–31.

Ha-Huy, T., Le Van, C., & Nguyen, T. T. M. N. (2018, May 24–25). Optimal growth with renewable resources: A two-sectors model. In *6th International Symposium on Environment Energy & Finance Issues, ISEFI 2018*, Paris, France.

Hancock, P. A., Ross, J. M., & Szalma, J. L. (2007). A meta-analysis of performance response under thermal stressors. *Human Factors: The Journal of the Human Factors and Ergonomics Society, 49*(5), 851. https://doi.org/10.1518/001872007X230226

Harrod, R. F. (1948). *Towards a dynamic economics*. Macmillan.

Hatfield, J. L., & Prueger, J. H. (2015). Temperature extremes: Effect on plant growth and development. *Weather and Climate Extremes, 10*, 4–10. https://doi.org/10.1016/j.wace.2015.08.001

Heal, G., & Park, J. (2014). *Feeling the heat: Temperature, physiology and the wealth of nations*. The Harvard Project on Climate Agreements Discussion Paper 14-60. Belfer Center, Harvard University, Cambridge, MA.

Heal, G., Park, J., & Zhong, N. (2017). *Labor productivity and temperature*. https://geoffreyheal.files.wordpress.com/2017/09/yale_paper_draft_august_25.pdf

Heidari, Z., Kamkar, B., & Sinaki, J. M. (2014). Determination of cardinal temperatures of milk thistle germination. *Advances in Plants and Agriculture Research, 1*(5), 1–7.

Helm, A. (2009). Global marketing classroom: Green marketing. *American Marketing Association Global SIG Newsletter, Global Interests*.

Hollinger, S. E., & Angel, J. R. (2009). Weather and crops. In E. Nafziger (Ed.), *Illinois agronomy handbook* (pp. 1–12). Illinois Agronomy Extensions Publications.

Hosseini, M., Mojab, M., & Zamani, G. R. (2017). Cardinal temperatures for seed germination of wild barley, barley grass and hoary cress. *Archives of Agronomy and Soil Science, 63*(3), 352–361. https://doi.org/10.1080/03650340.2016.1213385

Huizenga, C., Abbaszadeh, S., Zagreus, L., & Arens, E. (2006). Air quality and thermal comfort in office buildings: Results of a large indoor environmental quality survey. *Proceedings of Health Buildings, 3*, 393–397.

Hulme, M. (2016). Concept of climate change. In D. Richardson, N. Castree, M. F. Goodchild, A. Kobayashi, W. Liu, & R. A. Marston (Eds.), *The international encyclopedia of geography* (pp. 1–6). Wiley-Blackwell Association of American Geographers.

Ielpo, F., & Sevi, B. (2013). *Forecasting the density of oil futures returns using model-free implied volatility and high-frequency data*. Cahier de recherche no. 13.10.106, Centre de Recherche en Economie et Droit de l'Energie CREDEN Equipe ART Dev, Université Montpellier, Faculté d'Economie, Montpellier.

Kant, I. (1785/1993). *Grounding for the metaphysics of morals*. Hackett.

Karjalainen, S. (2007). Biological sex differences in thermal comfort and use of thermostats in everyday thermal environments. *Building and Environment, 42*(4), 1594. https://doi.org/10.1016/j.buildenv.2006.01.009

Kilian, L., & Zhou, X. (2017). *Modeling fluctuations in the global demand for commodities*. Discussion Paper No. DP12357. Centre for Economic Policy Research, London, UK. https://papers.ssrn.com/sol3/papers.cfm?abstract_id=3049914

Kirat, Y. (2018, May 24–25). Carbon curse in developed countries. In *6th International Symposium on Environment Energy & Finance Issues, ISEFI 2018*, Paris, France.

Klaassen, G. A. J., & Opschoor, J. B. (1991). Economics of sustainability or the sustainability of economics: Different paradigms. *Ecological Economics, 4*, 93–115. https://doi.org/10.1016/0921-8009(91)90024-9

Konnert, M., Fady, B., Gömöry, D., A'Hara, S., Wolter, F., Ducci, F., Koskela, J., Bozzano, M., Maaten, T., & Kowalczyk, J. (2015). *Use and transfer of forest reproductive material in Europe in the context of climate change*. European Forest Genetic Resources Programme. http://www.euforgen.org/fileadmin/templates/euforgen.org/upload/Publications/Thematic_publications/EUFORGEN_FRM_use_transfer.pdf

Kruse, R., & Wegener, Ch. (2017). *Time-varying persistence in real oil prices and its determinants*. Working paper. http://www.wisostat.uni-koeln.de/sites/statistik/abstracts/RKB-Time-Varying_Persistence_in_Real_Oil_Prices_and_its_Determinant.pdf

Lan, L., Lian, Z., Liu, W., & Liu, Y. (2007). Investigation of biological sex differences in thermal comfort for Chinese people. *European Journal of Applied Physiology, 102*(4), 471–480. https://doi.org/10.1007/s00421-007-0609-2

Lauterbach, C., Schmitt, B., Jordan, U., & Vajen, K. (2012). The potential of solar heat for industrial processes in Germany. *Renewable Sustainable Energy Review, 16*, 5121–5130. https://doi.org/10.1016/j.rser.2012.04.032

Lenzuni, P., Freda, D., & Del Gaudio, M. (2009). Classification of thermal environments for comfort assessment. *Annals of Occupational Hygiene, 53*(4), 325–332. https://doi.org/10.1093/annhyg/mep012

Leon, L. R. (2008). Thermoregulatory responses to environmental toxicants: The interaction of thermal stress and toxicant exposure. *Toxicology and Applied Pharmacology, 233*(1), 146–161. https://doi.org/10.1016/j.taap.2008.01.012

Luo, Q. (2011). Temperature thresholds for a range of crops from cereal crops to horticultural crops. *Climate Change, 109*(3–4), 583–598. https://doi.org/10.1007/s10584-011-0028-6

Marron, D. B., & Morris, A. C. (2016). *How to use carbon tax revenues*. Tax Policy Center Urban Institute & Brookings Institution. https://doi.org/10.2139/ssrn.2737990

Massol, O., Banal-Estanol, A., & Ruperez-Micola, A. (2018, May 24–25). Production intermittence in spot electricity markets. In *6th International Symposium on Environment Energy & Finance Issues, ISEFI 2018*, Paris, France.

McCormick, J. I., Goodger, R. A., & Chynoweth, R. J. (2014). Cardinal temperatures and vernalisation requirements for a selection of vegetables for seed production. *Agronomy New Zealand, 44*, 71–83.

Myhren, J. A., & Holmberg, S. (2008). Flow patterns and thermal comfort in a room with panel, floor and wall heating. *Energy and Buildings, 40*(4), 524. https://doi.org/10.1016/j.enbuild.2007.04.011

Naegler, T., Simon, S., Klein, M., & Gils, H. C. (2015). Quantification of the European industrial heat demand by branch and temperature level. *International Journal of Energy Research, 39*, 2019–2030. https://doi.org/10.1002/er.3436

Najm, S. (2018, May 24–25). Oil production strategies, institutions, and the green paradox. In *6th International Symposium on Environment Energy & Finance Issues, ISEFI 2018*, Paris, France.

Nature Climate Change. (2018). https://www.nature.com/nclimate/

Nordhaus, W. D. (1994). *Mapping the global commons: The economics of climate change*. MIT Press.

Nordhaus, W. D. (2008). *A question of balance: Weighting the options on global warming*. Yale University Press.

Oppenheimer, M., O'Neill, B. C., Webster, M., & Agrawal, S. (2011). Climate change: The limits of consensus. *Science, 317*(5844), 1505–1506. https://doi.org/10.1126/science.1144831

Orlov, S., Rovenskaya, E., Puaschunder, J. M., & Semmler, W. (2024). Green bonds, transition to a low-carbon economy, and intertemporal welfare allocation: Evidence from an extended DICE model. *AIMS Environmental Science, 11*, 4, 628-648. https://doi.org/10.3934/environsci.2024031

Orttung, R. (2009). *Improving energy relations with Russia: The roles of energy efficiency and alternative energy*. Jefferson Institute-PASOS Policy Paper.

Orttung, R. (2011a, August 17). *Natural gas*. Bucharest Summer University.

Orttung, R. (2011b, March 25). Russland und der Wandel der internationalen Erdgasmärkte. Die Bedeutung von Flüssiggas und Schiefergas [The impact of global LNG and shale gas developments on the Russian gas industry]. *Russland-Analysen*, 217.

Orttung, R. (2011c, October 25). *Shale gas and European energy security*. Uppsala University.

Orttung, R. (2011d, December 7–8). *Twenty years of energy in foreign policy: Two decades of post-Soviet independence: What have we learned?* The George Washington University.

Orttung, R. (2013a, January 24). *The Arctic's place in center-periphery coordination and conflict: Integrating local, regional, federal, and corporate interests*. Arctic Frontiers Conference, University of Tromso.

Orttung, R. (2013b, April 22). *What lies ahead: Russia's political and energy outlooks over the next decade*. Delta Phi Epsilon Professional Foreign Service Fraternity, The George Washington University.

Orttung, R. (2014a, March 20). *Center-periphery relations in the development of Russia's arctic energy resources: The challenges for Russia's politicized economic system*. Swedish Defense Research Agency (FOI).

Orttung, R. (2014b, October 20). *Russia's oil and social responsibility: Hydrocarbons, governance and social accountability in the Caspian Basin and beyond.* The George Washington University.

Orttung, R. (2017). Energy security. In M. D. Cavelty & T. Balzacq (Eds.), *Routledge handbook of security studies* (pp. 204–212). Routledge.

Orttung, R., Langenbach, S., & Wenger, A. (2012, December 11). *The role of hydrocarbons in maritime claim-making: Patterns of conflict and cooperation.* Post-Communist Politics Social Science Workshop, The George Washington University.

Orttung, R., & Perovic, J. (2010). Energy security. In V. Mauer & M. Dunn (Eds.), *Routledge handbook of security studies* (pp. 211–220). Routledge.

Orttung, R., & Wenger, A. (2012, January 9). *Energy's impact on international conflict and cooperation: Improving our understanding of the causal mechanisms linking markets and politics.* Institute for Security and Conflict Studies Energy Security Workshop, The George Washington University.

Papapetrou, M., Kosmadaki, G., Cipollina, A., La Commare, U., & Micale, G. (2018). Industrial waste heat: Estimation of the technically available resource in the EU per industrial sector, temperature level and country. *Applied Thermal Engineering, 138*, 207–218. https://doi.org/10.1016/j.applthermaleng.2018.04.043

Pardo, N., Vatopoulos, K., Krook-Riekkola, A., Moya, J. A., & Perez, A. (2012). Heat and cooling demand and market perspective. *European Commission Report 25382.* Joint Research Centre, Institute for Energy and Transport. Petten, Netherlands, European Union.

Park, J. (2016). *Will we adapt? Temperature shocks, labor productivity and adaptation to climate change in the United States (1986–2012).* Belfer Center, The Harvard Project on Climate Agreements, Harvard University. https://www.belfercenter.org/sites/default/files/legacy/files/dp81_jpark_hpca.pdf

Park, J. (2017). *Will we adapt? Labor productivity and adaptation to climate change.* Working paper. Harvard University. https://scholar.harvard.edu/files/jisungpark/files/will_we_adapt_-_jpark_4-3-17.pdf

Peeters, L., de Dear, R., Hensen, J., & d'Haeseleer, W. (2009). Thermal comfort in residential buildings: Comfort values and scales for building energy simulation. *Applied Energy, 86*(5), 772. https://doi.org/10.1016/j.apenergy.2008.07.011

Pope Francis. (2015). *Encyclical Letter Laudato Si of the Holy Father Francis on care for our common home.* http://w2.vatican.va/content/dam/francesco/pdf/encyclicals/documents/papa-francesco_20150524_enciclica-laudato-si_en.pdf

Potter, S., Cabbage, M., & McCarthy, L. (2017). *NASA, NOAA data show 2016 warmest year on record globally.* https://www.nasa.gov/press-release/nasa-noaa-data-show-2016-warmest-year-on-record-globally

Puaschunder, J. M. (2010). *On corporate and financial social responsibility*. Unpublished dissertation, University of Vienna.
Puaschunder, J. M. (2016a). Intergenerational climate change burden sharing: An economics of climate stability research agenda proposal. *Global Journal of Management and Business Research: Economics and Commerce, 16*(3), 31–38.
Puaschunder, J. M. (2016b). Mapping climate justice. In *Proceedings of the 2016 Young Scientists Summer Program Conference, International Institute for Applied Systems Analysis (IIASA)*, Laxenburg, Austria.
Puaschunder, J. M. (2016c). On eternal equity in the fin-de-millénaire: Rethinking capitalism for intergenerational justice. *Journal of Leadership, Accountability and Ethics, 13*(2), 11–24.
Puaschunder, J. M. (2016d). The call for global responsible intergenerational leadership in the corporate world: The quest for an integration of intergenerational equity in contemporary Corporate Social Responsibility (CSR) models. In D. Jamali (Ed.), *Comparative perspectives in global corporate social responsibility* (pp. 275–288). IGI Global Advances in Business Strategy and Competitive Advantage Book Series.
Puaschunder, J. M. (2017a). Climate in the 21st century: A macroeconomic model of fair global warming benefits distribution to grant climate justice around the world and over time. In *Proceedings of the 8th International RAIS Conference on Social Sciences and Humanities organized by Research Association for Interdisciplinary Studies (RAIS) at Georgetown University*, Washington, DC, USA, March 26–27, pp. 205–243.
Puaschunder, J. M. (2017b). Mapping climate in the 21st century. *Development, 59*(3), 211–216.
Puaschunder, J. M. (2017c). The climatorial imperative. *Agriculture Research and Technology, 7*(4), 1–2.
Puaschunder, J. M. (2018). *Intergenerational responsibility in the 21st century*. Vernon.
Puaschunder, J. M., & Schwarz, G. (2012). *The future is now: How joint decision making curbs hyperbolic discounting but blurs social responsibility in the intergenerational equity public policy domain*. Working paper. Harvard University Situations Law and Mind Sciences.
Ramsey, F. P. (1928). A mathematical theory of saving. *Economic Journal, 38*(152), 543–559. https://doi.org/10.2307/2224098
Ranasinghe, C. S., Kumarathunge, M. D. P., & Kiriwandeniya, K. G. S. (2017). Genotypic differences in cardinal temperatures for in vitro pollen germination and pollen tube growth of coconut hybrids. *Experimental Agriculture, 1*, 1–13.
Raveneau, M. P., Coste, F., Moreau-Valancogne, P., Lejeune-Hénaut, I., & Durr, C. (2011). Pea and bean germination and seedling responses to temperature and water potential. *Seed Science Research, 21*(3), 205–213.

Rawls, J. (1971). *A theory of justice*. Harvard University Press. https://doi.org/10.4159/9780674042605

Rodrigues, B. R. A., dos Santos, R. C., Nietsche, S., Mercadante-Simões, M. O., da Cunha, I. R., & Pereira, M. C. T. (2016). Determination of cardinal temperatures for sugar apple (Annona squamosa L.). *Agricultural Sciences, 40*(2), 145–154.

Rohles, F. H. (2007). Temperature & temperament: A psychologist looks at comfort. *ASHRAE Journal, 49*(2), 14–22.

Rovenskaya, E. (2008, January). Optimal economic growth under stochastic environmental impact: Sensitivity analysis. *Interim Report*. International Institute for Applied Systems Analysis.

Russell, J. A. (1957). *Industrial operations under extremes of weather*. Meteorological Monographs, American Meteorological Society.

Sachs, J. D. (2014). Climate change and intergenerational well-being. In L. Bernard & W. Semmler (Eds.), *The Oxford handbook of the macroeconomics of global warming* (pp. 248–259). Oxford University Press.

Savva, Ch. (2018, May 24–25). Anthropogenic greenhouse gas concentrations and global temperature: A threshold analysis. In *6th International Symposium on Environment Energy & Finance Issues, ISEFI 2018*, Paris, France.

Schlenker, W., & Roberts, M. J. (2009). Nonlinear temperature effects indicate severe damages to U.S. crop yields under climate change. *Proceedings of the National Academy of Sciences, 106*, 15594–15598. https://doi.org/10.1073/pnas.0906865106

Shahbaz, M., Papavassiliou, V. G., & Lahiani, A. (2018, May 24–25). Are we moving towards decarbonization of the global economy? Lessons from the distant past to the present. In *6th International Symposium on Environment Energy & Finance Issues, ISEFI 2018*, Paris, France.

Somanathan, E., Somanathan, R., Sudarshan, A., & Tewari, M. (2014). *The impact of temperature on productivity and labor supply: Evidence from Indian manufacturing*. Discussion Paper in Economics 14-10. Indian Statistical Institute, Economics and Planning Unit.

Stevens, A. (2016). *Temperature, wages, and agricultural labor productivity*. Working paper. https://are.berkeley.edu/sites/default/files/job-candidates/paper/stevens_jmp_jan16.pdf

Strobl, E., Elliott, R., & Cole, M. (2018). *Biodiversity and economic land use*. http://www.lse.ac.uk/GranthamInstitute/event/rss-matthew-cole/

Sudarshan, A., & Tewari, M. (2014). *The economic impacts of temperature on industrial productivity: Evidence from Indian manufacturing*. ICRIER Paper No. 278. https://www.researchgate.net/publication/283721979_Sudarshan_A_and_M_Tewari_2014_The_Economic_Impacts_of_Temperature_on_Industrial_Productivity_Evidence_from_Indian_Manufacturing_ICRIER_Paper_No_278

Szokolay, S. V. (2010). *Introduction to architectural science: The basis of sustainable design*. Architectural Press.

Talevi, M. (2018, May 24–25). The evolution of spatial spillovers in the diffusion of domestic solar photovoltaic panels. In *6th International Symposium on Environment Energy & Finance Issues, ISEFI 2018*, Paris, France.

The World Bank. (2015). *Green bonds attract private sector climate finance*. World Bank Brief.

The World Economic Forum 2015 Report. (2015). Davos: World Economic Forum.

Tsutsui, Y. (2012). Weather and individual happiness. *Weather, Climate and Society, 5*, 70–82. https://doi.org/10.1175/WCAS-D-11-00052.1

Wyon, D. P., Andersen, I., & Lundqvist, G. R. (2009). Spontaneous magnitude estimation of thermal discomfort during changes in the ambient temperature. *Journal of Hygiene, 70*(2), 203–221. https://doi.org/10.1017/S0022172400022269

Yildirim, K., Koyuncu, C., & Koyuncu, J. (2009). Does temperature affect labor productivity: Cross-country evidence. *Applied Economics and International Development, 9*(1), 1–21.

Zehtab-Salmasi, S. (2006). Study of cardinal temperatures of pumpkin (cucurbita pepo) seed germination. *Journal of Agronomy, 5*(1), 95–97. https://doi.org/10.3923/ja.2006.95.97

Zhang, P., Deschenes, O., Meng, K. C., & Zhang, J. (2017). *Temperature effects on productivity and factor reallocation: Evidence from a half million Chinese manufacturing plants*. IZA Working paper 11132. Institute of Labor Economics, Bonn, Germany.

CHAPTER 11

An Inquiry into the Nature and Causes of Climate Wealth of Nations: What Temperature Finance Gravitates Toward? Sketching a Climate-Finance Nexus and Outlook on Climate Change-Induced Finance Prospects

11.1 Introduction

Climate change accounts for one of the most debated international concerns. Never before in the history of humankind, the earth has been warming consistently in an unprecedentedly fast pace as outlined by Graph 11.1, based on United States National Aeronautics and Space Administration (NASA) data on the world land and ocean temperature from 1880 until 2023.[1]

Global warming also appears to be the most lasting external shock event coming down on society ever. The manifold impacts of a hotter temperature are described to be irreversible and widespreadly touching on humankind and natural habitat alike (Environmental Defense Fund, 2018). Climate instability and weather extremes are attributed as one of

[1] https://www.nasa.gov/stem-content/global-land-ocean-temperature-index-data-set/; https://data.giss.nasa.gov/gistemp/graphs/graph_data/Global_Mean_Estimates_based_on_Land_and_Ocean_Data/graph.txt

J. Puaschunder, *Governance & Climate Justice*,
https://doi.org/10.1007/978-3-031-87971-5_11

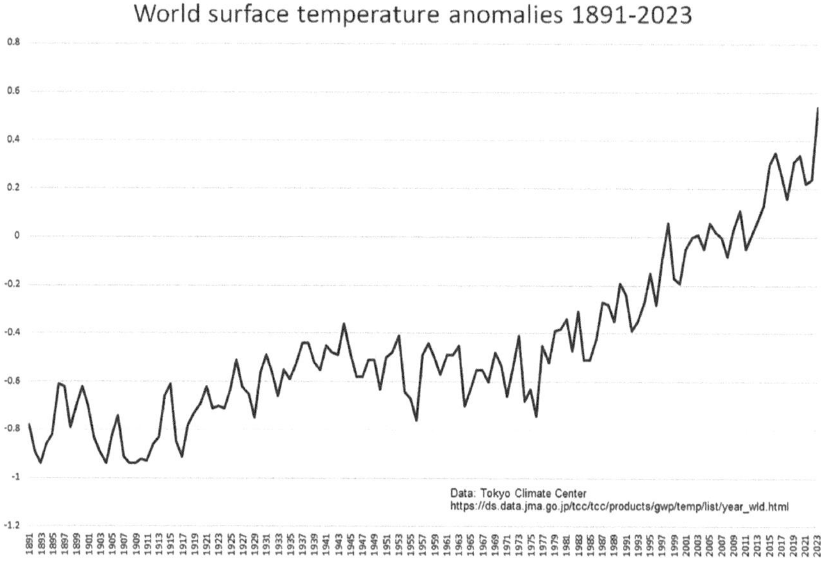

Graph 11.1 World temperature 1880–2023 derived from NASA world land and ocean temperature measurement (*Source* Author's own)

the most pressing emergent risks imposing dangers onto contemporary humankind (Centeno et al., 2013; Centeno & Tham, 2012). Climate stabilization accounts for one of the most contested public policy domains as what is at stake is novel, unprecedented and will shape the living conditions of all generations on earth to come (Puaschunder, 2017c, 2018c; Stiglitz, 1998). The impacts of climate change are believed to influence our daily lives already and the toll climate change takes on us and future world inhabitants is expected to increase exponentially with time (Environmental Defense Fund, 2018). Climate mitigation and adaptation therefore shape markets now and even more so in the future to come.

Money makes the world go round. Around the world, finance is the ultimate means of exchange and a necessary condition to sustain a pleasant living. Access to finance is directly related to freedom. Like no other good, financial assets are linked to social status throughout societal orders of world populations. Finance is the ultimate hammer to nail economies. Financial reserves are storages of economic power and policy leverage. While methods and means have changed, finance is—to this day—the longest serving stabilizer of prosperous societies. Within markets, financial assets move the allocation of goods and services. Finance therefore determines the shape, nature and form of any living community. It is a necessary motivator within economies and a psychological safety coverage for universal risk management. No society exists that does not use any form of finance. Any regime that tried to establish and order society without attention to finance collapsed or imploded.

Given the impetus and importance of climate change and finance around the globe and over time, surprisingly striking is that less attention is paid to how climate change will impact finance. While there is research on how finance can be used to curb climate change and orchestrate global warming mitigation and adaptation around the world (Greiner & Semmler, 2008; Marron & Morris, 2016; The World Bank, 2015), we hardly know anything about the reverse effect of climate change on financial flows and outcomes. How will a hotter temperature change the finance world? What financial flows will arise in light of climate change gains and losses being distributed unequally around the globe? Lastly, how does climate determine the future wealth of nations?

These are the questions that are arising and this paper seeks to answer. The paper is structured as follows: The theoretical part tries to connect the worlds of finance and temperature through price mechanisms, with attention to supply- and demand-side perspectives. The impact of a warming earth on financial markets will be prospected (Sect. 11.2). A temperature-finance nexus will be introduced (Sect. 11.3). A macroeconomic finance prospectus on a warming earth will be presented as results (Sect. 11.4) comprising climate change-induced market flows (Sects. 11.4.1 and 11.4.2), climate flexibility (Sects. 11.4.3 and 11.4.4) as well as climate change expectations driving commodity bubbles (Sects. 11.4.5 and 11.4.6). A discussion of the findings (Sect. 11.5) will lead to final conclusions and a future research endeavors description.

11.2 Theoretical Background

11.2.1 Climate Change

Never before in the history of humankind have environmental concerns in the wake of economic growth heralded governance predicaments as we face today (Nordhaus, 1994). Climate change presents societal, international and intergenerational fairness as a challenge for modern economies and contemporary democracies (Bürgenmeier, 1994). In today's climate change mitigation and adaptation efforts, high- and low-incomehouseholds, developed and underdeveloped countries and overlapping generations are affected differently. Based on insights from the current endeavor to finance climate stabilization action plans, the *Mapping Climate Justice Project* proposes a three-dimensional climate justice approach in order to find a universally fair climate strategy (Puaschunder, 2018a).

Mapping Climate Justice elucidates international climate regimes around the world based on geographic, technological, socio-economic and political factors. An interactive graphic solution highlighting different countries' climate stability efforts targets at finding ways to share the benefits and burdens of climate change equally *within society*, *between countries* and *over time*.[2] Comparing international climate mitigation and adaptation as well as unraveling complex interdependencies is aimed at helping to protect vulnerable communities from variegated climate change risks while opening ways for all to enjoy the upsides of a warming earth.

Climate justice accounts for the most challenging global governance goal. In the post-COP21 Paris agreement climate change mitigation and adaptation efforts, the financialization of the ambitious goals has leveraged into a blatant demand. In the weighting of the burden of global warming, the benefits of a warming earth have been neglected since recently.

Following the introduction of the gains from climate change (Puaschunder, 2017a, 2017d), the *Climate Justice in the 21st Century* endeavor proposes a model to distribute the benefits of a warming earth in a fair way based on which countries are losing and which countries are winning from a warming earth until the year 2100. A macroeconomic cost-benefit analysis thereby aids in finding the optimum solution on how to distribute climate change benefits and burden within society and over time.

When unidimensionally focusing on estimated GDP growth given a warmer temperature, over different calculated models assuming linear, prospect or hyperbolic gains and losses, the world will be gaining more than losing from a warming earth until the year 2100. Based on the *WL*

[2] https://juliampuaschunder.com/mapping-climate-justice/

index of 188 countries of the world, less countries ($n = 78$) will gain more from global warming until 2100 than more countries ($n = 111$) will lose from a warming earth.

Based on the overall *WLTT* index factored by GDP per inhabitant, global warming benefits are demanded to be redistributed in a fair way to offset climate change loser countries for climate change mitigation and adaptation efforts and in order to instigate a transition to renewable energy (Puaschunder, 2017a). Restructuring industrial systems needs a diverse, cross-sector approach so that green industrial policy frameworks can become a valuable tool for all countries to restructure economies toward better societal outcomes (Altenburg & Assmann, 2018).

Adding onto contemporary climate fund-raising strategies ranging from emissions trading schemes (ETS) and carbon tax policies, as well as financing climate justice through bonds as viable mitigation and adaptation strategies, climate justice is introduced to comprise fairness within a country but also among different nation-states and over time in a unique and unprecedented tax-and-bonds climate change gains and losses distribution (Puaschunder, 2017b, 2018a; The World Bank, 2015).

Thereby, climate change winning countries should be using taxation to raise revenues to offset the losses incurred by climate change. Climate change losers could raise revenues by issuing bonds that have to be paid back by taxing future generations. Regarding taxation, in the winning countries, foremost the gaining GDP sectors should be taxed.

Climate justice within a country should also pay tribute to the fact that low- and high-income households share the same burden proportional to their dispensable income, for instance, enabled through a progressive carbon taxation. Those who caused climate change could be regulated to bear a higher cost through carbon tax in combination with retroactive billing through a corporate inheritance tax that comes into play when a board member changes or there is a merger and acquisition or a hostile takeover. Such corporate inheritance tax could reap benefits from past wealth accumulation that has potentially exacerbated global warming (Rosenlund Soysal, 2018).

This paper builds on the established climate change winning and losing prospects but adds a novel perspective on climate flexibility as the range of different temperatures a country can offer to produce in. Climate flexibility will be captured around the globe, and production ranges also analyzed based on flexibility of agricultural crops. A climate-finance nexus will be introduced based on expectations of scarcity driving up prices and inflating value of goods that may become scarce in a global warming scenario. This theoretical model will be backtested on actual commodity market prices. The impact of most recent geopolitical tensions on commodity prices around the globe will be thematized and science diplomacy solutions to alleviate politically-induced resource bottlenecks proposed. Deriving respective policy recommendations for the wider climate change community in the discussion of the results is aimed at ensuring to share the burden but also the benefits of climate change within a society in an economically efficient, legally equitable and practically feasible way.

11.2.2 Climate Change Gains

The implementation of climate stability accounts for the most challenging contemporary global governance predicament that seems to pit world countries but also today's generation against future world inhabitants. In a trade-off of economic growth versus sustainability, a broad-based international coalition could establish climate stability. As a novel angle toward climate justice, attention is paid to global warming economic gains and losses being distributed unequally around the globe (Puaschunder, 2017b) in order instigate a well-balanced climate mitigation and adaptationpublic policy mix for redistribution guided by micro- and macroeconomic analysis results.

Based on the medium temperature of countries around the world and the optimum temperature for Gross Domestic Product (GDP) production, the world can be divided into climate change gaining and losing territories until the year 2100. Graph 11.2 holds those countries that have the most time ahead until optimum temperature for GDP production in green and those that are around peak condition in yellow.

The climate change winning territories are also outlined in Table 11.1, ranked by country name.

Having found that there are gains from a warming earth demands to transfer benefits into areas of the world that will be primarily losing from climate change. Having shed light on the gains of a warming earth allows for the redistribution of climate change benefits to those areas of the world that will be losing from a warming earth. In the implementation, climate change bonds and taxation strategies are recommended. In order to avoid governmental expenditure on climate change hindering economic growth (Barro, 1990), the climate transfers should

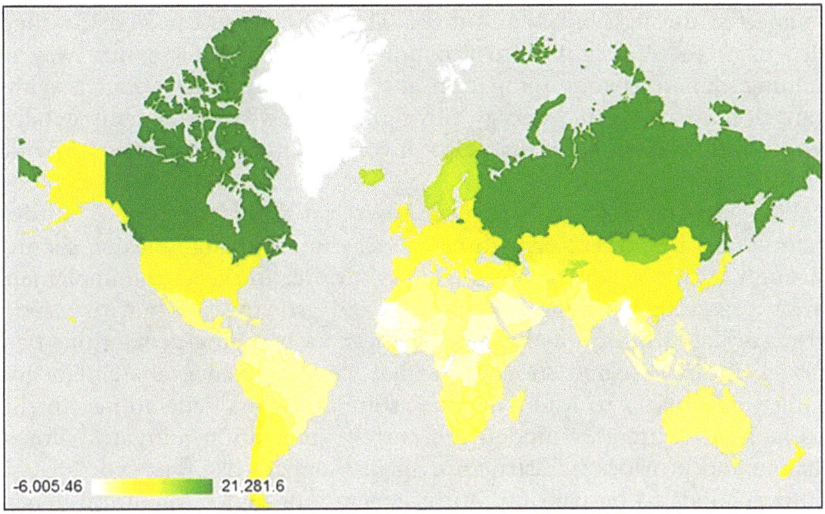

Graph 11.2 Climate change winning territories until the year 2100 (*Source* Author's own)

Table 11.1 Climate change winning territories by country ranked

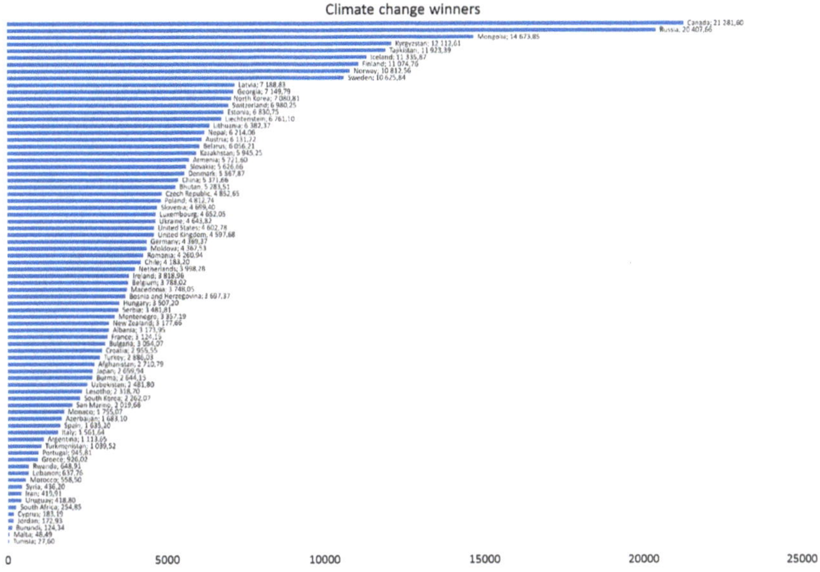

be enacted through bonds and taxes. The 'Climate in the 21st Century' idea takes these insights into account in order to offer a new way of funding climate change mitigation and adaptation policies, but also the transition to renewable energy through a broad-based climate stability bonds-and-taxation mix that also involves future generations (World Bank, 2015).

First, Jeffrey Sachs (2014) proposed an intergenerational burden sharing idea by presenting a three-model climate change burden sharing through fiscal policy with bond-issuing in order to reflect the implementation regarding contemporary finance and growth models with respect to maximizing utility of the model. In an overlapping-generations-type model, future research should elucidate climate change abatement and mitigation policies to lead to a fairer solution across generations. In this tax-and-bonds transfer model, the current generation mitigates climate change and provides infrastructure against climate risk financed through climate bonds to be paid by future generations. Since for future generations the currently created externalities from economic activities—the effects of CO_2 emissions—are removed, this entails that the current generations remain financially as well off as without mitigation, while improving environmental well-being of future generations. As Sachs (2014) shows, this intergenerational tax-and-transfer policy turns climate change mitigation and adaptation policy into a Pareto improving strategy. Shifting the costs for climate abatement to the recipients of the benefits of climate stability appears as novel, feasible and easily implementable solution to nudge many overlapping generations toward future-oriented loss aversion in the sustainability domain (Puaschunder, 2016).

One of the most prominent forms to create revenues for public long-term investment causes is tax. Taxation is codified in all major societies and is a hallmark of democracy (Puaschunder, 2015). Aimed at redistributing assets to provide public goods and ensure societal harmony, taxation improves societal welfare and fairness notions within the society. Tax compliance is a universal phenomenon based on cooperation in the wish for improving the social compound. Taxpayers voluntarily decide to what extent to pay or avoid tax that limits the personal freedom (Puaschunder, 2015). In a social dilemma, individual interests are in conflict with collective goals. From a myopic economic perspective, the optimal strategy of rational individuals would be to not cooperate and thus evade tax. Short-term, the single civilian tax contribution does not make a significant difference in the overall maintenance of public goods—if only a few taxpayers evade taxes, public goods will not disappear or be reduced. But if a considerable number of taxpayers do not contribute to tax over time, common goods are not guaranteed and ultimately everyone will suffer from suboptimal societal conditions (Dawes, 1980; Puaschunder, 2015; Stroebe & Frey, 1982). Contemporary economic research has focused on costs and risks of tax evasion (Tyler & De Cremer, 2006). Coercive means—such as audits and fines—were found to crowd-out tax morale and ultimately result in greater non-compliance as people feel controlled and not being trusted (Cialdini, 1996; Feld & Frey, 2002; Frey, 1992; Hasseldine, 1998). In the last decade, researchers have started to recognize the importance of incorporating morals and social dynamics in the economic theory on tax behavior (Andreoni, Erard, & Feinstein, 1998). When analyzing tax behavior, recently behavioral economics insights have drawn attention to social influences, such as trust and reciprocity (Puaschunder, 2015).

Behavioral economists widen the lens of incorporating sociological and socio-psychological notions of fairness stemming from social comparisons regarding tax burdens that could be positive drivers of tax compliance to overcome the 'burden of taxes' and associations of losses. The cases of voluntary, self-chosen tax ethics and situational influences on social tax compliance norms have just recently been covered by behavioral approaches toward public administration. In general, social comparisons determine social norms that define internalized standards on how to behave. Yet internalized social norms are based on comparisons with others that may determine tax morale (Frey, 1997; Mumford, 2001; Schmölders, 1960). Social norms elicit concurring behavior when taxpayers identify with the goals of a group and also if they feel being treated in a fair manner by that group. Social fairness considerations in a tax reference group may further taxpayer compliance. Fairness is believed to decrease egoistic utility maximization, leveraging trust and reciprocity as interesting social norms building factors (Kirchler, 2007). Social perceptions of fairness as underlying social norms are therefore potential tax ethics nudges. But psychological facets of fairness for the formation of social norms have been left out. If taxpayers believe that non-compliance is widespread and socially accepted, then it is more likely that they will not comply as well. Non-compliance may stem from the notion of unfairness in how the tax burden is weighted heavier on some parts of society.

The respective bonds-and-tax climate stability financing strategy therefore proposes to bear the burden of climate in a right, just and fair way around the globe. In the climate change winner countries, taxation should become the main driver over financing climate stability strategies. Foremost, the industries winning from a warming climate should be taxed. The Winner-Loser-index is based on the cardinal temperatures for all GDP contributing sectors. Based on the cardinal temperatures for the three GDP components—agriculture, industry and service—the taxation should be enacted for those sectors having most time ahead. The rational is that these sectors will be gaining the most from a warming earth and will therefore be flourishing in the near future.

The taxation models should aid to share the burden of climate change within society in a fair way. Regarding concrete climate taxation strategies, a carbon tax on top of the existing tax system should be used to reduce the burden of climate change and encourage economic growth through subsidies (Chancel & Piketty, 2015). Within a country, high- and low-income households should face the same burden of climate stabilization adjusted for their disposable income. First, climate justice within a country should pay tribute to the fact that low- and high-income households share the same burden proportional to their income, for instance, enabled through a progressive carbon taxation. Those who caused climate change could be regulated to bear a higher cost through carbon tax in combination with retroactive billing through a corporate inheritance tax when there is a merger or acquisition or a board member change. Besides progressive taxation schemes to imbue a sense of fairness in climate change burden sharing, corporate inheritance taxation is also a flexible means to reap past wealth accumulation, which potentially caused environmental damage. The burden of climate change mitigation and adaptation could therefore also be allocated in a fair way within society through contemporary corporate inheritance tax in order to reap benefits of past wealth accumulation.

In addition, developed and underdeveloped countries as well as various overlapping generations are affected differently. But in order to determine a fair corporate inheritance tax grantors and beneficiaries, one needs information on what countries and what sectors of the economy have a rising or declining economic prospect in light of climate change.

Finding the optimum balance between consumption tax adjusted for disposable income through a progressive tax scheme will aid to unravel drivers of tax compliance in the sustainability domain. If climate taxation is perceived as fair and just allocation of the climate burden, this could convince taxpayers to pay one's share. A novel 'service-and-client' atmosphere could promote taxpayers as cooperative citizens who are willing to comply if they feel their share as a fair contribution to the environment. Taxpayers as cooperative citizens would then be willing to comply voluntarily, following the greater goal to promote taxpayer collaboration and enhance tax morale in the environmental domain.

International comparisons of tax behavior also reveal tax norms being related to different stages of institutional development of the government, which is an essential consideration in sharing the climate change burden in a fair manner between countries. A completely novel approach is to shed light on the benefits of a warming earth in order to derive fair climate gains distribution strategies around the world (Puaschunder, 2017e). This paper will add on the basic climate change winning and losing prospects model with specific market outlooks. Concretely, given a global warming trend, novel concepts will be studied, such as climate flexibility and climate change-induced financial and human migration flows as well as market price expectations around scarcity (Cifarelli & Paesani, 2018; Dukhanina, Leveque, & Massol, 2018; Ekholm, 2018; Elgouacem & Legrand, 2018; Puaschunder, 2018b).

11.2.3 Climate Change Losses

Based on the medium temperature of countries around the world and the optimum temperature for Gross Domestic Product (GDP) production, the world can be divided into climate change gaining and losing territories until the year 2100. Graph 11.3 shows in red those countries that have passed the peak temperature for GDP production.

The climate change losing territories are also outlined in Table 11.2, ranked by country name.

Introducing financing climate change mitigation through bonds to be paid back by future generations through taxation is a means to raise funds for offsetting the losses of global warming. As a novel way to amend individual saving preferences in favor of future generations, Sachs (2014) proposes to mitigate climate change by debt to be repaid by tax revenues on labor income in the future. In a 2-period model, one generation works in period 1 and retires in period 2. Part of the disposable wage income is saved for consumption in the second period. CO_2 emission mitigation imposes immediate costs onto current generations and

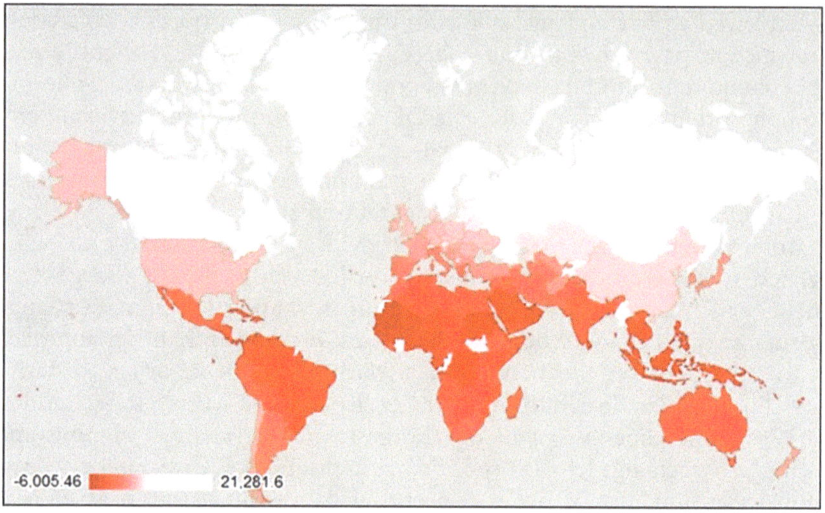

Graph 11.3 Climate change losing territories until the year 2100 (*Source* Author's own)

reduces wages. Greenhouse gas concentrations in period 2 are determined by the emissions in period 1. Wages of the young in the second period are reduced by climate change dependent on greenhouse gas levels. Disposable labor income of the young equals market wage net of taxes. Leaving the current generation with unchanged disposable income allocates the burdens of climate change mitigation across generations without the need to trade-off one generation's well-being for another's.

While today's young generation is left unharmed, the second-period young generation is made better off ecologically. The bonds solution should be pursued in climate change loser countries, in order to offset the costs for climate change in a more intergenerationally harmonious way. Since the majority of GDP contributing factors are losing in climate change loser countries, future generations should be serving as last resort to pay for climate stability. While future generations enjoy a favorable climate and averted environmental lock-ins, the current populace does not face drawbacks on already constraint economic growth (Puaschunder, 2017a, 2017b, 2017d). All generations are better off with mitigation through climate bonds as compared to the business-as-usual (BAU) non-mitigation scenario (Sachs, 2014).

Table 11.2 Climate change losing territories by country ranked

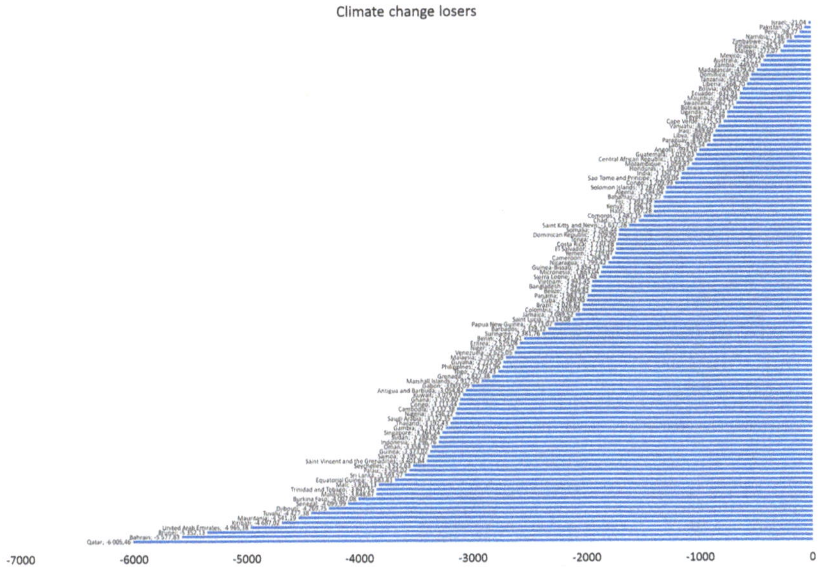

Source Author's own

Governments in loser countries should receive tax transfers in the present from the winning countries. Since here borrowing equals loans or issuing of bonds to be paid back by future generations, the government must pay back debt plus interest payments by raising taxes on later generations. Taxing future generations is justified as for the assumed preferences of future generations to avoid higher costs of climate change long-term damages and environmental irreversible lock-ins. Overall this tax-and-transfer mitigation policy thus appears as a Pareto-improving fair solution across the world and generations. While there seems to be a strategy on how to raise funds for the stabilization of the climate, not so much information exists on how the changing climate may affect the finance world.

11.2.4 Finance Industry

The financial industry provides financial services, which encompass a broad range of businesses that manage money, including credit unions, commercial and investment banks, credit-card companies, insurances, accountancy consulting, consumer-finance companies, stock brokerages, investment funds, individual managers and some government-sponsored enterprises (Asmundson, 2017).

The financial sector is a category of the economy made up of firms that provide financial services to commercial and retail customers. The term 'financial services' became more prevalent in the United States partly as a result of the Gramm-Leach-Bliley Act of the late 1990s, which enabled different types of companies operating in the US financial services industry at that time to merge. Foreign exchange services are provided by banks and specialist foreign exchange brokers around the world, including wire transfer and remittances services (World Bank Group Migration and Development Brief, 2016). Insurances are additional financial services offered as risk management tool. Financial export is a financial service provided by a domestic firm (regardless of ownership) to a foreign firm or individual. A large portion of this sector generates revenue from mortgages and loans.

Finance is a field that deals with the study of investments and is the science of money management. It includes the dynamics of assets and liabilities over time under conditions of different degrees of uncertainties and risks. Market participants aim to price assets based on their risk level, fundamental value and expected rate of return. Finance is used by individuals (personal finance), by governments (public finance), by businesses (corporate finance) and by a wide variety of other organizations, such as schools and non-profit organizations. Finance is one of the most important aspects of business management and includes analyses related to the use and acquisition of funds for the enterprise. Finance can be broken into three sub-categories: public finance, corporate finance and personal finance.

Macroeconomic public finance concerns center around the amount of money in an economy, inflation within economies, exchange rate differences between countries, retirement and social security planning, gold reserves and estate investments. Central Banks are public finance entities, acting as lenders of last resort with influences on monetary and credit conditions in the economy.

Corporate finance deals with the source funding and capital structure of corporations, shareholder wealth maximization, cash flow and capital budgeting, investment management and asset diversification. Commercial finance adds a broad range of financial services—such as stock brokerage, portfolio management, risk management strategies, hedge fund management, private equity and asset accumulation goals. The accounting industry is primarily focused on the reporting of historical financial information. Financial risk management strategies encompass risk estimations based on credit and market risk prospects.

Personal finance is concerned about the investment and accumulation of assets; insurance protection against unforeseen personal events, as well as events in the wider economies; transfer of family wealth across generations; tax planning; management of personal finances; effects of credit on individual financial standing, savings plan or financing for large purchases; and planning a secure financial future in an environment of economic instability.

Banks are financial intermediaries that aggregate the activities of many borrowers and lenders. A bank accepts deposits from lenders, on which it pays interest. The bank then lends these deposits to borrowers. Banks allow borrowers and lenders, of different sizes, to coordinate their activity. The owners of both bonds (either government bonds or corporate bonds) and stock (whether its preferred stock or common stock) may be institutional investors—financial institutions such as investment banks and pension funds or private individuals, called private investors or retail investors.

Financial economics is the branch of economics studying the interrelation of financial variables, such as prices, interest rates and shares. Financial economics concentrates on influences of real economic variables on financial ones. A country's economy is a social institution that organizes a society's production, distribution and consumption of goods and services, all of which must be financed.

Given the impetus and importance of climate change and finance around the globe and over time, surprisingly striking is that research is limited on how climate change will impact finance. How will a hotter temperature change the finance world? What financial flows will arise in light of climate change gains and losses being distributed unequally around the globe? And lastly, how will global warming determine the future wealth of nations?

This chapter tries to introduce three market mechanisms of climate change on finance based on (1) climate flexibility as the range of different temperatures offered by a nation-state and agricultural production commodities as well as (2) climate change-related finance, human and innovation prospects and (3) inflated prices in the prospect of rising natural resources scarcity induced by global warming.

11.2.4.1 Price Mechanisms

The financial industry is centered around prices. Price determines the value of a good based on supply and demand. Classical price theory started with Adam Smith (1776) who described at first that free interaction creates an orderly pattern guided by the invisible hand. Competition creates order out of activities in the market, where consumer and producer meet over determining the price of goods. Profit rates gravitate around natural prices. Market prices are invisible centers of gravity, and profit rates equalize due to the mobility of capital. High profit rates attract capital that brings the market prices down. Low profit rates will slow down supply in the movements of market prices. Prices move around centers of gravity among each other. The natural course is that actual market prices are above or below the natural price. The central price is the price to which all prices gravitate continually. According to Smith (1776), the price of production has a natural profit, which is achieved only through gravitation. The natural profit is the profit one deserves added to costs.

Exchange takes place in disorder, in which prices guide through the monetary expression of a commodity's quantitative worth. Market prices gravitate in a turbulent manner around prices of production. Profit is an objective measure subject to constant scrutiny by the firm's managers, the stock market, the banks and the public in general. The uniform price is assumed to be supremely responsive to the market demand and supply. According to Kalecki's (1943) theory of price, the price of individual firms depends on the relative size of firms due to union power of employees, industry prices and average costs and monopoly power that determine the mark-up structure.

According to Adam Smith (1776), the profit motive is the dominant factor in the regulation of economic growth. Actual decisions are always made in terms of current and expected market prices. Expected profitability at the core of any economic activity implies a mode of interaction between aggregate demand and supply. Profit thus regulates both—supply and demand. Profit determines an endogenous natural rate of growth. The interest rate thereby becomes the price of finance. Financial firms exist to make profit, and competition makes the profit rate of the regulating financial capitals gravitate around the general rate of profit.

Price is the quantitative worth of a commodity expressed through the medium of money. Direct prices represent a set of monetary prices that are proportional to labor values. The sum of direct prices equals the sum of market prices. The inclusion of depreciation lowers the dispersion for direct prices, and the inclusion of fixed capital lowers the dispersion for prices of production. Price mechanisms around scarcity feature demand-driven bubbles, hence inflated prices (Bai et al., 2018; Bonato, 2018; Ielpo & Sevi, 2013; Ji et al., 2018). For instance, during the great stagflation, the price of gold rose from $125 to $615 before subsiding to $375 by 1982. During the German hyperinflation of the 1920s, prices were actually in gold even though fiat money dominated the actual exchange. Profit drives capitalism. If profit fails, the shock lets capital begin to atrophy.

In general, there is a complexity of the calculation of prices via an input-output system that prevents a simple description of the relation of a commodity's prices to any other single parameter (Bienenfeld, 1988; Sraffa, 1960). Divergences in price expectations are mainly captured to stem from the organic composition of capital in each industry and the relative capital intensity in layers of goods, and the vertically integrated value composition of capital as an intricate network of relations between rate of profit and prices in the whole economic system (Bienenfeld, 1988).

But how temperature may change the perception of goods and scarcity in light of climate changing patterns of production which may influence pricing mechanism, we do not know (Ganepola et al., 2018; Ordu-Akkaya & Soytas, 2018). The described climate change winning and losing territory aspects and prevailing demands for a restructuring of gains into the losing areas are expected to create market distortions, which have not been described sufficiently. In determining price mechanisms, information on how climate influences prices is missing. Therefore, supply- and demand-side perspectives should be considered. Politics may also play a crucial role (Orttung, 2010a, 2010b, 2014, 2015, 2016; Orttung & Overland, 2011a, 2011b; Perovic et al., 2009).

Prices appear to be dependent on the prices of production, which depend on fixed capital, turnover, wages, return on capital, capacity utilization and capital stock (Cockshott & Cottrell, 1993; Leontief, 1997). Ricardo (1817/1951) argued that changes in relative prices are dominated by changes in relative vertically integrated labor coefficients. Marx (1867/1995) outlined that relative prices move directly with changes in vertically integrated labor coefficients and inversely with changes in relative vertically integrated labor productivity, which even holds for prices of production (Chilcote, 1997). Evaluating the sensitivity of different prices of production estimators to temperature changes can provide insights for market estimations about depreciation, fixed capital, turnover and capacity utilization into the estimation of prices of production.

All firms have market prices. Firms constantly try to reduce prices in order to get an edge over competitors. Competition is driven by competitors trying to sell cheaply to drive out competitors based on technological change lowering costs and cheapening commodities. Competition tends to equalize wage rates and profit rates, around which market prices tend to gravitate, but remain different from natural prices of production.

Market prices may gravitate around feasibility of production rather than direct prices. Deviations of market prices from prices of production mostly reflect imbalances in supply and demand conditions. Expectations about shrinking production possibilities may inflate prices to an irrational exuberance around scarcity. Competition between industries makes relative prices of individual commodities gravitate around prices of production reflecting equal profit rates. Weather plays a role in the cost of production—for instance, when considering the cost of obtaining oil and gas. During warm winters, pipelines and trade in oil and gas is cheaper than during cold winters given shipment going through icy oceans and on land pipeline structures.

John Maynard Keynes (1936) and Karl Marx (1867/1995) tie prices to periodic recurrence of boom and bust in *The General Theory* and *Das Kapital* (Sherman, 1967). According to Keynes (1936), firms do not set prices; competition sets prices. Firms cannot set any level, raise them at certain level or pressure competition. In the classical theory of money, price wages and levels of production determine the long-run growth rate. According to Keynes (1936), prices depend on confidence in the future, whereby a collapse of confidence precipitates price drops and capital flight.

In the Marxian tradition, one could argue that the inequality of climate change wins and losses around the globe opens a climate hegemony in the twenty-first century. Naturally imposed temperature constraints may implicitly determine prices. Scarcity around temperature flexibility may create market distortions and surreal demand inflating prices to climate-induced bubbles (Moore, 2009). With increasing scarcity, prices are expected to rise, which provides high returns to shareholders. In a world where some places may run out of production capabilities due to temperature, climate spectra may become the new center of gravity toward which production naturally gravitates.

According to Sraffa (1960), the price of commodities is a joint result of the price of the commodity itself and the costs of the commodity production as responding to changes in the prices of those commodities, which enter into its production. A firm's competitive edge may be determined by climate flexibility, which may derive from natural propensities of national territories and climate areas in the age of global warming.

In today's world, temperature-related production flexibility may regulate the market price. Firms with leeway to produce in different climatorial zones may thereby become active price-setters. Fluctuations in aggregate demand cause the economy to come to short-run equilibrium above or below full employment. Shrinking temperature windows for production of commodities may trigger expectations that determine prices. Expectations are a thought or a belief about anticipated gains or losses (Soros, 2003). Expectations are thereby anticipations of an expected future value in an uncertain world (Soros, 2009, 2013). Expectations are a central part of value calculations in economics. The subjective expected utility is derived from the value of an outcome and the probability that it will occur. Expectations elicitation is used in many domains, including health, wealth, survival and educational outcomes, but may be most prominent in financial realms. While we know that consumers with more optimistic stock market expectations tend to more likely hold riskier assets and acquire stocks in the near future, we still are in need to explore what information and factors drive financial market expectations in light of a warming globe.

Expectations play a vital role in economic calculus. Since firms have to produce for the future, any production is based on expectations of future outcomes in Keynes (1936). Expectations are formed in a dynamic process. Anticipations are skewed and dependent on other people. Expectations influence what happens in the process, as people do not correctly anticipate what happens. Soros' (2003) financial market theory holds that some people will realize that prices are not sustainable, which will cause a bubble bursts in financial markets. One knows that it will happen, but not when it will happen. According to Friedman (1962), surprises in markets matter more than expectations. Every deviation besides surprises is temporary. Information on the linguistic correlates of economic ups and downs could help 'bridge between rational behavior and societal collective action' (Soskice, 1978).

To every good and service there is a (1) fundamental value$f$, what a good is actually worth, when considering the direct market price and price of production, and (2) expectations$e$. When information in markets builds up expectations, $e > f$; hence, expectations start overpromising beyond the fundamental value of the good. People believe a good or service is more worth than its actual fundamental value is. If e gets largely inflated and majorly detached from f, first market experts and investment leaders start realizing the unsustainable inflation of a good's or services' expected value. The first market actors start losing faith in the product and service, hence try to sell it. If the amount of sale bids reaches a certain threshold, by the law of supply and demand, the price will eventually decline, and then, if bleeding into a run and more investors to follow, eventually collapse. This is the nature of the collective soul of booms and busts as outlined in Graph 11.4.

While we have information about the mathematical calculus of the expected value of a discrete random variable being the probability-weighted average of all possible values, mainly estimated as probabilities by frequencies, we lack an understanding of the economic correlates in the wake of socio-psychological phenomena regarding expectations of scarcity and fear of losses surrounding climate change. To this day, we have no fully explanatory theory on how expectations are built up in markets and how the over- or underperformance creates social norms around markets that may trigger herd behavior or inflate swarm phenomena. In general, we can assert the higher a person's expectation and the lower the performance, the higher the frustration and trigger of cessation of activity and quitting behavior—yet what this means for financial market decisions in the eye of a changing climate shrinking our natural prospects of productivity possibilities, we have no concrete information about.

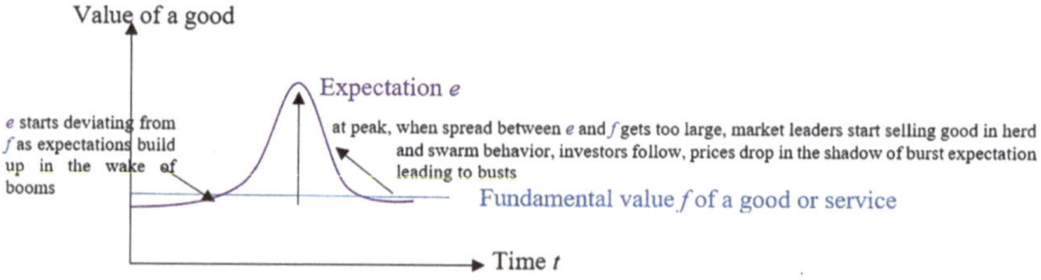

Graph 11.4 The collective soul of booms and busts (*Source* Author's own)

Behavioral economics found people to be risk-averse and their loss prospects loom larger than gains, which may influence their inflated willing-to-pay in light of loss prospects surrounding scarcity of goods (Kahneman & Tversky, 1979). How scarcity prospects reflect on energy prices is a currently growing research field in finance with applications in the natural gas market and oil sector (Kilian & Zhou, 2018; Klein et al., 2018; Kruse & Wegener, 2017; Legnazzi, 2018; Mahadeo et al., 2018; Markowitz et al., 2015).

As people develop favorable and unfavorable expectations that guide their choices and behavior on a daily basis, we need to unravel how climate change-related expectations are created and how they influence economic market actors' decisions bleeding into the collective soul of booms and busts. In the need for sustainable finance, we therefore should unravel the impact of expectations regarding how a changing climate determines our economic prospects and hence market actions (Schütze et al., 2018; Schweiger & Stepanov, 2018). Central Banks and market prospects play a crucial role in shaping expectations and correcting overshooting or undershootingin market predictions (Puaschunder, 2019).

11.2.5 Research Contribution

The following part will connect the contemporary research on climate change gains with the financial prospects in light of a warming globe. The chapter will thereby release unprecedented information on finance prospects and what finance market changes to expect given climate change. Lastly, the chapter will also provide recommendations for market actors on how to adapt to climate change and reap benefits from a warming globe.

11.3 Method

11.3.1 Overall Model Assumptions

11.3.1.1 Study 1: Climate Change-Induced Market Flows

The previously presented model is built on GDP optimal cardinal temperature for production. The GDP pillars of agriculture, industry and service are the foundation of the model. The model divided the world nation-states into climate change winning and losing territories. This is in accordance with International Monetary Fund (IMF) research concluding that in low-temperature countries, an increase in temperature tends to boost economic activity, while in higher temperature countries, an increase in temperature will dampen economic activity (Acevedo et al., 2018). The findings of uneven economic effects are robust over different models and backed by regional cluster analysis and geographic case studies (Burke et al., 2015; Somanathan et al., 2014). The IMF model now sophisticates these findings with insights about the effect of temperature on productivity, capital and labor.

In order to consolidate these findings and derive implications for economic and financial market performance, three market-focused hypotheses will test whether climate change winning and losing territories attract different levels of human capital inflow, finance inflow and Internet connectivity.

Overall geographical mobility is expected to increase, and thus should be supported, especially when massive shifts of the industry structure result from green industrial policies. This may be the case, for example, when shifting from fossil to renewable energy—fossil energy plants are usually located next to energy-intensive industries and cities, whereas renewable power plants are located where the respective power resources are abundant. Shifting energy systems will therefore affect spatial distribution, and workers are expected to move physically (Esposito et al., 2018).

Hypothesis H1 investigates if countries that economically benefit from climate change feature a significant foreign human capital inflow.

Hypothesis H2 investigates if countries that economically benefit from climate change feature a significant foreign capital inflow that does not correspond with remittance outflow from the FDI receiving territories into the FDI sponsoring countries.

Hypothesis H3 investigates if countries that economically benefit from climate change feature a significant higher level of internet connectivity.

11.3.1.2 Study 2: Climate Flexibility

In a model extension of the original calculation of climate change winners and losers around the world based on GDP prospects, the optimum cardinal temperatures for specific agriculture products, industry representations and service sector providers were retrieved from literature sources.

A future market prospect in light of global temperature rising was then given based on the mean temperature of the world now and as estimated in the year 2100. Prospective market shifts in the GDP composition but also agricultural products, industry sector representatives and service providers were provided. Underlying to all these future prospects is the idea that GDP production and its manifold components are enabled at best at a certain temperature, and therefore specific markets will follow and gravitate toward these temperatures.

In light of global warming, an influx of labor and mobility of corporations is predicted. These insights fortify the idea that the world will become more globalized and location-flexible due to climate change. Those corporations and individuals realizing this and seeking the gains of a warming globe—maybe beyond their own territories—are prospected to become market winners. The importance of climate flexibility for productivity is expected to rise with the progression of global warming.

11.3.1.3 Study 3: Climate Change Expectations Driving Commodity Bubbles

Since the turn of the millennium, the United Nations (UN) detected an unprecedented agriculture products commodities bubble trend, especially in countries around the equator belt, which are the worst off regarding climate change due to a natural high base mean temperature (United Nations Report on the World Social Situation, 2011). There also seems to be a recent and unprecedented credit risk and bubble behavior in the corporate energy sector—such as oil-supply shocks and stock prices—that may be related to the demand for a transition to renewable energy (Figuerola-Ferretti & Cervera, 2018; Kaabia et al., 2018; Lips, 2018). In the oil energy sector, price spikes have been found around procurement shortages, which are expected to be occurring in all climate-related extinction cases (Roberts et al., 2018). Research also already pits food versus fuel and finds evidence of a contemporary predicament in nation-state's investment allocations in either one resource (Filip et al., 2018; Trinks et al., 2017). There is also already evidence for sector competition for energy resources in a carbon-constrained hoteling model (Renz, 2018). This may only be a first sign of agricultural products becoming more precious and valuable for economic production given a longer durability in a warming climate (Wenger et al., 2009). Energy prices and mitigation policies also impact Foreign Direct Investment (FDI) locations

(Saussay & Sato, 2018; The World Economic Forum 2015 Report, 2015; World Investment Report, 2015).

The following financial market model now estimates prices centered on agriculture, industry and service sector transitions. Based on evidence from the energy sector, the relation of a price of a good and time to extinction of a resource is not linear but hyperbolically tilted toward the end before extinction (Roberts et al., 2018). When a resource is about to extinct, the market value of this resource will rise shortly before the good goes away from the market forever. For instance, if we run out of a natural resource, no alternatives may exist, so scarcity will drive up the price, shortly before the good extinguishes (Roberts et al., 2018; Yarlik et al., 2018; Zhang, 2018). Derived from this insight, the model will be based on the maximum temperature for agricultural goods and industry production as well as time for extinction of natural resources and cooling and storage maximum temperatures in order to determine not only market transition and shifting trends but also price spikes and peaks based on the world and regional temperatures.

11.4 Operationalization

11.4.1 Data Study 1: Climate Change-Induced Market Flows

In order to operationalize and consolidate the underlying premise that countries that economically benefit from climate change feature better market prospects, worldwide Foreign Direct Investment (FDI) flows[3] (Hypothesis H1), remittances[4] (Hypothesis H2) and Internet connectivity[5] (Hypothesis H3) were retrieved online form the World Bank database for the year 2015 as well as the *Global Connectivity Index* homepage as an indicator of market relevance and innovation potential.

11.4.2 Results Study 1: Climate Change-Induced Market Flows

11.4.2.1 Climate Change Winners and Losers Model

The relation of climate change gains and losses and market prospects was investigated. Based on the climate change winners and losers model,

[3] http://data.worldbank.org/indicator/BX.KLT.DINV.CD.WD

[4] http://data.worldbank.org/indicator/BX.TRF.PWKR.CD.DT

[5] http://www.huawei.com/minisite/gci/en/country-rankings.html

gaining and losing from climate change was related to human migration, financial inflows as well as innovation.

Human migration: When investigating the relation of human migration and climate change winners and losers in 187 countries of the world, a highly significant correlation ($r_{\text{Pearson}}(187) = 0.184$, $p < 0.006$) is found between being climate change winner and human migration inflows. A highly significant independent t-test with $t(185) = -2.401$, $p = 0.017$ of 110 climate change loser and 77 winner countries consolidates the finding.

Financial inflows: When investigating the relation of financial inflows and climate change winners and losers including 181 countries of the world, a significant correlation ($r_{\text{Pearson}}(181) = 0.180$, $p < 0.016$) is found between being climate change winner and FDI inflows. A highly significant independent t-test with $t(179) = 2.680$, $p = 0.008$ of 110 climate change loser and 77 winner countries consolidates the finding. As a cross-validation of lasting financial inflows into climate change winner countries, a non-significant correlation and non-significant independent t-tests are reported.

Innovation: When investigating the innovation potential currently prevalent in climate change winners and losers including 79 countries of the world, a significant correlation ($r_{\text{Pearson}}(79) = 0.507$, $p < 0.000$) is found between being climate change winner and global connectivity as measured by the 2018 *Global Connectivity Index*.[6] A highly significant independent t-test with $t(77) = 6.306$, $p = 0.000$ of 33 climate change loser and 46 winner countries consolidates the finding.

11.4.3 Data Study 2: Climate Flexibility

Climate change was recently found to affect countries differently. Given various mean temperatures and GDP compositions, countries around the globe will be affected differently by a warming climate. Given different climate change gain and loss prospects, climate flexibility appears as future trade asset. Climate flexibility is defined as leeway countries have in coping with a changing climate due to a range of climate zones prevalent in their territory. Climate flexibility can be grounded on the relative latitude and altitude of countries around the globe. The wider the range of

[6] http://www.huawei.com/minisite/gci/en/country-rankings.html

Table 11.3 Temperature range (in Celsius) by country determining climate flexibility

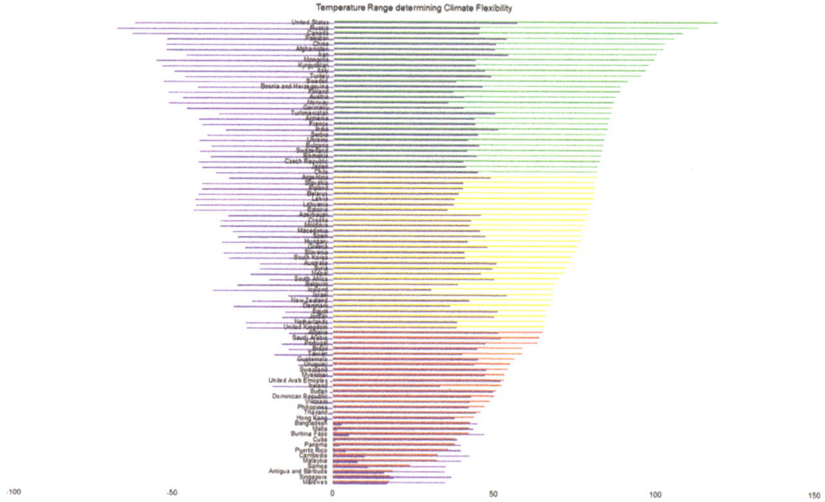

latitude and altitude within a nation-state, the more climate flexibility and favorable economic degrees of freedom for multiple production peaks are assumed to be prevalent. When dividing the world based on temperature ranges that are prevalent, calculated based on the *List of Countries by Extreme Temperatures*,[7] the following temperature ranges are found, as exhibited in Table 11.3.

11.4.4 Results Study 2: Climate Flexibility

Table 11.3 exhibits 87 countries' temperature range in comparison to each other. One-third of the sample features countries with the widest temperature range and therefore most flexible in green color, one-third of the sample features medium temperature range of climate flexibility in yellow color and one-third of the sample with lowest temperature range and relative climate inflexibility in red color.

The United States (US) leads on climate flexibility with a temperature range of $r = 118.9$°C based on [−62.2 having been the lowest

[7] https://en.wikipedia.org/wiki/List_of_countries_by_extreme_temperatures

temperature ever recorded in the USA and 56.7°C the highest temperature ever recorded in the USA], followed by Russia r = 113.2°C based on [−67.8; 45.4°C], Canada r = 108°C based on [−63; 45°C], Pakistan r = 105.5°C based on [−52; 53.5°C], China r = 102.6°C based on [−52.3; 50.3°C], Afghanistan r = 102.6°C based on [−52.2; 49.9°C], Iran r = 100°C based on [−46; 54°C], Mongolia r = 99.3°C based on [−55.3; 44°C], Kyrgyzstan r = 97.6°C based on [−53.6; 44°C], Italy r = 96.6°C based on [−49.6; 47°C], Turkey r = 95.2°C based on [−46.4; 48.8°C], Sweden r = 91°C based on [−53; 38°C], Finland r = 88.7°C based on [−51.5; 37.2°C], Bosnia and Herzegovina r = 88.7°C based on [−42.5; 46.2°C], Austria r = 87.6°C based on [−47.1; 40.5°C], Norway r = 87°C based on [−51.4; 35.6°C], Germany r = 86.2°C based on [−45.9; 40.3°C], Turkmenistan r = 86.1°C based on [−36; 50.1°C], Armenia r = 85.7°C based on [−42; 43.7°C], France r = 85.1°C based on [−41; 44.1°C], India r = 84.9°C based on [−33.9; 51°C], Serbia r = 84.4°C based on [−39.5; 44.9°C], Ukraine r = 83.9°C based on [−41.9; 42°C], Bulgaria r = 83.5°C based on [−38.3; 45.2°C], Switzerland r = 83.3°C based on [−41.8; 41.5°C], Romania r = 83°C based on [−38.5; 44.5°C], Czech Republic r = 82.6°C based on [−42.2; 40.4°C], Japan r = 82.1°C based on [−41; 41°C] and Chile r = 81.9°C based on [−37; 44.9°C].

The countries in the medium climate flexibility range are Argentina with r = 81.7°C based on [−32.8; 48.9°C], Slovakia r = 81.3°C based on [−41; 40.3°C], Poland r = 81.2°C based on [−41; 40.2°C], Belaruse r = 81.1°C based on [−42.2; 38.9°C], Latvia r = 81°C based on [−43.2; 37.8°C], Lithuania r = 80.4°C based on [−42.9; 37.5°C], Estonia r = 79.1°C based on [−43.5; 35.6°C], Azerbaijan r = 79°C based on [−33; 46°C], Croatia r = 78.3°C based on [−35.5; 42.8°C], Moldova r = 77.9°C based on [−35.5; 42.4°C], Spain r = 77.2°C based on [−30; 47.2°C], Macedonia r = 77.2°C based on [−31.5; 45.7°C], Hungary r = 76.9°C based on [−35; 41.9°C], Greece r = 75.8°C based on [−27.8; 48°C], Slovenia r = 75.3°C based on [−34.5; 40.8°C], South Korea r = 73.8°C based on [−32.8; 41°C], Australia r = 73.7°C based on [−23; 50.7°C], Syria r = 72.4°C based on [−23; 49.4°C], Nepal r = 72°C based on [−26; 46°C], South Africa r = 70.1°C based on [−20.1; 50°C], Iceland r = 68.4°C based on [−37.9; 30.5°C], Israel r = 68.2°C based on [−14.2; 54°C], New Zealand r = 68°C based on [−25.6; 42.4°C], Denmark r = 67.6°C based on [−31.2; 36.4°C], the Netherlands r = 66°C based on [−27.4; 38.6°C], Jordan r = 66°C based on [−16; 50°C], Egypt r = 66°C based on [−15; 51°C] and the United Kingdom r = 65.7°C based on [−27.2; 38.5°C].

The countries with the least climate flexibility are Algeria r = 65.1°C based on [−13.8; 51.3°C], Saudi Arabia r = 64°C based on [−12; 52°C], Portugal r = 63.4°C based on [−16; 47.4°C], Brazil r = 58.7°C based on [−14; 44.7°C], Taiwan r = 58.6°C based on [−18.4; 40.2°C], Guatemala r = 56.5°C based on [−11.5; 45°C], Uruguay r = 55°C based on [−11; 44°C], Swaziland r = 54.4°C based on [−6.7; 47.7°C], Myanmar r = 53.2°C based on [−6; 47.2°C], United Arab Emirates r = 53.1°C based on [−1; 52.1°C], Ireland r = 52.4°C based on [−19.1; 33.3°C], Sudan r = 50.7°C based on [−1; 49.7°C], Dominican Republic r = 50°C based on [−7; 43°C], Vietnam r = 48.8°C based on [−6.1; 42.7°C], Philippines r = 47.2°C based on [−5; 42.2°C], Thailand r = 46°C based on [−1.4; 44.6°C], Hong Kong r = 43.9°C based on [−6; 37.9°C], Bangladesh r = 42.5°C based on [2.6; 45.1°C], Malta r = 42.4°C based on [1.4; 43.8°C], Cuba r = 38.2°C based on [0.6; 38.8°C], Panama r = 38°C based on [2; 40°C], Puerto Rico r = 36°C based on [4; 40°C], Cambodia r = 32.6°C based on [10; 42.6°C], Malaysia r = 32.3°C based on [7.8; 40.1°C], Samoa r = 24.2°C based on [11.1; 35.3°C], Antigua and Barbuda r = 18.8°C based on [16.1; 34.9°C], Singapore r = 18°C based on [19; 37°C], Maldives r = 17.7°C based on [17.2; 34.9°C] and Burkina Faso r = 6.2°C based on [41; 47.2°C].

Climate flexibility is prospected to open a gate for productivity. Countries with climate flexibility are likely productive grounds for agriculture, industry and service production at different temperature levels.

11.4.4.1 Climate Flexibility Model

The relation of climate flexibility and market prospects was investigated. Based on the determined climate flexibility, the range of temperature variability—as retrieved and calculated based on the *List of Countries by Extreme Temperatures*[8]—was related to human migration.

Human migration: When investigating the relation of human migration and climate flexibility in 86 countries of the world, a highly significant correlation ($r_{\text{Pearson}}(86) = 0.319$, $p < 0.002$), is found between being

[8] https://en.wikipedia.org/wiki/List_of_countries_by_extreme_temperatures

climate flexible and human migration inflows. No significant relation is reported for financial inflows as well as innovation.

11.4.5 Data Study 3: Climate Change Expectations Driving Commodity Bubbles

The world is divided in nation-states. Recent research has defined climate change winners and losers by GDP prospects in a changing climate. Stepping aside from national territories, climate change winning industries and sectors were unraveled most recently.

11.4.6 Results Study 3: Climate Change Expectations Driving Commodity Bubbles

For the GDP comprising pillars of agriculture, industry and service sectors, the peak temperatures for production were derived from a meta-analysis of literature sources. For agriculture, crop durability—meaning how long crops can prosper in a changing climate—was determined based on literature resources. Graph 11.5 holds crops and their germination temperature, peak performance temperature and maximum possible temperature.

The temperature ranges are outlined in Graph 11.6, which holds the crop durability. Longest temperature range crops are indicated by green color, medium temperature range in yellow color and shortest temperature range in red color.

Beans, pak choi, peas, cabbage, carrots, red beet and milk thistle will have the widest application range—as outlined in green color—regarding temperature. Onions, ryegrass, Chinese cabbage, Asian radish, corn, parsnips and cold-season crops such as oats, rye and wheat have a medium flexibility range—as exhibited in yellow color—when it comes to temperature. Soybean, mustard, pumpkins, rice, coconut, sorghum, peanut and groundnut as well as all the cool-season crops such as broccoli, Brussels sprouts, cabbage, chives, collards, corn salad, garlic, horseradish, kale, kohlrabi, leeks, onions, parsley, peas, radishes, rhubarb, rutabaga, spinach and turnips will run out of time first as outlined in red coloring of the smallest temperature range. The latter's cultivation is expected to become the most expensive first, followed by the medium and then the longest temperature-ranged crops.

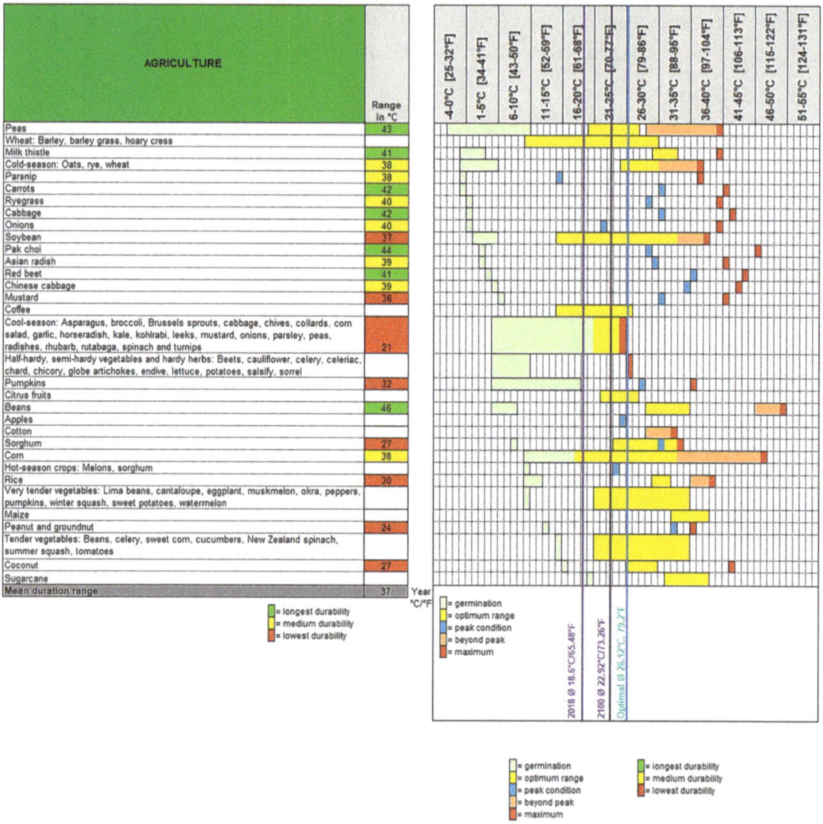

Graph 11.5 Agriculture sector temperature-based condition in relation to world mean temperature (*Source* Author's own)

Given the assumption of price increases around scarcity, a model for crops' extinction and price mechanism is sketched out (Figuerola-Ferretti & Cervera, 2018; Kaabia et al., 2018; Lips, 2018). In the oil energy sector, price spikes have been found around procurement shortages, which are expected to be occurring in all climate-related extinction cases (Roberts et al., 2018).

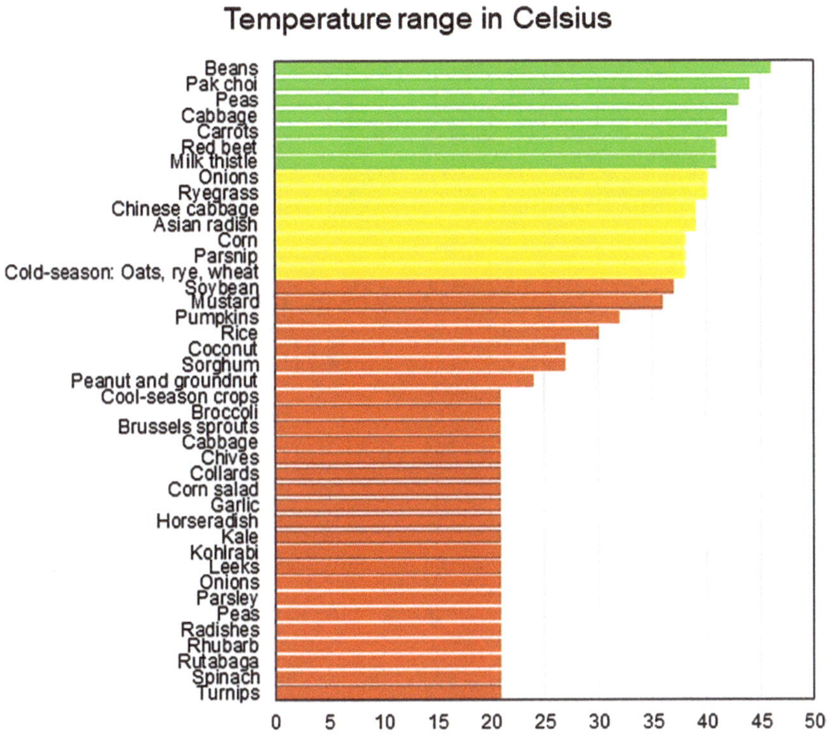

Graph 11.6 Temperature ranges for outlining agriculture durability (*Source* Author's own)

Time series of commodity price indices from June 1989 to 2017, retrieved from Index Mundi,[9] allowed to determine the price development of commodity prices for food and beverages. Since the turn of the millennium, temperature records have skyrocketed, which fueled vast communication about global warming. Price expectations are therefore assumed to have spiraled a commodities bubble for food and beverages. As outlined in Graph 11.7, commodity prices for food as represented by barley, Canadian wheat, maize/corn, rice, soft red winter wheat and sorghum have risen extraordinarily since 2000.

[9] https://www.indexmundi.com/commodities/?commodity=wheat&months=360

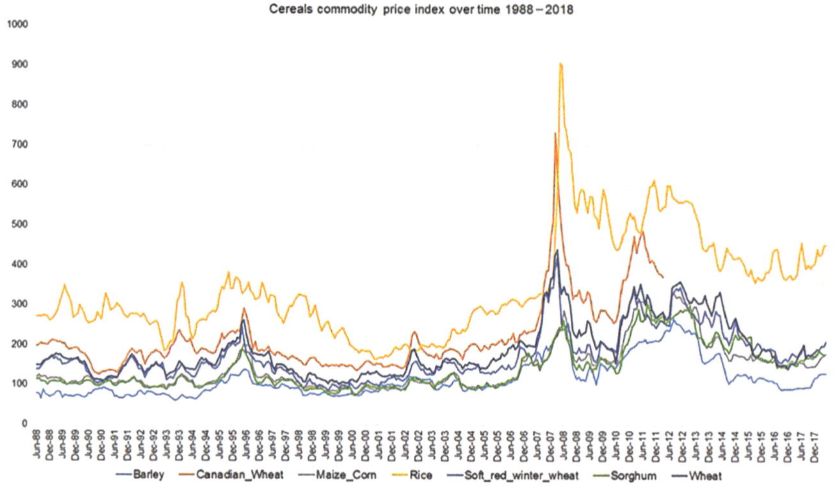

Graph 11.7 Commodity price index retrieved from Index Mundi for food (*Source* Author's own. https://www.indexmundi.com/commodities/?commodity=wheat&months=360)

As outlined in Graph 11.8, commodity prices for beverages, as represented by Arabica coffee as well as tea, have risen extraordinarily since 2000.

Highly significant independent t-tests comparing the price index per crop prior to 2000 and thereafter consolidate the extraordinary price rise in commodities for barley with $t(358) = -11.984$, $p = 0.000$ of 139 data points in the previous millennium and 221 observations in the current millennium based on monthly price indices measurement; Canadian wheat with $t(284) = -7.502$, $p = 0.000$ of 139 data points in the previous millennium and 147 observations in the current millennium based on monthly price indices measurement; maize/corn with $t(358) = -8.82$, $p = 0.000$ of 139 data points in the previous millennium and 221 observations in the current millennium based on monthly price indices measurement; rice with $t(358) = -7.331$, $p = 0.000$ of 139 data points in the previous millennium and 221 observations in the current millennium based on monthly price indices measurement; soft red winter wheat with $t(358) = -8.786$, $p = 0.000$ of 139 data points in the previous millennium and 221 observations in the current

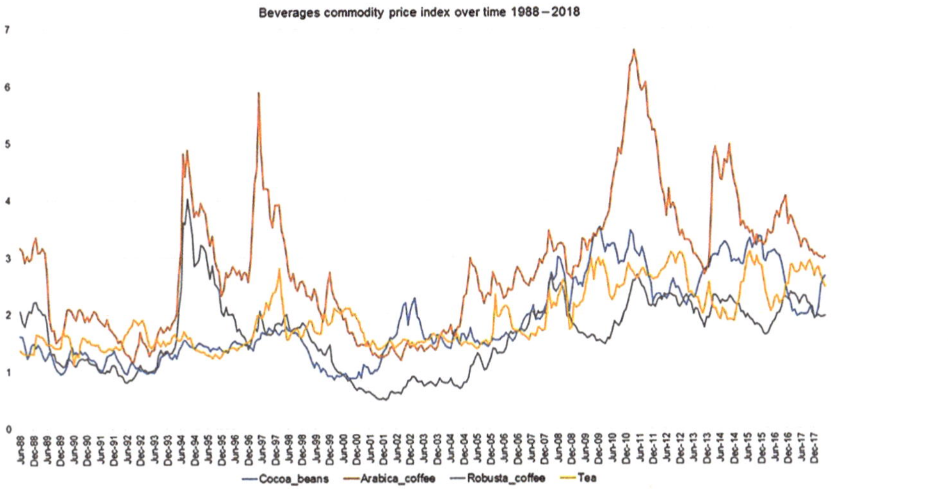

Graph 11.8 Commodity price index retrieved from Index Mundi for food (*Source* Author's own. https://www.indexmundi.com/commodities/?commodity=wheat&months=360)

millennium based on monthly price indices measurement; sorghum with $t(357) = -10.352$, $p = 0.000$ of 139 data points in the previous millennium and 220 observations in the current millennium based on monthly price indices measurement; wheat with $t(355) = -9.209$, $p = 0.000$ of 139 data points in the previous millennium and 218 observations in the current millennium based on monthly price indices measurement; cocoa beans with $t(358) = -13.702$, $p = 0.000$ of 139 data points in the previous millennium and 221 observations in the current millennium based on monthly price indices measurement; Arabica coffee with $t(358) = -3.631$, $p = 0.000$ of 139 data points in the previous millennium and 221 observations in the current millennium based on monthly price indices measurement; and tea with $t(358) = -11.493$, $p = 0.000$ of 139 data points in the previous millennium and 221 observations in the current millennium based on monthly price indices measurement. Out of all commodity indices for food and beverage reported in the Index Mundi database[10] with sufficient data points, Robusta coffee was the only

[10] https://www.indexmundi.com/commodities/?commodity=wheat&months=360

good with non-significant price increases for the period after the turn of the millennium. To cross-validate the findings, price indices for gasoline, medical care, fast food, soda, jewelry, new cars and computers[11] were eyeballed, and no stringent pattern of increased prices could be found for these goods and services since turn of the millennium.

Given the backtesting results consolidating inflated commodity prices since the turn of the millennium, future prospects of price deviations based on the extinction temperature are drawn as stylized, hypothetical estimation. Graph 11.9 holds prospective extinction-based market price predictions as deviations from normal prices. As one can see, the world with medium temperature of 18.6°C is not at the efficiency frontier yet (Foley, 2007). In addition, the mean temperature for continents was derived from the respective literature, and critical mean continental temperature in Central and South America, Africa and Australia was put into perspective.[12]

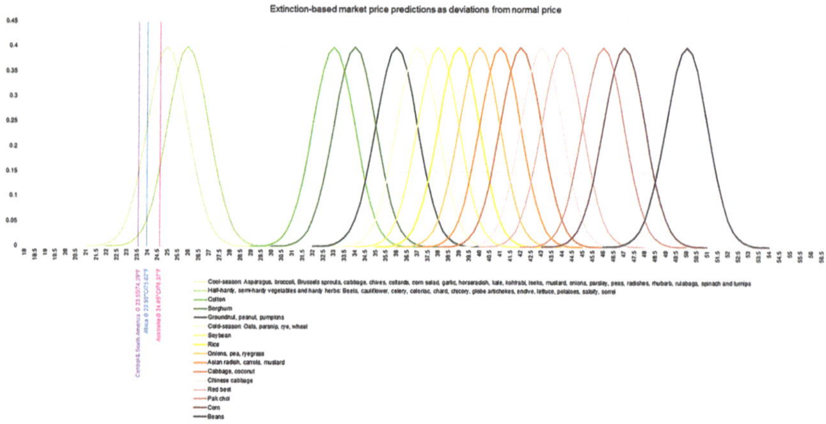

Graph 11.9 Extinction-based market price predictions as deviations from normal prices (*Source* Author's own)

[11] https://people.duke.edu/~rnau/411infla.htm

[12] https://simple.wikipedia.org/wiki/List_of_countries_by_continents

11.4.7 Geo-politics of food prices

The impact of climate change on commodity prices was investigated (Puaschunder, under review). Global warming was found to be a more significant international influence on food prices than war activities and financial crises. Empirical research identified the causal relationship between global warming, wartime activities and financial crises on food commodity prices (Puaschunder, under review).

The macroeconomic investigation spanned data over a 26-year period from 1994 to 2019. Climate change, war activities and financial crises data was regressed on commodity prices (Puaschunder, under review). Climate change was captured in the National Aeronautics and Space Administration (NASA) Land-Ocean Temperature Index. War destruction was estimated in the sum of deaths in armed conflicts around the world. Financial crises were measured by the Metrick-Schmelzing Banking-Crisis Intervention database. Commodity prices were operationalized by the Index Mundi Commodity Food and Beverage Price Index.

The overall causal relation of climate change and rising commodity prices was discovered (Puaschunder, under review). Overall, worldwide war activities and financial crises are not systemically related to worldwide commodity price changes. This finding underlines the long-term and substantial but somewhat chronic food price inflation caused by global warming (Puaschunder, under review).

In a more targeted investigation covering the time period from the beginning of 1990 to the end of 2024, the impact of financial crises on food prices operationalized by the real, adjusted for inflation, United Nations Food and Agriculture Organization (FAO) Food Price Index showed a significant long-term price elevation of global food prices after the systemic internal 2008/09 World Financial Crisis and the 2020 COVID-19 external shock (Puaschunder, under review). In an additional targeted investigation of data ranging from the beginning of 1990 to the end of 2024, the impact of the annexation of the Crimea and the outbreak of Russia's hostile takeover attempt of the Ukraine on cereal prices operationalized by the real, adjusted for inflation, United Nations Food and Agriculture Organization (FAO) Cereal Price Index showed a significant long-term price elevation after both political events (Puaschunder, under review). While the investigation covered a global outlook, a more regional and temporal focus as well as case studies and disparate impact analyses of food price changes are recommended to follow.

11.5 Results and Discussion

The presented results may shed light on climate change from a different angle and grant hope but also alleviate the contemporary reported predicament of economic growth versus sustainability (Täuber et al., 2015; Zhou et al., 2018). Behavioral economics found people to be risk-averse, and their loss prospects loom larger than gains (Kahneman & Tversky, 1979; Tversky & Shafir, 1992). Changing individual temperature conditions make people afraid, and there is an immediate loss of the status quo visible. But in reality, the world by itself will be winning from a warming globe for quite some time, just needs to relocate and transfer skills and production into those areas that will become newly arable and pleasant to stay and work in.

Looking forward to world peak condition with regard to a favorable climate for GDP production, the future wealth of nations may be determined by countries' access to global markets (e.g. via infrastructure such as airports and ports, trade relations such as general agreements and trade unions and practiced customs such as openness to take in migrants or a favorable easy-to-obtain language like English or Spanish) as well as natural propensities to temperature flexibility by being either a widespread country (like Canada or Russia) or one featuring different temperature zones due to tectonic landscape formation and altitude (e.g. like Italy, New Zealand, Luxembourg or Switzerland featuring mountains).

Future favorable market operations appear to be (1) investments in insurances specializing on climate change and capitalizing the collective fear that has been built around the change in temperature, which actually also features positive aspects, (2) mergers and conglomerates of multiple entities to diversify from climate-related risks and shortages with higher economies of scale (Emmert, 2018) and (3) surge price strategies as ticket prices that can offer a discount during shoulder seasons disrupted from unfavorable climate and high prices during peak seasons (Emmert, 2018). The winter sports industry appears to be heading toward a direction of high luxury price and low customer base due to snow shortages and shortened winter seasons (Emmert, 2018). Mountain ownership by insurance agencies is another trend that is seen in the winter industry (Emmert, 2018). Resorts have immensely high insurance costs, both for body injury and weather. If a resort can farm its insurance in-house, it can save in the long run. This trend may also be extended onto other industries that may become more entwined with insurances, as climate change causes weather unpredictability. Consumer lock-in strategies and weather-related loss exemptions for satisfaction guarantee clauses may be additional trends we are heading toward in the age of global warming. Revenue off weather-dependent sales are potential other areas that are prospected to see economic growth in times of weather uncertainty caused by climate change. For instance, in the skiing industry, at present, half of the revenue is made up with food, lodging, rentals and lesson sales (Emmert, 2018).

Overall, the market-driven results of price mechanisms but also FDI and market streams underline the importance of research on the effect of climate change on market mechanisms. This first preliminary introduction of the idea of climate flexibility as future determinant of competitive advantage and wealth of nations opens a door for future research on climate impacts on international trade. The impact of climate on markets also implicitly calls for a further investigation of industrial policies that rest on the fundamental insight that market prices are not often sufficient in allocating scarce resources in a humane society (Altenburg & Assmann, 2018). For instance, preferences of future generations cannot be fully reflected in market prices (Altenburg & Rodrik, 2018).

As a caveat to the presented results, it deserves mentioning that price expectations—to this day—are hard to predict, and the actual temporal flow of these recurrent patterns is hard to measure quantitatively. While there is a pattern of the economy, problems of agents remain to distinguish between noise and information, and between shocks and change. In the ever-changing nature of capitalism, having evolved from industrialization to post-World War booms to globalization and the new media revolution but also slowbalisation trends and geo-political tensions, economic cycles have become more fuzzy and unpredictable in most recent decades (Held & McGrew, 2007).

In addition to standard economic prediction impediments, commodities and farming depend on different inputs, including equipment, fertilizers, and distribution. If costs of these resources increase or their procurement is at risk, the cost of production increases. Overall risks that drive up prices include transportation costs determined by oil and gas prices, which depend on weather conditions. Such price increases therefore are also influenced by global warming trends, which determine weather conditions. Supply chain bottlenecks and livestock illness can also increase production costs in the agriculture sector. Moreover, inflation impacts the production costs associated with land construction and maintenance. Increased costs of raw materials raises production costs, which exacerbates affordability challenges in the overall market.

Investment in agriculture and commodities is also volatile due to political frictions. In general, global political crises impact commodity prices (Toledo & Duncan, 2024). Commodities, such as food and beverages, already had elevated price levels since 2000 and experienced a significant rise during the 2008/09 World Financial Recession, which never adjusted to previous levels, as investors can use commodities as a fairly financially unregulated speculative good. The commodities price inflation further exacerbated during the COVID pandemic having driven up transportation costs and creating bottlenecks (Cox, 2021). The political crisis between Russia and Ukraine then led to bottlenecks in access to fertilizers and seeds, implying drastic food price spikes worldwide (Krugman, 2023). The countries in conflict are major exporters of commodities and fertilizers to the world, but also oil and natural gas prices set the tone for fertilizer production throughout Europe, which further exacerbated the already elevated price levels (Krugman, 2023). The Russia–Ukraine war including different nations to support both parties at dispute has further exacerbated the detected trends on food price inflation, even around the world.

Overall, the Russia–Ukraine conflict has direct and indirect impacts on food grain prices in the U.S. and Europe. While food commodity prices have been relatively stable in the U.S. since the outbreak of the Ukraine crisis, Europe experienced more food price volatility, likely due to the relative closeness to the countries at war and less agricultural production in European highly specialized service sector industry countries (Krugman, 2023). Direct impact is measured in terms of high Brent crude oil prices due to the war and the dependency of the United States and other European nations on the Russian crude oil supply for food procurement. The indirect impact is the spillover effect caused by inflated oil and natural gas prices, which erupted cost-push inflation in the country, especially in terms of food grain inflation. As the U.S. is dependent on fertilizer imports from the two countries engaged in a war, this causes a cost-push inflation in agriculture, as fertilizer is the basic input in agriculture.

In its entirety, the political East-West crisis has led to a global rise in prices and thus global inflation (Ozili, 2022). For instance, the geopolitical tensions were found to be highly correlated with Euro area food prices, the world food price index, world oil price index, world dairy price index, and world cereal price index during the month of the Ukraine invasion (Ozili, 2022). Empirical evidence for the impact of the dependency on oil and fertilizer needs serves to elucidate the various factors that build up inflationary pressures within the economy. Geopolitical conflicts were found to be associated with food inflation, even in countries not directly involved in the tensions.

War therefore has an impact on food procurement on a global scale. Concrete downsides of food inflation include malnutrition and poor health as well as a degraded quality of life. As food and nutrition are essential for all human life, understanding the effect of war on food prices is also important for international development and global welfare to combat a substantial disparate impact on marginalized groups. Food inflation in particular affects livelihoods and well-being as for the essential nature of food for living.

Commodities' price inflation can be combated with several policies. Direct interference in markets can include price controls and price caps at a lower price level. Countries should also diversify import opportunities to help navigate limited climate flexibility. Subsidizing and compensating farmers for weather dependencies are additional means to stabilize prices. Food reserves help overcome bottlenecks and supply shortages. Alternative energy sources, including renewables, are additional means to curb price volatility. Self-sustaining production could strategically incentivize farmers to produce less price volatile goods that are less dependent on weather conditions and negative price spillovers by sudden political shocks. Inflation effects on consumers can be alleviated by direct economic stimulus and interest rate adjustments. In addition, macroeconomic tools such as open market operations and reducing the money in circulation, e.g., through interest rate raising and bank reserve requirements, help ease inflationary pressure. Strategic reserves, contingency planning and communication are additional measures to aid with food inflation a country can choose. On the international level, diplomacy and energy-strategic ties can help stabilize commodity markets and manage supply chain disruptions during geopolitical tensions. In addition, diplomatic trade policies can help overcome bottlenecks through strategic resource alliances and policy action. Agricultural investments aid in developing crops and food supply chain productivity. Consumer education and strategic waste management through recycling can target demand and foster efficient food consumption. Consumer protection should focus on marginalized groups, on whom commodities' price spikes have the most detrimental impact. Overall, a comprehensive approach should combine the outlined strategies in order to address food inflation effectively in light of weather uncertainty.

Future research may investigate climate change impacts on the 'demand' side of economic pricing behavior. As studies have shown, demand preferences are sticky as habits usually form over a long time and therefore do not change quickly. Future studies may unravel how climate changes preferences for specific goods and services in the long run. Thereby, especially the role of critical life events and external climate shocks in changing preferences immediately and definitively should be scrutinized by behavioral insights specialists targeting at finding innovative ways on how to align the demand side with the supply side of climate policies from an interdisciplinary perspective (Puaschunder, 2020). Future studies may also focus on how to teach the gains of a warming earth in applied business cases (Gonglewski & Helm, 2010; Helm, 2009). All these presented insights gained may help administer the transition to new technologies and renewable energy as while stabilizing economic prices and reaping the economic gains of a warming globe to the benefit of all.

References

Acevedo, S., Mrkaic, M., Novta, N., Poplawski-Ribeiro, M., Pugacheva, E., & Topalova, P. (2018, April 12). *How can low income countries cope? The effects of weather shocks on economic activity*. Presentation delivered at The New School Economics of Climate Change Project, The New School, New York.

Altenburg, T., & Assmann, C. (2018). *United Nations Environment Programme Report on Green Industrial Policy: Concept, policies, country experiences*. United Nations: United Nations Environment Programme & German Development Institute/Deutsches Institut für Entwicklungspolitik.

Altenburg, T., & Rodrik, D. (2018). Green industrial policy: Accelerating structural change towards wealthy green economies. In T. Altenburg & C. Assmann (Eds.), *United Nations Environment Programme Report on Green Industrial Policy: Concept, policies, country experiences* (pp. 2–20). United Nations: United Nations Environment Programme & German Development Institute/Deutsches Institut für Entwicklungspolitik.

Andreoni, J., Erard, B., & Feinstein, J. S. (1998). Tax compliance. *Journal of Economic Literature, 36*(2), 818–860.

Asmundson, I. (2017). *Financial services: Getting the goods*. IMF entry. http://www.imf.org/external/pubs/ft/fandd/basics/finserv.htm

Bai, S., Koong, K., & Valenzuela, E. (2018, May 24–25). The dynamic impact of oil price shocks on stock returns: Evidence from U.S. energy companies in the value chain. In *6th International Symposium on Environment Energy & Finance Issues, ISEFI 2018*, Paris, France.

Barro, R. (1990). Government spending in a simple model of endogenous growth. *Journal of Political Economy, 98*, 103–125. https://doi.org/10.1086/261726

Bienenfeld, M. (1988). Regularity in price changes as an effect of changes in distribution. *Cambridge Journal of Economics, 12*, 247–255.

Bonato, M. (2018, May 24–25). Realized correlations, betas and volatility spillover in the agricultural commodity market: What has changed? In *6th International Symposium on Environment Energy & Finance Issues, ISEFI 2018*, Paris, France.

Bürgenmeier, B. (1994). Environmental policy: Beyond the economic dimension. In B. Bürgenmeier (Ed.), *Economy, environment, and technology: A socio-economic approach* (pp. 166–175). Armonk Sharpe.

Burke, M., Hsiang, S. M., & Miguel, E. (2015). Global non-linear effect of temperature on economic production. *Nature, 527*, 235–239. https://doi.org/10.1038/nature15725

Centeno, M. A., Creager, A. N., Elga, A., Felton, E., Katz, St. N., Massey, W. A., et al. (2013). *Global systemic risk: Proposal for a research community.* Working paper. Princeton Institute for International and Regional Studies, Princeton University, Princeton, NJ.

Centeno, M. A., & Tham, A. (2012).*The emergence of risk in the global system.* Working paper. Princeton University, Princeton, NJ.

Chancel, L., & Piketty, T. (2015). *Carbon and inequality: From Kyoto to Paris.* Paris School of Economics.

Chilcote, E. B. (1997). *Interindustry structure, relative prices, and productivity: An input-output study of the U.S. and O.E.C.D countries.* Doctoral thesis, The New School for Social Research Department of Economics, New York, NY.

Cialdini, R. B. (1996). Social influence and the triple tumor structure of organizational dishonesty. In D. M. Messick & A. E. Tenbrunsel (Eds.), *Codes of conduct* (pp. 44–58). Sage.

Cifarelli, G., & Paesani, P. (2018, May 24–25). Fundamentalists, chartists and hedgers: A non-linear dynamic model of futures oil pricing: Theory and empirical evidence. In *6th International Symposium on Environment Energy & Finance Issues, ISEFI 2018*, Paris, France.

Cockshott, W. P., & Cottrell, A. (1993). *Towards a new socialism.* Spokesman.

Cox, J. (2021). Consumer prices jump 5% in May, fastest pace since the summer of 2008. *CNBC News.* https://www.cnbc.com/2021/06/10/cpi-may-2021.html

Dawes, R. M. (1980). Social dilemmas. *Annual Review of Psychology, 31*(1), 169–193. https://doi.org/10.1146/annurev.ps.31.020180.001125

Dukhanina, E., Leveque, F., & Massol, O. (2018). *Policy measures targeting a more integrated gas market: Impact on prices and arbitrage activity.* https://www.eeg.tuwien.ac.at/conference/iaee2017/files/paper/649_Dukhanina_fullpaper_2017-06-30_21-58.pdf

Ekholm, T. (2018, May 24–25). Climatic cost-benefit analysis under uncertainty and learning on climate sensitivity and damages. In *6th International*

Symposium on Environment Energy & Finance Issues, ISEFI 2018, Paris, France.
Elgouacem, A., & Legrand, N. (2018, May 24–25). The delaying effect of storage on investment. Evidence from the crude oil sector. In *6th International Symposium on Environment Energy & Finance Issues, ISEFI 2018*, Paris, France.
Emmert, H. (2018). *The melting point: The economics and environmental impact of the winter sports industry*. Unpublished working paper. Economics of the Environment, The New School, New York.
Environmental Defense Fund. (2018). *How climate change plunders the planet: A warming earth disturbs weather, people, animals and much more*. https://www.edf.org/climate/how-climate-change-plunders-planet
Esposito, M., Haider, A., Semmler, W., & Samaan, D. (2018). Enhancing job creation through the green transformation. In T. Altenburg & C. Assmann (Eds.), *United Nations Environment Programme Report on Green Industrial Policy: Concept, policies, country experiences* (pp. 50–68). United Nations: United Nations Environment Programme & German Development Institute/Deutsches Institut für Entwicklungspolitik.
Feld, L. P., & Frey, B. S. (2002). Trust breeds trust: How taxpayers are treated. *Economics of Governance, 3*(2), 87–99. https://doi.org/10.1007/s101010100032
Figuerola-Ferretti, I., & Cervera, I. (2018). *Recent credit risk and bubble behavior in the corporate energy sector*. https://repositorio.comillas.edu/jspui/handle/11531/26571
Filip, O., Janda, K., Kristoufek, L., & Zilberman, D. (2018). *Food versus fuel: An updated and expanded evidence*. CAMA working paper 73. https://papers.ssrn.com/sol3/papers.cfm?abstract_id=3070704
Foley, D. (2007). *The economic fundamentals of global warming*. Working paper 07-12-044. Santa Fe Institute, Santa Fe, NM. https://www.santafe.edu/research/results/working-papers/the-economic-fundamentals-of-global-warming
Frey, B. S. (1992). Tertium datur: Pricing, regulation and intrinsic motivation. *Kyklos, 45*(2), 161–184. https://doi.org/10.1111/j.1467-6435.1992.tb02112.x
Frey, B. S. (1997). *Not just for the money: An economic theory of personal motivation*. Edward Elgar.
Friedman, M. (1962). *Price theory: A provisional text*. Aldine.
Ganepola, C., Garrett, I., & Cho, S. (2018, May 24–25). Oil prices, discount rate news and cash flow news. In *6th International Symposium on Environment Energy & Finance Issues, ISEFI 2018*, Paris, France.

Gonglewski, M., & Helm, A. (2010). An examination of business case methodology: Pedagogical synergies from two disciplines. *Global Business Languages, 15*(3), 17–31.

Greiner, A., & Semmler, W. (2008). *The global environment, natural resources, and economic growth*. Oxford University Press. https://doi.org/10.1093/acprof:oso/9780195328233.001.0001

Hasseldine, J. D. (1998). Prospect theory and tax reporting decisions: Implications for tax administrators. *International Bureau of Fiscal Documentation, 52*(11), 501–505.

Held, D., & McGrew, A. G. (2007). *A new world economic order? Global markets and state power: Beyond globalization/anti-globalization: Beyond the great divide*. Polity.

Helm, A. (2009). Global marketing classroom: Green marketing. *American Marketing Association Global SIG Newsletter, Global Interests*.

Ielpo, F., & Sevi, B. (2013). *Forecasting the density of oil futures returns using model-free implied volatility and high-frequency data*. Montpellier: Centre de Recherche en Economie et Droit de l'Energie CREDEN Equipe ART Dev, Université Montpellier, Faculté d'Economie, Cahier de recherche n° 13.10.106.

Ji, Q., Liu, B. Y., Zhao, W. L., & Fan, Y. (2018, May 24–25). Modelling dynamic dependence and risk spillover between all oil price shocks and stock market returns in the BRICS. In *6th International Symposium on Environment Energy & Finance Issues, ISEFI 2018*, Paris, France.

Kaabia, O., Dhaoui, A., Abid, I., & Guesmi, K. (2018). *Oil supply-demand shocks and stock prices*. https://www.researchgate.net/publication/324687727_Oil_supply-demand_shocks_and_stock_prices

Kahneman, D., & Tversky, A. (1979). Prospect theory: An analysis of decision under risk. *Econometrica, 47*(2), 263–291. https://doi.org/10.2307/1914185

Kalecki, M. (1943). *Studies in economic dynamics*. Allen and Unwin.

Keynes, J. M. (1936). *The general theory of employment, interest and money*. Macmillan.

Kilian, L., & Zhou, X. (2018). *The propagation of regional shocks in housing markets: Evidence from oil price shocks in Canada*. CEPR Discussion Paper DP12845. Centre for Economic Policy Research, London, UK. https://papers.ssrn.com/sol3/papers.cfm?abstract_id=3159148

Kirchler, E. M. (2007). *The economic psychology of tax behaviour*. Cambridge University Press. https://doi.org/10.1017/CBO9780511628238

Klein, T., von Mettenheim, H. J., Walther, Th., & Wegener, Ch. (2018, May 24–25). Forecasting high frequency realized volatility of oil with exogenous regressors. In *6th International Symposium on Environment Energy & Finance Issues, ISEFI 2018*, Paris, France.

Krugman, P. (2023, August 8). Why are groceries so expensive? There's no one simple cause for higher food prices. *The New York Times*. https://www.nytimes.com/2023/08/08/opinion/inflation-us-groceries-economy.html
Kruse, R., & Wegener, Ch. (2017). *Time-varying persistence in real oil prices and its determinants*. Working paper. http://www.wisostat.uni-koeln.de/sites/statistik/abstracts/RKB-Time-Varying_Persistence_in_Real_Oil_Prices_and_its_Determinant.pdf
Legnazzi, C. (2018, May 24–25). Forward-looking VaR and CVaR: An application to the natural gas market. In *6th International Symposium on Environment Energy & Finance Issues, ISEFI 2018*, Paris, France.
Leontief, W. (1997). *Input-output economics*. Oxford University Press.
Lips, J. (2018). *Debt and the oil industry: Analysis on the firm and production level*. https://papers.ssrn.com/sol3/papers.cfm?abstract_id=3026063
Mahadeo, S. M. R., Heinleinb, R., & Legrenzib, G. D. (2018, May 24–25). New contagion applications for the energy-finance nexus: A higher moment approach. In *6th International Symposium on Environment Energy & Finance Issues, ISEFI 2018*, Paris, France.
Markowitz, E. M., Grasso, M., & Jamieson, D. (2015). Climate ethics at a multidisciplinary crossroad: Four directions for future scholarship. *Climate Change, 130*(3), 465–474. https://doi.org/10.1007/s10584-015-1404-4
Marron, D. B., & Morris, A. C. (2016). *How to use carbon tax revenues*. Tax Policy Center Urban Institute & Brookings Institution. https://doi.org/10.2139/ssrn.2737990
Marx, K. (1867/1995). *Capital: A critique of political economy*. Progress.
Moore, J. W. (2009, October 9–10). *Ecology and the accumulation of capital: A brief environmental history of neoliberalism*. Paper presented at the Workshop on Food, Energy, Environment: Crisis in the Modern World-System. Fernand Braudel Center, Binghampton University.
Mumford, A. (2001). *Taxing culture*. Ashgate.
Nordhaus, W. D. (1994). *Mapping the global commons: The economics of climate change*. MIT Press.
Ordu-Akkaya, B. M., & Soytas, U. (2018, May 24–25). The impact of foreign portfolio investment on financialization of commodities. In *6th International Symposium on Environment Energy & Finance Issues, ISEFI 2018*, Paris, France.
Orttung, R. (2010a, August 30). *Politics and profits in Russia's foreign oil and gas conflicts*. Center for Security Studies Colloquium, Swiss Federal Institute of Technology.
Orttung, R. (2010b, May 11). *Five myths about energy security*. Energy Forum 2010, Swiss Federal Institute of Technology.

Orttung, R. (2014, January 25). *Russia's place in the world energy system: Fueling the future: Deconstructing the role of energy in contemporary international relations*. University of Toronto.

Orttung, R. (2015). Promoting sustainability in Russia's artic: Integrating local, regional, federal, and corporate interest. In S. Oxenstierna (Ed.), *The challenges for Russia's politicized economic system* (pp. 202–219). Routledge.

Orttung, R. (2016). *Sustaining Russia's arctic cities: Resource politics, migration, and climate change*. Berghahn. https://doi.org/10.2307/j.ctvswx6s0

Orttung, R., & Overland, I. (2011a). A limited toolbox: Explaining the constraints on Russia's foreign energy policy. *Journal of Eurasian Studies, 2*(1), 74–85. https://doi.org/10.1016/j.euras.2010.10.006

Orttung, R., & Overland, I. (2011b). Russia and the formation of a gas cartel. *Problems of Post-Communism, 58*(3), 53–66. https://doi.org/10.2753/PPC1075-8216580305

Ozili, P. K. (2024). Global economic consequences of Russian invasion of Ukraine. In *Dealing with regional conflicts of global importance* (pp. 195–223). IGI Global.

Perovic, J., Orttung, R. W., & Wenger, A. (2009). *Russian energy power and foreign relations*. Routledge. https://doi.org/10.4324/9780203880098

Puaschunder, J. M. (2015). Trust and reciprocity drive common goods allocation norms. In *Proceedings of the Cambridge Business & Economics Conference*. Cambridge University.

Puaschunder, J. M. (2016). Intergenerational climate change burden sharing: An economics of climate stability research agenda proposal. *Global Journal of Management and Business Research: Economics and Commerce, 16*(3), 31–38.

Puaschunder, J. M. (2017a, March 26–27). Climate in the 21st century: A macroeconomic model of fair global warming benefits distribution to grant climate justice around the world and over time. In *Proceedings of the 8th International RAIS Conference on Social Sciences and Humanities organized by Research Association for Interdisciplinary Studies (RAIS) at Georgetown University*, Washington, DC, USA, pp. 205–243.

Puaschunder, J. M. (2017b). Mapping climate in the 21st century. *Development, 59*(3), 211–216.

Puaschunder, J. M. (2017c). Socio-psychological motives of socially responsible investors. *Advances in Financial Economics, 19*(1), 209–247. https://doi.org/10.1108/S1569-373220160000019008

Puaschunder, J. M. (2017d). Sunny side up! From climate change burden sharing to fair global warming benefits distribution: Groundwork on the metaphysics of the gains of global warming and the climatorial imperative. In *Proceedings of the Administrative Sciences Association of Canada Annual Conference*, HEC Montreal, Quebec, Canada.

Puaschunder, J. M. (2017e). The call for global responsible inter-generational leadership: The quest of an integration of inter-generational equity in corporate social responsibility (CSR) models. In *Natural resources management: Concepts, methodologies, tools, and applications: Information resources management association*. IGI Global.

Puaschunder, J. M. (2018a). Climate policies with burden sharing: The economics of climate financing. In J. Kose, A. Makhija & St.P. Ferris (2018). *Advances in Financial Economics, 19* (pp. 1–13). Emerald.

Puaschunder, J. M. (2018b). Gifts without borders: Intergenerational glue connecting over distance and time as pure international development in the age of migration. In J. M. Puaschunder (Ed.), *Intergenerational responsibility in the 21st century* (pp. 143–174). Vernon.

Puaschunder, J. M. (2018c). Nachhaltigkeit und Investment: *Psychologische Aspekte von nachhaltigkeitsorientiertem Investitionsverhalten.* In C. T. Schmitt & E. Bamberg (Eds.), *Psychologie und Nachhaltigkeit: Konzeptionelle Grundlagen, Anwendungsbeispiele und Zukunftsperspektiven* (pp. 127–134). Springer.

Puaschunder, J. M. (2019). Value at looking back: Towards an Empirical validation of the role of reflexivity in econo-historic backtesting: Economic market prediction corrections correlate with future market performance. *Oxford Journal of Finance and Risk Perspectives, 8*, 223–239.

Puaschunder, J. M. (2020). *Behavioral economics and finance leadership: Nudging and winking to make better choices.* Springer Nature.

Renz, A. (2018, May 24–25). Sector competition for energy resources in a carbon constrained hoteling model. In *6th International Symposium on Environment Energy & Finance Issues, ISEFI 2018*, Paris, France.

Ricardo, D. (1817/1951). On the principles of political economy and taxation. In P. Sraffa (Ed.), *The works and correspondence of David Ricardo* (pp. 1951–1973). Cambridge University Press.

Roberts, G., Barbier, E., & van't Veld, K. V. (2018). *The beginning of the end of the age of oil.* https://cenrep.ncsu.edu/cenrep/wp-content/uploads/2016/08/Roberts.pdf

Rosenlund Soysal, E. (2018, May 24–25). Share repurchase under diverging beliefs about carbon risk. In *6th International Symposium on Environment Energy & Finance Issues, ISEFI 2018*, Paris, France.

Sachs, J. D. (2014). Climate change and intergenerational well-being. In L. Bernard & W. Semmler (Eds.), *The Oxford handbook of the macroeconomics of global warming* (pp. 248–259). Oxford University Press.

Saussay, A., & Sato, M. (2018). *The impacts of energy prices on industrial foreign investment location: Evidence from global firm level data.* https://personal.lse.ac.uk/satom/publication/currentss/

Schmölders, G. (1960). *Das Irrationale in der öffentlichen Finanzwirtschaft*. Suhrkamp.

Schütze, F., Aleksovski, D., & Mozetic, I. (2018). *Stock market reactions to international climate negotiations*. https://bibbase.org/network/publication/schtze-aleksovski-mozeti-stockmarketreactionstointernationalclimatenegotiations-2018

Schweiger, H., & Stepanov, A. (2018). *Energy subsidies, energy intensity and management practices*. https://www.ebrd.com/cs/Satellite?c=Content&cid=1395236502077&d=Mobile&pagename=EBRD%2FContent%2FContentLayout

Sherman, H. J. (1967). Marx and the business cycle. *Science & Society, 31*(4), 486–504.

Smith, A. (1776). *An inquiry into the nature and causes of the wealth of nations*. Strahan.

Somanathan, E., Somanathan, R., Sudarshan, A., & Tewari, M. (2014). *The impact of temperature on productivity and labor supply: Evidence from Indian manufacturing*. Discussion Paper in Economics 14–10. Indian Statistical Institute, Economics and Planning Unit.

Soros, G. (2003). *The alchemy of finance*. Wiley Finance.

Soros, G. (2009). *The crash of 2008 and with is means*. Public Affairs.

Soros, G. (2013). Fallibility, reflexivity and the human uncertainty principle. *Journal of Economic Methodology, 20*, 309–329. https://doi.org/10.1080/1350178X.2013.859415

Soskice, D. (1978). Strike waves and wage explosions, 1968–1970: An economic interpretation. In C. Crouch & A. Pizzorno (Eds.), *The resurgence of class conflict in Western Europe since 1968* (pp. 221–245). Palgrave Macmillan. https://doi.org/10.1007/978-1-349-03025-5_9

Sraffa, P. (1960). *Production of commodities by means of commodities*. Cambridge University Press.

Stiglitz, J. (1998). The private uses of public interests: Incentives and institutions. *Journal of Economic Perspectives, 12*, 3–22. https://doi.org/10.1257/jep.12.2.3

Stroebe, W., & Frey, B. S. (1982). Self-interest and collective action: The economics and psychology of public goods. *British Journal of Social Psychology, 21*(2), 121–137. https://doi.org/10.1111/j.2044-8309.1982.tb00521.x

Täuber, S., van Zomeren, M., & Kutlaca, M. (2015). Should the moral core of climate issues be emphasized or downplayed in public discourse? Three ways to successfully manage the double-edged sword of moral communication. *Climate Change, 130*(3), 453–464. https://doi.org/10.1007/s10584-014-1200-6

The World Bank. (2015). *Green bonds attract private sector climate finance*. World Bank Brief.

The World Economic Forum 2015 Report. (2015). Davos: World Economic Forum.

Trinks, A., Mulder, M., & Scholtens, B. (2017). *Greenhouse gas emissions intensity and the cost of capital*. https://pdfs.semanticscholar.org/c8f5/d6b192df64cde5f578f100d7fb2a6b40515b.pdf

Toledo, P., & Duncan, R. (2024). Forecasting food price inflation during global crises. *Journal of Forecasting, 43*(4), 1087–1113.

Tversky, A., & Shafir, E. (1992). Choice under conflict: The dynamics of deferred decision. *Psychological Science, 3*(6), 358–361. https://doi.org/10.1111/j.1467-9280.1992.tb00047.x

Tyler, T. R., & De Cremer, D. (2006). How do we promote cooperation in groups, organizations, and societies? In P. A. M. van Lange (Ed.), *Bridging social psychology: Benefits of transdisciplinary approaches* (pp. 427–433). Lawrence Erlbaum.

United Nations Report on the World Social Situation. (2011). *The global social crisis*. Report ST/ESA/334. New York: United Nations Department of Economic and Social Affairs.

Wenger, A., Orttung, R. W., & Perovic, J. (2009). *Energy and the transformation of international relations: Toward a new producer-consumer framework*. Oxford University Press.

World Bank 2015 Report. (2015). Washington, DC: World Bank.

World Bank Group Migration and Development Brief 26. (2016, April). *Migration and remittances: Recent development and outlook*. Washington, DC: International Bank for Reconstruction and Development, World Bank Group.

World Investment Report. (2015). *Reforming international investment governance*. United Nations Conference on Trade and Development Report. United Nations.

Yarlik, M. V., Filis, G., Lloyd, T., & Degiannakis, St. (2018, May 24–25). Forecasting oil price volatility: The role of mixed-frequency data (MIDAS) model. In *6th International Symposium on Environment Energy & Finance Issues, ISEFI 2018*, Paris, France.

Zhang, D. (2018, May 24–25). Upgrading China's energy structure: The role of financial markets. In *6th International Symposium on Environment Energy & Finance Issues, ISEFI 2018*, Paris, France.

Zhou, S. L., Smulders, S., & Gerlagh, R. (2018, May 24–25). Closing the loop in a circular economy: Saving resources or suffocating innovations? In *6th International Symposium on Environment Energy & Finance Issues, ISEFI 2018*, Paris, France.

CHAPTER 12

Future Climate Wealth of Nations' Winners and Losers

12.1 Introduction

Never before in the history of humankind have environmental concerns in the wake of economic growth heralded governance predicaments as we face today (Centeno et al., 2013; Chichilnisky, 2007, 2010; Nordhaus, 1994). Climate change presents societal, international and intergenerational fairness as challenge for modern economies and contemporary democracies (Bürgenmeier, 1994; Chichilnisky & Bal, 2019). In today's climate change mitigation and adaptation efforts, high- and low-income households, developed and underdeveloped countries and overlapping generations are affected differently (Chichilnisky & Bal, 2019). Based on insights from the current endeavor to finance the post-COP21 Paris agreement action plan, the *Mapping Climate Justice Project* proposes a three-dimensional climate justice approach in order to find a universally fair climate strategy (Puaschunder, 2018c).

Mapping Climate Justice elucidates international climate regimes around the world based on geographic, technological, socio-economic and political factors. An interactive graphic solution highlighting different countries' climate stability efforts targets at finding ways to share the benefits and burdens of climate change equally *within society*, *between countries* and *over time*.[1]

[1] https://juliampuaschunder.com/mapping-climate-justice/

J. Puaschunder, *Governance & Climate Justice*,
https://doi.org/10.1007/978-3-031-87971-5_12

Comparing international climate mitigation and adaptation as well as unraveling complex interdependencies is aimed at helping to protect vulnerable communities from variegated climate change risks while opening ways for all to benefit from the upsides of a warming earth (Chichilnisky & Bal, 2019; Chichilnisky & Sheeran, 2018). In this chapter, economic and legal analyses unravel how to allocate the burden of climate protection in a fair manner and ensure humankind to feel a fair solution to climate justice is pursued. Climate flexibility will be introduced as the range of temperatures becoming the source of future wealth of nations. The more varied temperature zones a country offers for agriculture, industry and service production, the more likely a territory will be able to provide trade opportunities. Finally, scarcity in light of running out temperature ranges is discussed as a source of potential price spikes as exhibited in inflated commodity prices. The article ends with a discussion of the presented preliminary model and future perspectives of model extensions.

12.2 Theory

Climate justice within *countries*: In order to finance climate change mitigation and adaptation efforts, a diversified taxation scheme is proposed. To find a fair and just distribution of the burden of climate change, a taxation mix of (1) consumption tax, (2) progressive tax and (3) inheritance tax is recommended. Consumption tax can curb harmful emissions and directly nudge behavior toward sustainability. Yet to place a fair share of the burden of climate change mitigation upon society, these taxes have to be adjusted to the individual disposable income in order to not heavier charge low-income households. Retroactive taxation of past wealth accumulation at the expense of environmental damage can be enacted through a corporate inheritance tax. Industries should be taxed, when a merger or acquisition or a board member change occurs, in order to reap benefits from past wealth accumulation that potentially caused carbon emissions.

Climate justice between *countries*: Following the introduction of the gains from climate change (Puaschunder, 2017a, 2017d), the *Climate Justice in the 21st Century* endeavor proposes a model to distribute the benefits of a warming earth in a fair way. Based on legal subsumptions, argumentations of the those countries having better means of protection and conservation of a stable climate, will lead to the pledge of climate change winners having to bear a higher weight of climate stabilization efforts. A macroeconomic cost-benefit analysis thereby aids to find the optimum solution on how to distribute climate change benefits and burden within society and over time.

Contemporary Gross Domestic Product (GDP) serves as basis for economic estimations about the productivity of the agriculture, industry and service sectors around the world in a changing climate. When unidimensionally focusing on estimated GDP growth given a warmer temperature, over all calculated models assuming linear, prospect or hyperbolic gains and losses, the world will be gaining more than losing from a warming earth until the year 2100. Based on the *WL* index of 188 countries of the world, less countries ($n = 78$) will gain more from global warming until 2100 than more countries ($n = 111$) will lose from a warming earth (Puaschunder, 2018a). Graph 12.1 (author's own) holds climate change winners (green and yellow) and losers (orange and red) until 2100 around the world. Green countries are those with most time ahead until reaching a GDP production peak condition by climate, while yellow countries have some time ahead. Orange and red countries will have run out of time by 2100 for GDP production peak. Winning and losing from a warming earth is significantly positively correlated with the Paris COP 21 emissions country percentage of Greenhouse Gas (GHG) for ratification, leading to the conclusion that the countries with the longest time horizon regarding a warming earth lack motivation to mitigate global climate change (Puaschunder, 2017b).

12.2.1 Climate Change Gains

Based on the medium temperature of countries around the world and the optimum temperature for Gross Domestic Product (GDP) production, the world can be divided into climate change gaining and losing territories until the year 2100. Graph 12.2 (author's own) holds those countries that have the most time ahead until optimum temperature for GDP production in green, and those that are around peak condition in yellow.

The climate change winning territories are also outlined in Table 12.1 (author's own) ranked by country name.

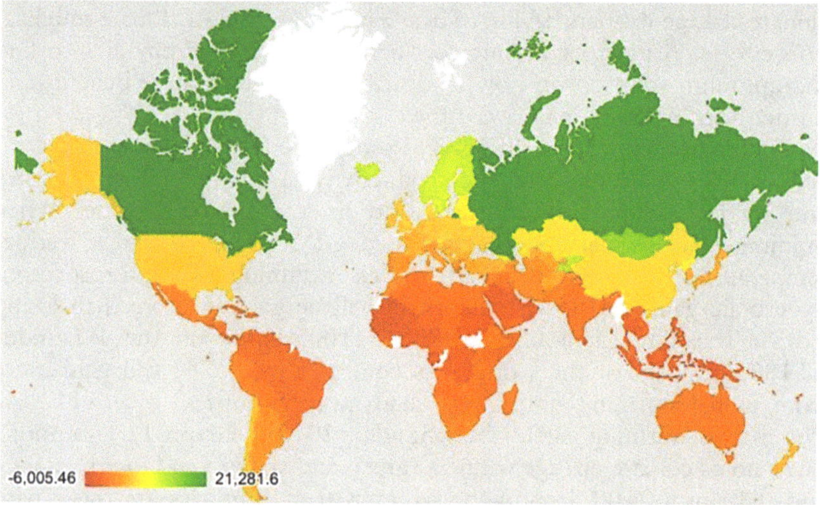

Graph 12.1 World climate change winners (green and yellow) and losers (orange and red) until 2100 (*Source* Author's own)

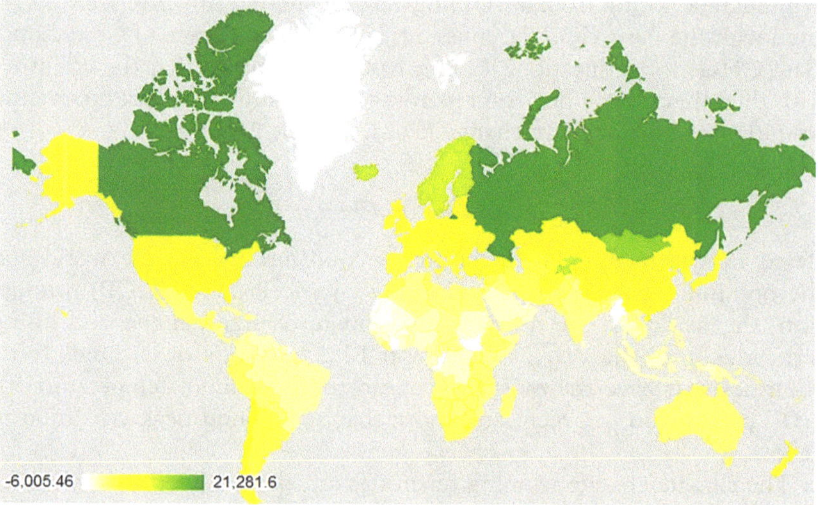

Graph 12.2 Climate change winning territories until the year 2100 (*Source* Author's own)

Table 12.1 Climate change winning territories by country ranked

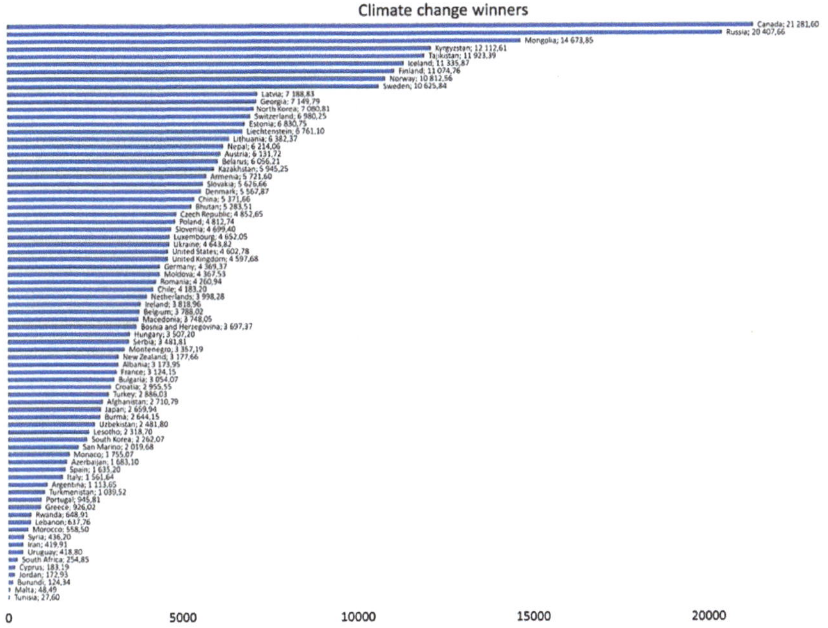

Source Author's own

12.2.2 Climate Change Losses

Based on the medium temperature of countries around the world and the optimum temperature for Gross Domestic Product (GDP) production, the world can be divided into climate change gaining and losing territories until the year 2100. Graph 12.3 (author's own) holds those countries that have passed the peak temperature for GDP production in red.

The climate change losing territories are also outlined in Table 12.2 (author's own) ranked by country name.

The results target at underlining the importance of climate change benefits transfer to offset the losses incurred in the global warming-burdened areas of the world. The philosophical justification is grounded in the climatorial imperative—advocating for the need for fairness in the distribution of the global earth benefits among nations based on Kant's imperative (1788/2003) to only engage in actions one wants to experience being done to oneself. Passive neglect of action on climate mitigation

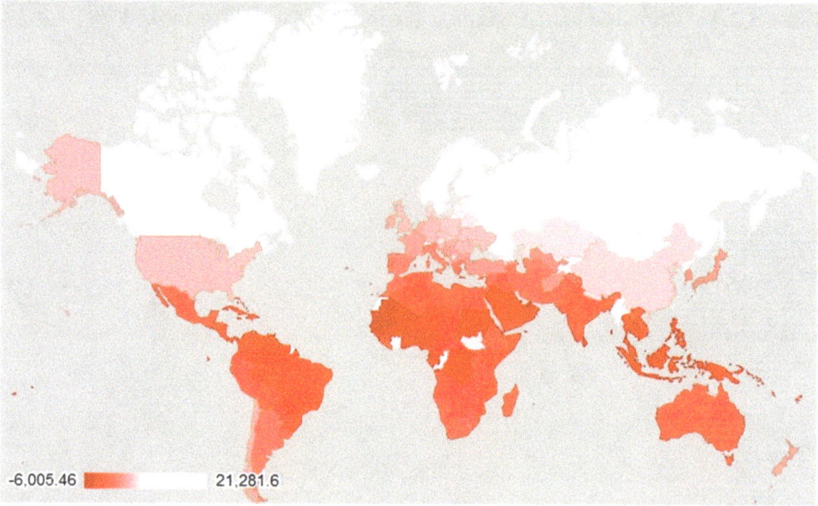

Graph 12.3 Climate change losing territories until the year 2100 (*Source* Author's own)

is viewed as an active injustice to others (Chichilnisky, 1996; Chichilnisky et al., 1998; Puaschunder, 2017b).

12.2.3 Climate Justice Between *Generations: Tax-and-Transfer-Bonds Strategy*

Having found that there are gains from a warming earth demands to transfer benefits into areas of the world that will be primarily losing from climate change (Chichilnisky, 1996, 2016; Chichilnisky et al., 1998; Chichilnisky & Heal, 2000). Having shed light on the gains of a warming earth allows for the redistribution of climate change benefits to those areas of the world that will be losing from climate change. In the implementation, climate change bonds as well as taxation strategies are recommended (Chichilnisky, 1996, 2016).

In order to avoid governmental expenditure on climate change hindering economic growth (Barro, 1990; Chichilnisky, 2007, 2010, 2016), the climate transfers should be enacted through bonds and taxes. The 'Climate in the 21st Century' idea takes these insights into account in order to offer a new way of funding climate change mitigation and

Table 12.2 Climate change losing territories by country ranked

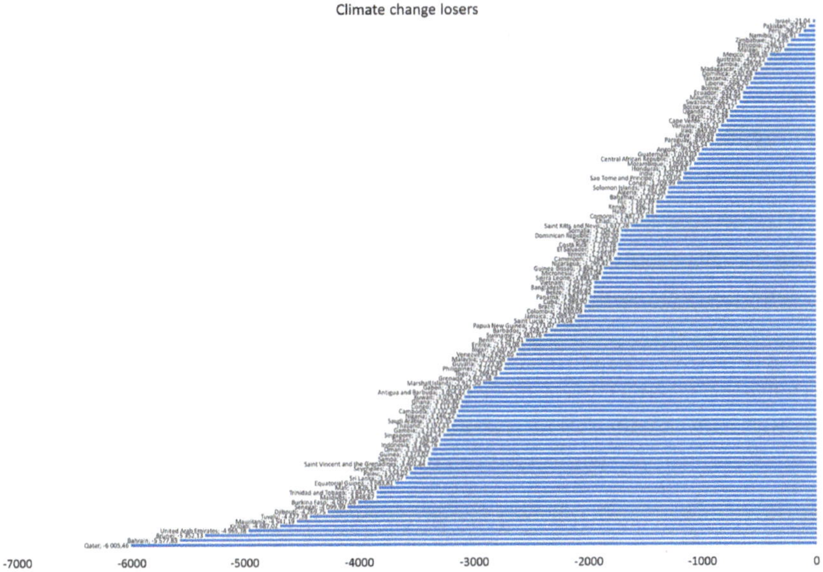

Source Author's own

adaptation policies as well as the transition to renewable energy through a broad-based climate stability bonds-and-taxation mix that also involves future generations (World Bank 2015 Report, 2015).

In order to finance climate change abatement, climate bonds could subsidize the current world industry for transitioning to green solutions. Sharing the costs of climate stabilization between and across generations is a Pareto-optimal strategy to immediately instigate climate action without curbing today's economic growth potentials (Chichilnisky et al., 1998; Chichilnisky & Heal, 2000; Chichilnisky & Sheeran, 2018).

Jeffrey Sachs (2014) proposed an intergenerational burden-sharing idea by presenting a model featuring fiscal policy with bond issuing. In an overlapping-generations-type model, future research should elucidate climate change abatement and mitigation policies to lead to a fairer solution across generations. The implementation pays tribute to contemporary finance and growth models with respect to maximizing utility of the model. The current generation thereby mitigates climate change and provides infrastructure against climate risk financed through climate bonds to be paid by future generations. Since for future generations the currently created externalities from economic activities—the effects of CO_2 emissions—are removed, this entails that the current generations remain financially as well off as without mitigation while improving environmental well-being of future generations. As Sachs (2014) shows, this intergenerational tax-and-transfer policy turns climate change mitigation and adaptation policy into a Pareto-improving strategy. Shifting the costs for climate abatement to the recipients of the benefits of climate stability appears as novel, feasible and easily implementable solution to nudge many overlapping generations toward future-oriented loss aversion in the sustainability domain (Puaschunder, 2016).

The respective bonds-and-tax climate stability financing strategy therefore proposes to bear the burden of climate in a right, just and fair way around the globe (Chichilnisky, 1996, 2007, 2010, 2016). In the climate change winner countries, taxation should become the main driver over financing climate stability strategies. Foremost, the industries winning from a warming climate should be taxed. The Winner-Loser index is based on the cardinal temperatures for all GDP-contributing sectors. Based on the cardinal temperatures for the three GDP components, agriculture, industry and service, the taxation should be enacted for those sectors having most time ahead. The rational is that these sectors will be gaining the most from a warming earth and will therefore be flourishing.

The taxation models should aid to share the burden of climate change within society in a fair way. Regarding concrete climate taxation strategies, a carbon tax on top of the existing tax system should be used to reduce the burden of climate change and encourage economic growth through subsidies (Chancel & Piketty, 2015). Within a country, high- and low-income households should face the same burden of climate stabilization adjusted for their disposable income. First, climate justice within a country should pay tribute to the fact that low- and high-income households share the same burden proportional to their dispensable income, for instance, enabled through a progressive carbon taxation. Those who caused climate change could be regulated to bear a higher cost through carbon tax in combination with retroactive billing through inheritance tax. But also developed and underdeveloped countries as well as various overlapping generations are affected differently. Besides progressive taxation schemes to imbue a sense of fairness in climate change burden sharing, inheritance taxation is also a flexible means to reap past wealth accumulation, which potentially caused environmental damage. The burden of climate change mitigation and adaptation could therefore also be allocated in a fair way within society through contemporary corporate inheritance tax in order to reap benefits of past wealth accumulation. But in order to determine fair corporate inheritance tax grantors and beneficiaries, one needs information on what countries and what sectors of the economy have a rising or declining economic prospect in light of climate change (Chichilnisky, 1996, 2007).

In addition, finding the optimum balance between consumption tax adjusted for disposable income through a progressive tax scheme will aid to unravel drivers of tax compliance in the sustainability domain. If climate taxation is perceived as fair and just allocation of the climate burden, this could convince tax payers to pay one's share. A novel 'service-and-client' atmosphere could promote taxpayers as cooperative citizens who are willing to comply if they feel their share as fair contribution to the environment (Puaschunder, 2015). Taxpayers as cooperative citizens would then be willing to comply voluntarily following the greater goal to promote taxpayer collaboration and enhance tax morale in the environmental domain. International comparisons of tax behavior also reveal tax norms being related to different stages of institutional development of the government, which is an essential consideration in sharing the climate change burden in a fair manner between countries. A completely novel approach is to shed light on the benefits of a warming earth in order to derive fair climate gains distribution strategies around the world (Puaschunder, 2017a, 2017c, 2017e).

In climate change winner countries weighted by GDP per capita (Graph 12.4, author's own, Green & Yellow), taxation should become the main climate stability financialization strategy. Foremost, the industries winning from a warming climate should be taxed. Regarding concrete climate taxation strategies, a carbon tax on top of the existing tax system should be used to reduce the burden of climate change and encourage economic growth through subsidies. Within a country, high- and low-income households should face the same burden of climate stabilization adjusted for their disposable income. Finding the optimum balance between consumption tax adjusted for disposable income through a progressive tax scheme will aid to unravel drivers of tax compliance in the sustainability domain.

Governments in global warming loser countries weighted by GDP per capita (Graph 12.4, author's own, Orange & Red) should receive tax transfers in the present from the winning countries. The climate change

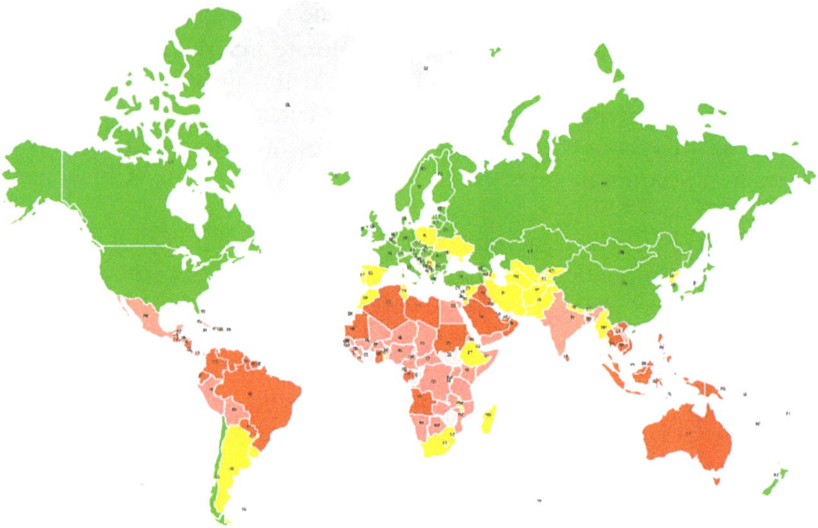

Graph 12.4 Climate stabilization transfers grantors (green and yellow) and beneficiaries (orange and red) until 2100 (*Source* Author's own)

loser countries should also borrow by loans or by issuing of bonds to be paid back by future generations. Taxing future generations is justified as for the assumed preferences of future generations to avoid higher costs of climate change long-term damages and environmental irreversible lock-ins.

To conclude, climate change winning countries are advised to use taxation of the gaining GDP sectors to raise revenues to offset the losses incurred by climate change (Chichilnisky, 2010, 2016). Climate change losers should issue bonds to be paid back by taxing future generations. Tax-and-bonds transfers could be used to subsidize industry actors for choosing clean energy in order to shift the general race-to-the-bottom price-cutting behavior to a race-to-the-top hunt for subsidies for employing clean energy.

12.2.4 Future Climate Wealth of Nations

This chapter builds on the established climate change winning and losing prospects but adds a novel perspective on climate flexibility as the range of different temperatures a country can offer to produce in. Climate flexibility is captured around the globe, and production ranges analyzed based on flexibility of agricultural crops production. A climate-finance nexus is introduced based on expectations of scarcity driving up prices and inflating value of goods that may become scarce in a global warming scenario (Puaschunder, 2019). Lastly, this theoretical model will be backtested on actual commodity market prices.

12.3 Method

12.3.1 Overall Model Assumptions

12.3.1.1 Study 1: Climate Change-Induced Market Flows

Most recently, a trend has been become apparent to demand a shift in focus from the poor to the rich in climate economics (Otto, Kim, Dubrovsky, & Lucht, 2019). Around 2016 with a model was built on GDP optimal cardinal temperature for production (Puaschunder, 2016). The GDP pillars of agriculture, industry and service are the foundation of the model. The model divides the world nation-states into climate change winning and losing territories. This is in accordance with research of the International Monetary Fund (IMF) research division. Contemporary IMF research concludes that in low-temperature countries, an increase in temperature boosts economic activity, while in higher temperature countries, an increase in temperature will dampen economic activity (Acevedo et al., 2018; Chichilnisky, 1996, 2007, 2010, 2016; Chichilnisky & Bal, 2019). The findings of uneven economic effects of climate change around the globe are robust over the different models and backed by regional cluster analysis and geographic case studies (Burke et al., 2015; Chichilnisky, 2010, 2016; Somanathan, Somanathan, Sudarshan, & Tewari, 2014; Topalova, in speech, 2018). The IMF model now sophisticates these findings with insights about the effect of temperature on productivity, capital and labor (Topalova, in speech, 2018).

In order to consolidate the finding of unequal climate change gains and losses and derive implications for economic and financial market performance, three market-focused hypotheses tested whether climate change winning and losing territories attract different levels of human capital inflow, finance inflow and Internet connectivity.

Overall geographical mobility is expected to increase, especially when massive shifts of the industry structure result from green industrial policies. This may be the case, for example, when shifting from fossil to renewable energy: Fossil energy plants are usually located next to energy-intensive industries and cities, whereas renewable power plants are located where the respective power resources are abundant. Shifting energy systems will therefore affect spatial distribution and workers are therefore expected to move physically (Chichilnisky, 2007, 2010; Chichilnisky et al., 1998; Chichilnisky & Bal, 2019).

Hypothesis H1 investigates if countries that economically benefit from climate change feature a significant foreign human capital inflow.

Hypothesis H2 investigates if countries that economically benefit from climate change feature a significant foreign capital inflow that does not correspond with remittance outflow from the FDI receiving territories into the FDI sponsoring countries.

Hypothesis H3 investigates if countries that economically benefit from climate change feature a significant higher level of Internet connectivity.

12.3.1.2 Study 2: Climate Flexibility

A model extension of the original calculation of climate change winners and losers around the world based on GDP prospects focuses on agriculture, industry and service sectors representing economic growth factors. Methodologically, the optimum cardinal temperatures for specific agriculture products, industry representations and service sector providers were retrieved from literature sources. A future market prospect in light of global temperature rising was then given based on the mean temperature of the world now and as estimated in the year 2100. Prospective market shifts in the GDP composition but also agricultural products, industry sector representatives and service providers were provided. Underlying to all these future prospects is the idea that GDP production and its manifold components are enabled at best at a certain temperature, and therefore specific markets will follow and gravitate toward these temperatures.

In light of global warming, an influx of labor and mobility of corporations are predicted. These insights fortify the idea that the world will become more globalized and location-flexible due to climate change. Those corporations and individuals realizing this and seeking the gains of a warming globe—maybe beyond their own inhabited territories—are prospected to become market winners.

12.3.1.3 Study 3: Climate Change Expectations Driving Commodity Bubbles

Since the turn of the millennium, the United Nations (UN) detected an unprecedented agriculture products commodities bubble trend—especially in countries around the equator belt, which are the worst off regarding climate change due to a natural high base mean temperature (United Nations Report on the World Social Situation, 2011). There also seems to be a recent and unprecedented credit risk and bubble behavior in the corporate energy sector—such as oil-supply shocks and stock prices—that may be related to the demand for a transition to renewable energy (Figuerola-Ferretti & Cervera, 2018; Kaabia et al., 2018; Lips, 2018). In the oil energy sector, price spikes have been found around procurement shortages, which are expected to be occurring in all climate-related extinction cases (Roberts et al., 2018). Research also already pits food versus fuel and finds evidence of a contemporary predicament in nation-state's investment allocations in either one resource (Filip et al., 2018; Trinks et al., 2017). There is also already evidence for sector competition for energy resources in a carbon-constrained hoteling model (Renz, 2018). This may only be a first sign of agricultural products becoming more precious and valuable for economic production given their expected lower durability in a warming climate (Wenger et al., 2009). Energy prices and mitigation policies also impact on Foreign Direct Investment (FDI) locations (Saussay & Sato, 2018; The World Economic Forum 2015 Report, 2015; World Investment Report, 2015).

The following financial market model now estimates prices centered around agriculture, industry and service sector transitions. Based on evidence from the energy sector, the relation of a price of a good and time to extinction of a resource is not linear but hyperbolically tilted toward the end before extinction (Roberts et al., 2018). When a resource is about to become extinct, the market value of this resource will in general rise shortly before the good goes away from the market forever. For instance, if we run out of a natural resource, no alternatives may exist, so scarcity will drive up the price shortly before the good extinguishes (Roberts et al., 2018; Yarlik et al., 2018; Zhang, 2018). Derived from this insight, the model will be based on the maximum temperature for agricultural goods and industry production as well as time for extinction of natural resources, cooling and storage maximum temperatures in order to determine market transition and shifting trends but also price spikes and peaks based on the world and regional temperatures.

12.4 Operationalization

12.4.1 *Data Study 1: Climate Change-Induced Market Flows*

In order to operationalize and consolidate the underlying premise that countries that economically benefit from climate change feature better market prospects, worldwide Foreign Direct Investment (FDI) flows[2] (Hypothesis H1), remittances[3] (Hypothesis H2) and Internet connectivity[4] (Hypothesis H3) were retrieved online form the World Bank database for the year 2015 as well as the *Global Connectivity Index* homepage as an indicator of market relevance and innovation potential.

12.4.2 *Results Study 1: Climate Change-Induced Market Flows*

12.4.2.1 Climate Change Winners and Losers Model

The relation of climate change gains and losses and market prospects was investigated. Based on the climate change winners and losers model (Puaschunder, 2018c), gaining and losing from climate change was related to human migration, financial inflows as well as innovation.

Human migration: When investigating the relation of human migration and climate change winners and losers in 187 countries of the world, a highly significant correlation ($r_{\text{Pearson}}(187) = 0.184$, $p < 0.006$) is found between being climate change winner and human migration inflows. A highly significant independent t-test with $t(185) = -2.401$, $p = 0.017$ of 110 climate change loser and 77 winner countries consolidates the finding.

Financial inflows: When investigating the relation of financial inflows and climate change winners and losers including 181 countries of the world, a significant correlation ($r_{\text{Pearson}}(181) = 0.180$, $p < 0.016$) is found between being a climate change winner and FDI inflows. A highly significant independent t-test with $t(179) = 2.680$, $p = 0.008$ of 110

[2] http://data.worldbank.org/indicator/BX.KLT.DINV.CD.WD.

[3] http://data.worldbank.org/indicator/BX.TRF.PWKR.CD.DT.

[4] http://www.huawei.com/minisite/gci/en/country-rankings.html.

climate change loser and 77 winner countries consolidates the finding. As a cross-validation of lasting financial inflows into climate change winner countries, a non-significant correlation and non-significant independent t-tests are reported.

Innovation: When investigating the innovation potential currently prevalent in climate change winners and losers including 79 countries of the world, a significant correlation ($r_{\text{Pearson}}(79) = 0.507$, $p < 0.000$) is found between being climate change winner and global connectivity as measured by the 2018 *Global Connectivity Index*.[5] A highly significant independent t-test with $t(77) = 6.306$, $p = 0.000$ of 33 climate change loser and 46 winner countries consolidates the finding.

12.4.3 Data Study 2: Climate Flexibility

Climate change was recently found to affect countries differently. Given different mean temperatures and GDP compositions, countries around the globe will be affected differently by a warming climate (Puaschunder, 2018c). In the determination into climate change gain and loss prospects, climate flexibility needs to be integrated. Climate flexibility is defined as the leeway countries have in coping with a changing climate due to a range of climate zones prevalent in their territory. Climate flexibility can be grounded on the relative latitude and altitude of countries around the globe. The wider the range of latitude and altitude within a nation-state, the more climate flexibility and favorable economic degrees of freedom for multiple production peaks are assumed. When dividing the world given temperature ranges prevalent calculated based on the *List of Countries by Extreme Temperatures*,[6] the following temperature ranges are found, as exhibited in Table 12.3 (author's own).

12.4.4 Results Study 2: Climate Flexibility

Table 12.3 (author's own) exhibits 87 countries' temperature range in comparison to each other. One-third of the sample features countries with the widest temperature range and therefore most flexible in green color, one-third of the sample features medium temperature range of

[5] http://www.huawei.com/minisite/gci/en/country-rankings.html

[6] https://en.wikipedia.org/wiki/List_of_countries_by_extreme_temperatures.

Table 12.3 Temperature range (in Celsius) by country determining climate flexibility

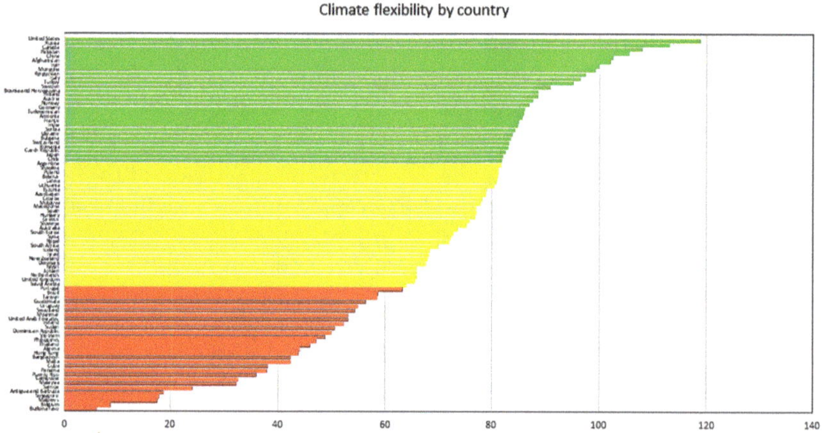

Source Author's own

climate flexibility in yellow color and one-third of the sample with lowest temperature range and relative climate inflexibility in red color.

The USA leads on climate flexibility with a temperature range of $r = 118.9$°C based on [−62.2 having been the lowest temperature ever recorded in the US and 56.7°C the highest temperature ever recorded in the US], followed by Russia $r = 113.2$°C based on [−67.8; 45.4°C], Canada $r = 108$°C based on [−63; 45°C], Pakistan $r = 105.5$°C based on [−52; 53.5°C], China $r = 102.6$°C based on [−52.3; 50.3°C], Afghanistan $r = 102.1$°C based on [−52.2; 49.9°C], Iran $r = 100$°C based on [−46; 54°C], Mongolia $r = 99.3$°C based on [−55.3; 44°C], Kyrgyzstan $r = 97.6$°C based on [−53.6; 44°C], Italy $r = 96.6$°C based on [−49.6; 47°C], Turkey $r = 95.2$°C based on [−46.4; 48.8°C], Sweden $r = 91$°C based on [−53; 38°C], Bosnia and Herzegovina $r = 88.7$°C based on [−42.5; 46.2°C], Finland $r = 88.7$°C based on [−51.5; 37.2°C], Austria $r = 87.6$°C based on [−47.1; 40.5°C], Norway $r = 87$°C based on [−51.4; 35.6°C], Germany $r = 86.2$°C based on [−45.9; 40.3°C], Turkmenistan $r = 86.1$°C based on [−36; 50.1°C], Armenia $r = 85.7$°C based on [−42; 43.7°C], France $r = 85.1$°C based on [−41; 44.1°C], India $r = 84.9$°C based on [−33.9; 51°C], Serbia $r = 84.4$°C based on [−39.5; 44.9°C], Ukraine $r = 83.9$°C based on [−41.9;

42°C], Bulgaria r = 83.5°C based on [−38.3; 45.2°C], Switzerland r = 83.3°C based on [−41.8; 41.5°C], Romania r = 83°C based on [−38.5; 44.5°C], Czech Republic r = 82.6°C based on [−42.2; 40.4°C], Japan r = 82°C based on [−41; 41°C] and Chile r = 81.9°C based on [−37; 44.9°C].

The countries in the medium climate flexibility range are Argentina with r = 81.7°C based on [−32.8; 48.9°C], Slovakia r = 81.3°C based on [−41; 40.3°C], Poland r = 81.2°C based on [−41; 40.2°C], Belarus r = 81.1°C based on [−42.2; 38.9°C], Latvia r = 81°C based on [−43.2; 37.8°C], Lithuania r = 80.4°C based on [−42.9; 37.5°C], Estonia r = 79.1°C based on [−43.5; 35.6°C], Azerbaijan r = 79°C based on [−33; 46°C], Croatia r = 78.3°C based on [−35.5; 42.8°C], Moldova r = 77.9°C based on [−35.5; 42.4°C], Macedonia r = 77.2°C based on [−31.5; 45.7°C], Spain r = 77.2°C based on [−30; 47.2°C], Hungary r = 76.9°C based on [−35; 41.9°C], Greece r = 75.8°C based on [−27.8; 48°C], Slovenia r = 75.3°C based on [−34.5; 40.8°C], Australia r = 73.7°C based on [−23; 50.7°C], South Korea r = 72.8°C based on [−32.8; 40°C], Syria r = 72.4°C based on [−23; 49.4°C], Nepal r = 72°C based on [−26; 46°C], South Africa r = 70.1°C based on [−20.1; 50°C], Iceland r = 68.4°C based on [−37.9; 30.5°C], Israel r = 68.2°C based on [−14.2; 54°C], New Zealand r = 68°C based on [−25.6; 42.4°C], Denmark r = 67.6°C based on [−31.2; 36.4°C], Egypt r = 66°C based on [−15; 51°C], Jordan r = 66°C based on [−16; 50°C], the Netherlands r = 66°C based on [−27.4; 38.6°C], United Kingdom r = 65.7°C based on [−27.2; 38.5°C] and Saudi Arabia r = 64°C based on [−12; 52°C].

The countries in the least climate flexibility are Portugal r = 63.4°C based on [−16; 47.4°C], Brazil r = 58.7°C based on [−14; 44.7°C], Taiwan r = 58.6°C based on [−18.4; 40.2°C], Guatemala r = 56.5°C based on [−11.5; 45°C], Uruguay r = 55°C based on [−11; 44°C], Swaziland r = 54.4°C based on [−6.7; 47.7°C], Myanmar r = 53.2°C based on [−6; 47.2°C], United Arab Emirates r = 53.1°C based on [−1; 52.1°C], Ireland r = 52.4°C based on [−19.1; 33.3°C], Sudan r = 50.7°C based on [−1; 49.7°C], Dominican Republic r = 50°C based on [−7; 43°C], Vietnam r = 48.8°C based on [−6.1; 42.7°C], Philippines r = 47.2°C based on [−5; 42.2°C], Thailand r = 46°C based on [−1.4; 44.6°C], Algeria r = 44.1°C based on [7.2; 51.3°C], Hong Kong r = 43.9°C based on [−6; 37.9°C], Bangladesh r = 42.5°C based on [2.6; 45.1°C], Malta r = 42.4°C based on [1.4; 43.8°C], Cuba r = 38.2°C based on [0.6; 38.8°C], Panama $r = 38$°C based on [2; 40°C], Puerto Rico $r = 36$°C based on [4; 40°C], Cambodia $r = 32.6$°C based on [10; 42.6°C], Malaysia $r = 32.3$°C based on [7.8; 40.1°C], Samoa $r = 24.2$°C based on [11.1; 35.3°C], Antigua and Barbuda $r = 18.8$°C based on [16.1; 34.9°C], Singapore $r = 18$°C based on [19; 37°C], Maldives $r = 17.7$°C based on [17.2; 34.9°C], Belgium $r = 8.7$°C based on [30.1; 38.8°C] and Burkina Faso $r = 6.2$°C based on [41; 47.2°C].

Climate flexibility is prospected to open a gate for productivity. Countries with climate flexibility are prone to be productive grounds for agriculture, industry and service production at all different temperature levels.

12.4.4.1 Climate Flexibility Model

The relation of climate flexibility and market prospects was investigated. Based on the determined climate flexibility, the range of temperature variability—as retrieved and calculated based on the *List of Countries by Extreme Temperatures*[7]—was related to human migration.

Human migration: When investigating the relation of human migration and climate flexibility in 86 countries of the world, a highly significant correlation ($r_{\text{Pearson}}(86) = 0.319$, $p < 0.002$), is found between being climate flexible and human migration inflows. No significant relation is reported for financial inflows as well as innovation.

12.4.5 Future Climate Wealth of Nations' Winners and Losers

The insights gained from the climate change winners and losers model and the climate flexibility model demand to be blended into a Future Climate Wealth of Nations' Winners and Losers model. This new model will pay attention to the GDP prospects of countries around the globe in a changing temperature as well as the climate flexibility based on individual country's latitude and altitude naturally determining extreme temperatures. The novel Future Climate Wealth of Nations' Winners and Losers ($W_x WL_{TT}$) Index derives from the factoring of the Climate Change Winner and Loser (WL_{TT}) Index, as defined by the climate flexibility index (W_x). The Future Climate Wealth of Nations' Winners and Losers

[7] https://en.wikipedia.org/wiki/List_of_countries_by_extreme_temperatures.

($W_x WL_{TT}$) Index is calculated based on formula (12.1).

$$W_x WL_{TT} = W_x \times WL_{TT} \tag{12.1}$$

If applying the new $W_x WL_{TT}$ Index to a world dataset, Graph 12.5 (author's own) reveals as dark green colored countries those nation-states that can expect to be climate flexible climate change winning territories, whereas the light green colored countries are the climate inflexible climate change losing territories (light green).

Limitations of this model may stem from the different time perspectives as climate change winners and losers are projections in 2100 and climate flexibility is measured on past temperature recordings. In addition, the temperature extremes should be controlled for outliers and over time. Temperature fluctuations should be separated from special temperature differences.

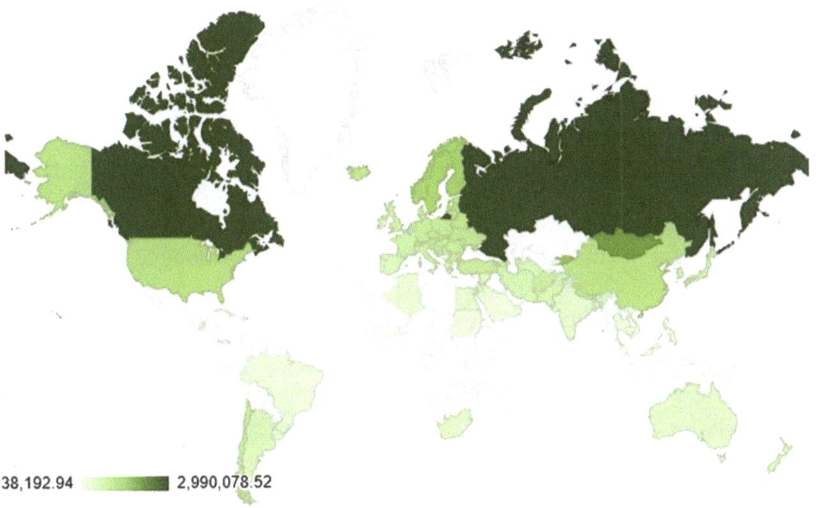

Graph 12.5 Future climate wealth of nations' winners and losers (*Source* Author's own)

12.4.6 *Data Study 3: Climate Change Expectations Driving Commodity Bubbles*

Recent research has defined climate change winners and losers by GDP prospects in a changing climate (Puaschunder, 2018c). Now by stepping aside from national territories, climate change winning industries and sectors were unraveled.

12.4.7 *Results Study 3: Climate Change Expectations Driving Commodity Bubbles*

For the GDP comprising pillars of agriculture, industry and service sectors, the peak temperatures for production were derived from a meta-analysis of literature sources. For agriculture, crop durability—meaning how long crops can prosper in a changing climate—was determined based on literature resources. Graph 12.6 (author's own) holds crops and their germination temperature, peak performance temperature and maximum possible temperature.

The temperature ranges are outlined in Graph 12.7 (author's own), which holds the crop durability. Longest temperature range crops are indicated by green color, medium temperature range in yellow color and shortest temperature range in red color.

Beans, pak choi, peas, cabbage, carrots, red beet and milk thistle will have the widest application range—as outlined in green color—regarding temperature. Onions, ryegrass, Chinese cabbage, Asian radish, corn, parsnips and cold-season crops such as oats, rye and wheat have a medium flexibility range—as exhibited in yellow color—when it comes

Graph 12.6 Agriculture sector temperature-based condition in relation to world mean temperature (*Source* Author's own)

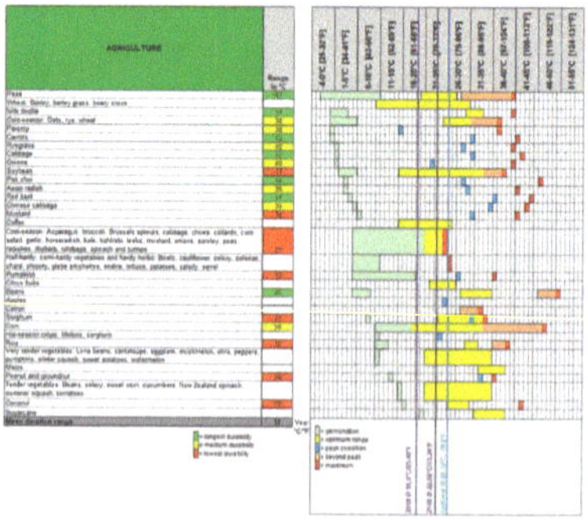

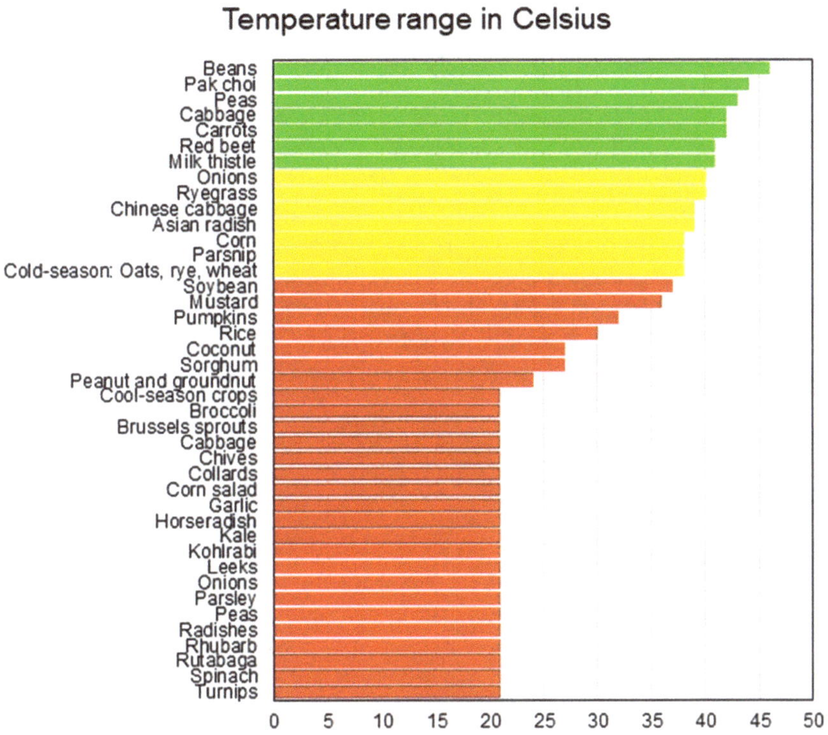

Graph 12.7 Temperature ranges for outlining agriculture durability (*Source* Author's own)

to temperature. Soybean, mustard, pumpkins, rice, coconut, sorghum, peanut and groundnut but also all the cool-season crops such as broccoli, Brussels sprouts, cabbage, chives, collards, corn salad, garlic, horseradish, kale, kohlrabi, leeks, onions, parsley, peas, radishes, rhubarb, rutabaga, spinach and turnips will run out of time first, as outlined in red coloring, as they represent the smallest temperature range. The latter's cultivation is expected to become the most expensive first, then the medium and then the longest temperature-ranged crops.

Given the assumption of price increases around scarcity, a model for crops' extinction and price mechanism is sketched out (Figuerola-Ferretti & Cervera, 2018; Kaabia et al., 2018; Lips, 2018). In the oil energy sector, price spikes have been found around procurement shortages, which are expected to be occurring in all climate-related extinction cases (Roberts et al., 2018).

Time series of commodity price indices from June 1989 to 2017 retrieved from Index Mundi[8]allowed to determine the price development of commodity prices for food and beverages. Since the turn of the millennium, temperature records have skyrocketed, which fueled vast communication about global warming. Price expectations are therefore assumed to have spiraled a commodities bubble for food and beverages. As outlined in Graph 12.8 (author's own), commodity prices for food as represented by barley, Canadian wheat, maize/corn, rice, soft red winter wheat and sorghum have risen extraordinarily since 2000.

As outlined in Graph 12.9 (author's own), commodity prices for beverages as represented by Arabica coffee as well as tea have risen extraordinarily since 2000.

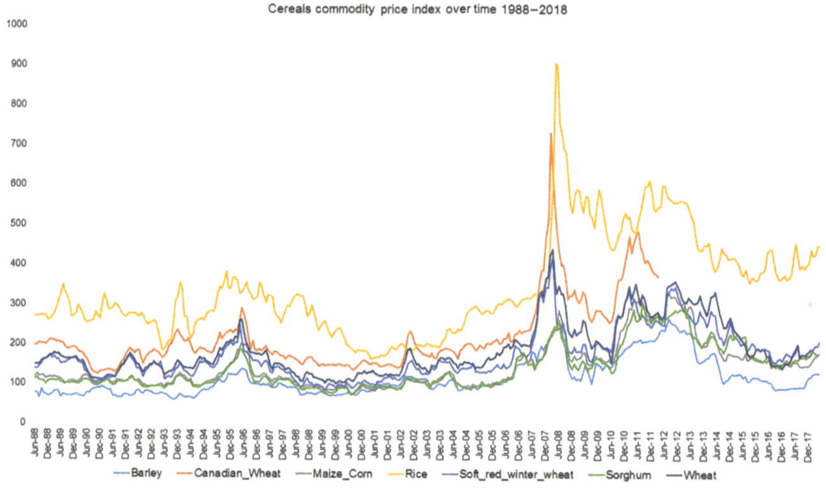

Graph 12.8 Commodity price index retrieved from Index Mundi for food (*Source* Author's own. https://www.indexmundi.com/commodities/?commodity=wheat&months=360)

[8] https://www.indexmundi.com/commodities/?commodity=wheat&months=360.

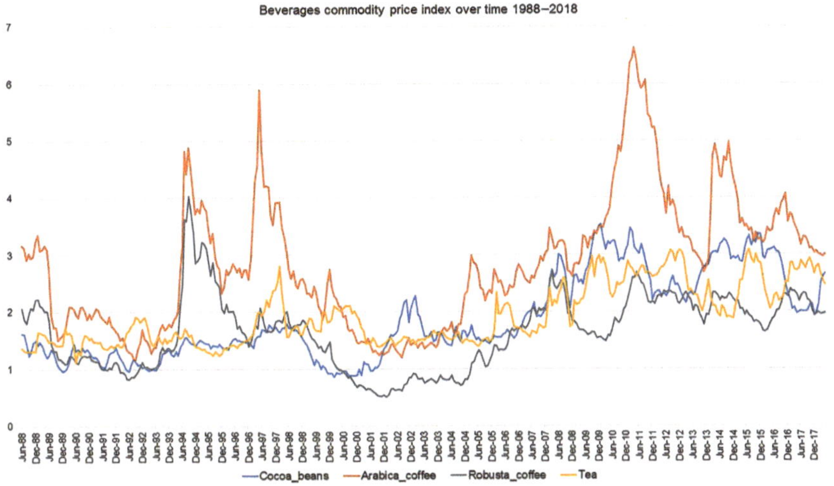

Graph 12.9 Commodity price index retrieved from Index Mundi for food (*Source* Author's own. https://www.indexmundi.com/commodities/?commodity=wheat&months=360)

Highly significant independent t-tests that compare the price index per crop prior to 2000 and thereafter consolidate the extraordinary price rise in commodities for barley with $t(358) = -11.984$, $p = 0.000$ of 139 data points in the previous millennium and 221 observations in the current millennium based on monthly price indices measurement; Canadian wheat with $t(284) = -7.502$, $p = 0.000$ of 139 data points in the previous millennium and 147 observations in the current millennium based on monthly price indices measurement; maize/corn with $t(358) = -8.82$, $p = 0.000$ of 139 data points in the previous millennium and 221 observations in the current millennium based on monthly price indices measurement; rice with $t(358) = -7.331$, $p = 0.000$ of 139 data points in the previous millennium and 221 observations in the current millennium based on monthly price indices measurement; soft red winter wheat with $t(358) = -8.786$, $p = 0.000$ of 139 data points in the previous millennium and 221 observations in the current millennium based on monthly price indices measurement; sorghum with $t(357) = -10.352$, $p = 0.000$ of 139 data points in the previous millennium and 220 observations in the current millennium based on monthly price indices measurement; wheat with $t(355) = -9.209$, $p = 0.000$ of 139 data points in the previous millennium and 218 observations in the current millennium based on monthly price indices measurement; cocoa bean with $t(358) = -13.702$, $p = 0.000$ of 139 data points in the previous millennium and 221 observations in the current millennium based on monthly price indices measurement; Arabica coffee with $t(358) = -3.631$, $p = 0.000$ of 139 data points in the previous millennium and 221 observations in the current millennium based on monthly price indices measurement; and tea with $t(358) = -11.493$, $p = 0.000$ of 139 data points in the previous millennium and 221 observations in the current millennium based on monthly price indices measurement. Out of all the commodity indices for food and beverage reported in the Index Mundi database[9] with sufficient data points, Robusta coffee was the only commodity with non-significant price increases for the period after the turn of the millennium. To cross-validate the findings, price indices for gasoline, medical care, fast food, soda, jewelry, new cars and computers[10] were eyeballed, and no stringent pattern of increased prices could be found for these goods and services since turn of the millennium.

Climate change, wartime and financial turmoil were scrutinized to have an impact on driving up food and nutrition costs. For a worldwide commodities index, the major influence over time since 1994 was found to be climate change (Puaschunder, forthcoming). For the more specific cases of the annexation of the Crimea and Russia's attempt to seize Ukraine, an influence on cereals food prices was found (Puaschunder, forthcoming). In regard to financial crises, the 2008/09 World Financial Recession as well as the COVID-19 economic downturn were found to be events that elevated food prices. Understanding the direct price impact but also the implicit consequences of climate change, political tensions as well as financial crises on food intake opportunities, serves as an example of the widespread effect of environmental conditions, political unrest, as well as financial insecurity within society. The findings aim at elucidating how environmental conditions, external shocks but also system-inherent crises can affect the livelihood and well-being around the world. Studying the direct impact of inflationary pressures is meant to offer ways how to combat rising prices in the food industry at the root cause. At the

9 https://www.indexmundi.com/commodities/?commodity=wheat&months=360

10 https://people.duke.edu/~rnau/411infla.htm.

same time also understanding the indirect implications of world events on nutrition and health offers ways to alleviate inequalities that are particularly hard on marginalized groups. The results find a long-term gradual impact of climate change being related to food price rises around the world but also specific major financial events and political upheaval having an influence of specific food commodities. In their entirety, the results have important ramifications for providing access to food and health around the world. With respect to policy recommendations, different measures are suggested to address food inflation. Policies should focus on diversifying the sources of fertilizer and crude oil imports, which can reduce food price shocks to some extent. Subsidizing more to farmers may help keep domestic prices stable even when input prices increase. The government can also consider setting up strategic food reserves to stabilize food prices and shortages occurring during any such exogenous uncertainty. For long-term policies, policymakers should emphasize diversifying energy dependency both in terms of a few nations and one source of energy. The focus should be to develop alternative sources of energy within the country, which will help reduce such price volatility. To avoid problems such as this, the government should focus on self-sustaining production, i.e., incentivizing farmers to produce those commodities that are price volatile and more import-depending. Such policies have the potential to reduce the price spillover created by sudden exogenous shocks in demand or supply. A forward-looking policy recommendation involves judicious investment in renewable energy. Accelerating investments in renewables enhances energy security and diminishes reliance on geopolitically-sensitive fossil fuels such as gasoline. Governments and policymakers can also implement several measures on the following grounds to maintain the FED's required level of inflation, such as monetary policy adjusting interest rates or controlling the money supply to manage inflation. Central banks can raise interest rates to reduce the money supply to make borrowing more expensive. Fiscal policy involves the government implementing fiscal measures such as taxation and spending policies to address inflation. Supply-side policies include the government, which focuses on increasing the supply of food products to meet rising demand and stabilize prices. Governmental price controls can impose temporary price caps on essential food items to prevent excessive price hikes. Strategic food reserves help stabilize prices during periods of scarcity or disruption in the food supply. Trade policies can promote free and fair trade in agricultural products by reducing tariffs, quotas, and other trade barriers. Agricultural investment and research can help improve farming techniques, develop high-yielding crop varieties, and increase overall food supply chain productivity. Circular economy consumer education and food waste reduction aid with food storage, meal planning, and food waste, all of which contribute to lower demand and more efficient use of food resources. Overall, a comprehensive approach that combines multiple strategies is necessary to address the complex issue of food inflation effectively. The findings consolidate that key system-inherent and external shock events can trigger financial turmoil that materializes in long-term elevated food prices, which warrants for attention to access to food and healthy nutrition during economic downturns. For marginalized communities, direct inequality alleviation strategies may include food stamps as transfer payments and governmental stimulus instead of cash allowances. Since different industries use different business operating models and have different inputs and processes, some industries are more significantly affected by inflation than others are. Further qualitative and quantitative research examining how other geographic areas affect food inflation and determining the best policy to serve food consumer prices would make a significant contribution to this field during our turbulent times. All these endeavors are targeted at ensuring fair access to food and healthy nutrition around the world and within all parts of society. Given the backtesting results consolidating inflated commodity prices since the turn of the millennium, future prospects of price deviations based on the extinction temperature are drawn as stylized, hypothetical estimation. Graph 12.10 (author's own) holds prospective extinction-based market price predictions as deviations from normal prices. As one can see, the world with medium temperature of 18.6°C is not at the efficiency frontier yet (Foley, 2007). In addition, the mean temperature for continents was derived from the literature, and critical mean continental temperature in Central and South America, Africa and Australia put into perspective.[11]

12.5 Results and Discussion

Climate change is reality. The implementation of climate stability accounts for the most challenging contemporary global governance predicament that seems to pit world countries but also today's generation against

[11] https://simple.wikipedia.org/wiki/List_of_countries_by_continents.

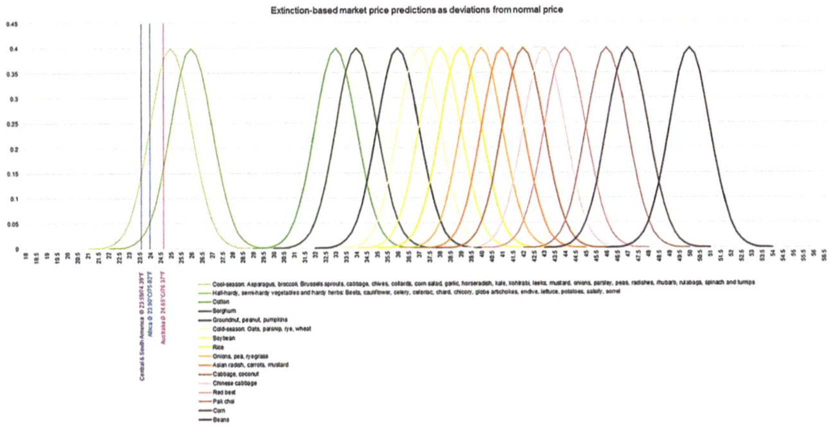

Graph 12.10 Extinction-based market price predictions as deviations from normal prices (*Source* Author's own)

future world inhabitants. In a trade-off of economic growth versus sustainability, a broad-based international coalition could establish climate stability. As a novel angle toward climate justice, the attention to global warming gains and losses being distributed unequally around the globe (Puaschunder, 2017b). The presented framework of climate change gains and losses allows to propose to search for a well-balanced climate mitigation and adaptation public policy mix guided by micro- and macroeconomic analysis results.

A three-dimensional Mapping Climate Justice approach was introduced to share the burden of climate change in a right, just and fair way around the globe. Climate justice within a country should pay tribute to the fact that low- and high-income households carry the same burden proportional to their dispensable income. Climate change burden sharing between countries ensures those countries benefiting more from a warmer environment also bear a higher responsibility regarding climate change mitigation and adaptation efforts. Climate justice over time can innovatively be enabled by sharing the gains in climate change benefitting territories via taxation to redistribute funds into climate change loser countries that issue climate stability bonds. All these recommendations are aimed at ensuring to share the burden but also the benefits of climate change within society in an economically efficient, legally equitable and practically feasible way.

Future wealth of nations was prospected by the concept of climate flexibility defined as the range of temperature variation of a country. In a changing climate, temperature range flexibility is a future asset for international trade of commodities but also for production flexibility leading to comparative advantages of countries (Ricardo, 1817/1951). A broad spectrum of climate zones has never been defined as asset and comparative edge in free trade, but climate change will require territories being more flexible in terms of changing economic production. The more climate variation a nation-state possesses, this paper argued, the more degrees of freedom a country has in terms of GDP production capabilities in a differing climate. Already now, the degree of climate flexibility was outlined to be related to human migration inflow and is predicted to determine future climate wealth of nations in a climate changing world. Future climate change-induced market changes were featured based on scarcity of agriculture production prospects.

Climate change-induced market flows were outlined to be already existent in terms of human capital, FDI inflows and Internet connectivity as predictor of technological superiority being prevalent in those areas of the world that gain from global warming. Climate flexibility was introduced as a future market asset for offering a range of economic possibilities as a basis for future climate wealth of nations. Climate flexibility as comparative advantage was already shown to be positively related to human migration. Migration and monetary flows are argued to vividly outline the competitive advantages of the climate change winner countries. The first introduction of climate-induced financial flows serves as unique justification for climate benefits transfers around the globe and demands for recognition of climate refugees under the Geneva Convention in order to alleviate the negative consequences of climate hegemony in the twenty-first century.

Behavioral economics found people to be risk-averse, and their loss prospects loom larger than gains (Kahneman & Tversky, 1979; Tversky & Shafir, 1992). Changing individual temperature conditions makes people afraid, and there is an immediate loss of the status quo visible. But in reality, the world by itself will be winning from a warming globe for quite some time; it just needs to relocate and transfer skills and production into those areas that will become newly arable and pleasant to stay and work in (Puaschunder, 2018c). Climate change-induced expectations of climate scarcity were associated with unnatural price spikes in the commodities market.

The presented results may shed light on climate change from a different angle and grant hope but also alleviate the contemporary reported predicament of economic growth versus sustainability (Täuber et al., 2015; Zhou et al., 2018). Looking forward to world peak condition with regard to a favorable climate for GDP production, the future wealth of nations may be determined by countries' access to global markets (e.g. via infrastructure such as airports and ports, trade relations such as general agreements and trade unions, and practiced customs such as openness to take in migrants or favorable easy-to-obtain languages like English or Spanish) as well as natural propensities to temperature flexibility by being either a widespread country (like Canada or Russia) or one featuring different temperature zones due to tectonic landscape formation and altitude (e.g. Italy, New Zealand, Luxembourg or Switzerland featuring mountains).

Future favorable market operations appear to be (1) investments in insurances specializing on climate change and capitalizing the collective fear that has been built around the change in temperature, which actually also features positive aspects (Puaschunder, 2018c), (2) mergers and conglomerates of multiple entities to diversify from climate-related risks and shortages with higher economies of scale (Emmert, 2018) and (3) surge price strategies as ticket prices that can offer a discount during shoulder seasons disrupted due to unfavorable climate and high prices during peak seasons. The winter sports industry appears to be heading toward a direction of high luxury price and low customer base due to snow shortages and shortened winter seasons (Emmert, 2018; Puaschunder, 2018c). Mountain ownership by insurance agencies is another trend that is seen in the winter industry (Emmert, 2018). Resorts have immensely high insurance costs, both for bodily injury and weather. If a resort can farm its insurance in-house, it can save in the long run. This trend may also be extended onto other industries that may become more entwined with insurances as climate change causes weather unpredictability. Consumer lock-in strategies but also weather-related loss exemptions for satisfaction guarantee clauses may be additional trends we are heading toward in the age of global warming. Weather-independent sales are potential other areas that are prospected to see economic growth in times of weather uncertainty caused by climate change. For instance, in the skiing industry, by now half of the revenue is earned from food, lodging, rentals and lesson sales (Emmert, 2018).

12.6 Discussion and Future Outlook

Overall, the market-driven results of price mechanisms but also FDI and market streams underline the importance of research on the effect of climate change on market mechanisms. This first preliminary introduction of the idea of climate flexibility as future determinant of competitive advantages and wealth of nations opens a door for future research on climate impacts on international trade (Ricardo, 1817/1951). Capturing the influence of climate on markets also implicitly calls for a further investigation of industrial policies that rest on the fundamental insight that market prices are not often sufficient in allocating scarce resources in a humane society (Altenburg & Assmann, 2018). For instance, preferences of future generations cannot be fully reflected in market prices (Altenburg & Rodrik, 2018; Chichilnisky & Bal, 2019; Chichilnisky & Heal, 2000).

As a caveat to the presented results, it deserves mentioning that price expectations—to this day—are hard to predict and the actual temporal flow of these recurrent patterns is hard to measure quantitatively. While there is a pattern of the economy, problem of agents remain to distinguish between noise and information, and between shocks and change. In the ever-changing nature of capitalism, having evolved from industrialization to post-World War booms to globalization and the new media and AI revolution, the cycles have become more fuzzy and unpredictable (Held & McGrew, 2007).

Future research may investigate climate change impacts on the 'demand' side of economic pricing behavior. As studies have shown, demand preferences are sticky as habits usually form over a long time and therefore do not change quickly. Future studies may unravel how climate changes preferences for specific goods and services in the long run. Thereby, especially the role of critical life events and external climate shocks in changing preferences immediately and definitively should be scrutinized (Puaschunder, 2020). Behavioral insight specialists may find innovative ways on how to align the demand side with the supply side of climate policies from an interdisciplinary perspective. Future studies may also focus on how to teach the gains of a warming earth in applied business cases (Gonglewski & Helm, 2010; Helm, 2009). All these presented insights gained may help administer the transition to new technologies and renewable energy as well as reap the economic benefits of a warming globe. In addition, climate transfer taxation could be proposed to reap the benefits from climate flexibility and redistribute these gains to territories that hit climate flexibility shortage. In the Climate Wealth of Nations' Winners and Losers model, the temperature extremes should be controlled for outliers. Over time temperature fluctuations should be separated from special temperature differences in order to find more sophisticated market predictions on the dominance of climate trade. Future public policy recommendations should start focusing away from the climate losses to redistributing the gains of a warming earth so that they can be enjoyed equally within societies, around the world and over time.

Appendix

Method

Overall Model Assumptions

Economic output was measured under projected conditions of a warming earth. For 189 world countries, the Gross Domestic Product (GDP) for agriculture, industry and service sector composition was retrieved from the Central Intelligence Agency (CIA) World Factbook (https://www.cia.gov/the-world-factbook/). *Agricultural output* is the component of the GDP of a nation that describes the process of producing food, feed, fiber and other goods by the systematic raising of plants and animals. *Industry* is the GDP segment of the economy concerned with production of goods including fuels and fertilizers, mining and extraction sectors. The *service* sector is the non-material equivalent of a good. Service provision is defined as an economic activity that does not result in ownership of physical goods. It is a process that creates benefits by facilitating a change in customers, a change in their physical possessions, or a change in their intangible assets. Service output is a component of the GDP of a nation that also includes—but is not limited to—farm- and factory-related activities.

Contemporary climate change projections estimate a mean temperature rise of approximately 4.24°C. The cardinal or optimal temperature per sector was related to the prospected temperature in 2100 per country. The cardinal age or optimal temperature for agriculture is 28.5°C for growing wheat and rice (http://agritech.tnau.ac.in/agriculture/agri_agrometeorology_temp.html). The cardinal temperature for the industry sector comprising of construction, automotive, airframe, locomotive, petroleum, coal and chemicals was found to be 21°C (Russell, 1957). The cardinal temperature for the service sector lies at 25°C (Somanathan et al., 2014).

From the current world temperature mean per country and the estimate of a 4.25°C climate change until 2100, the closeness to cardinal temperature per sector was calculated for each country by using the following formula (12.2):

$$T_A = C_A - T_e, \tag{12.2}$$

whereby T_A represents the closeness to the optimum cardinal temperature for agriculture calculated by the cardinal temperature for agriculture C_A, which equals to 28.5°C, subtracted by T_e as the temperature estimated for the year 2100 per country.

For calculating each country's time to optimum economic output in the industry sector, Eq. (12.3) was used:

$$T_I = C_I - T_e, \tag{12.3}$$

whereby T_I represents the closeness to the optimum cardinal temperature for the industry sector calculated by the cardinal temperature for the industry sector C_I, which equals to 21°C, subtracted by T_e as the temperature estimated for the year 2100 per country.

For calculating each country's time to optimum economic output in the service sector, Eq. (12.4) was used:

$$T_S = C_S - T_e, \tag{12.4}$$

whereby T_S represents the closeness to the optimum cardinal temperature for the service sector calculated by the cardinal temperature for service sector C_S, which equals to 25 degrees Celsius, subtracted by T_e as the temperature estimated for the year 2100 per country.

The result for the distance to the optimum temperature for each sector for each country was then multiplied by the GDP contribution percentage of the sector, using the following formulas (12.5), (12.6) and (12.7):

For the agricultural sector, formula (12.5) reads

$$WL_A = T_A \times GDP_A, \tag{12.5}$$

whereby WL_A stands for the climate change winner and loser index for agriculture per country comprised of the distance to the optimum cardinal temperature per GDP sector T_A multiplied by the percentage of agriculture contributing to the overall GDP indicated by GDP_A.

For the industry sector, formula (12.6) reads

$$WL_I = T_I \times GDP_I, \tag{12.6}$$

whereby WL_A stands for the climate change winner and loser index for industry per country comprised of the distance to the optimum cardinal temperature per GDP sector T_I multiplied by the percentage of industry contributing to the overall GDP indicated by GDP_I.

For the service sector, formula (12.7) reads

$$WL_S = T_S \times GDP_S, \tag{12.7}$$

whereby WL_A describes the climate change winner and loser index for service per country comprised of the distance to the optimum cardinal temperature per GDP sector T_S multiplied by the percentage of the service sector contributing to the overall GDP indicated by GDP_S.

The overall Winner–Loser WL_T index was calculated per GDP sector leading the sector-specific Winner-Loser indices WL_A for the agriculture sector, WL_I for the industry sector and WL_S for the service sector, which were then added up into the Winner-Loser total index WL_T based on the following formula (12.8):

$$WL_T = WL_A + WL_I + WL_S, \tag{12.8}$$

whereby WL_A denotes the Winner-Loser index for the agriculture sector, WL_I the index for the industry sector and WL_S the index for the service sector. All indices were calculated per country for the year 2100 under a business-as-usual projection.

Overall, a positive *WL* index result would indicate a long distance to the optimum from now until the year 2100, whereas a negative index would be considered as negative prospect. Basically, the more positive the index, the longer time the country could expect to be having a favorable climate for agriculture, industries or service production until peak condition; and the more negative the index, the sooner the country would (have) run out of efficiency time. In sum, the Winner-Loser index is an indicator of how much cool time a country still has ahead from now in prospect of an optimum GDP productivity temperature and the assumption that the earth is warming.

Model Variants
In order to derive the overall GDP gains and losses until the year 2100 and paying attention to literature controversies over the mathematical representation of wins and losses, three model variants of linear model, prospect model and hyperbolic model are calculated (Burke et al., 2015).

Linear Model
Model 1 assumes linear gain and loss functions to capture the earth's warming-related GDP prospects based on the following formulas: (12.9), (12.10) and (12.11). The winners' index *W* is calculated by multiplying the distance to the optimum cardinal temperature per *GDP* sector T_A for agriculture and GDP_A for GDP contribution to agriculture, T_I for industry sector and GDP_I for GDP contribution to industry, and T_S for service sector and GDP_S for GDP contribution to service using the formulas (12.2), (12.3) and (12.4), multiplied by the respective GDP per sector contributions as exhibited in formulas (12.9), (12.10) and (12.11):

$$W_A = {T_A}^1 \times GDP_A \tag{12.9}$$

$$W_I = {T_I}^1 \times GDP_I \tag{12.10}$$

$$W_S = {T_S}^1 \times GDP_S \tag{12.11}$$

For the losers' index *L*, a linear growth is calculated by formulas (12.12), (12.13) and (12.14) for each respective sector:

$$L_A = -{T_A}^1 \times GDP_A \tag{12.12}$$

$$L_I = -{T_I}^1 \times GDP_I \tag{12.13}$$

$$L_S = -{T_S}^1 \times GDP_S, \tag{12.14}$$

per country and per GDP sector. The overall Winner-Loser *WL* index was calculated per GDP sector leading to the sector-specific Winner-Loser indices WL_A for the agriculture sector, WL_I for the industry sector and

WL_S for the service sector, which were then added up into the Winner-Loser total index WL_T based on formula (12.8).

Concave Gains and Convex Losses Prospect Model

Model 2 assumes concave gain and convex loss functions to capture the earth's warming-related GDP prospects based on the following formulas: (12.15), (12.16) and (12.17). The winners' index W is calculated by multiplying the distance to the optimum cardinal temperature per GDP sector T_A for agriculture, T_I for industry sector and T_S for service sector using the formulas (12.2), (12.3) and (12.4), multiplied by the respective GDP per sector contributions as exhibited in formulas (12.15), (12.16) and (12.17):

$$W_A = {T_A}^{0.5} \times GDP_A \tag{12.15}$$

$$W_I = {T_I}^{0.5} \times GDP_I \tag{12.16}$$

$$W_S = {T_S}^{0.5} \times GDP_S \tag{12.17}$$

For the losers' index L, a convex growth is calculated by formulas (12.18), (12.19) and (12.20) for each respective sector:

$$L_A = -{T_A}^2 \times GDP_A \tag{12.18}$$

$$L_I = -{T_I}^2 \times GDP_I \tag{12.19}$$

$$L_S = -{T_S}^2 \times GDP_S, \tag{12.20}$$

per country and per GDP sector. The overall Winner-Loser WL index was calculated per GDP sector leading to the sector-specific Winner-Loser indices WL_A for the agriculture sector, WL_I for the industry sector and WL_S for the service sector, which were then added up into the Winner-Loser total index WL_T based on formula (12.8).

Hyperbolic Model

Model 3 assumes hyperbolic gain and loss functions, with the cardinal temperature equaling the maximum point to capture the earth's warming-related GDP prospects based on the following formulas: (12.21), (12.22) and (12.23). The winners' index W is calculated by multiplying the distance to the optimum cardinal temperature per GDP sector T_A for agriculture, T_I for industry sector and T_S for service sector using the formulas (12.2), (12.3) and (12.4), multiplied by the respective GDP per sector contributions as exhibited in formulas (12.21), (12.22) and (12.23):

$$W_A = {T_A}^2 \times GDP_A \quad (12.21)$$

$$W_I = {T_I}^2 \times GDP_I \quad (12.22)$$

$$W_S = {T_S}^2 \times GDP_S \quad (12.23)$$

For the losers' index L, a hyperbolic growth is calculated by formulas (12.24), (12.25) and (12.26) for each respective sector:

$$L_A = -{T_A}^2 \times GDP_A \quad (12.24)$$

$$L_I = -{T_I}^2 \times GDP_I \quad (12.25)$$

$$L_S = -{T_S}^2 \times GDP_S, \quad (12.26)$$

per country and per GDP sector. The overall Winner-Loser WL index was calculated per GDP sector leading to the sector-specific Winner-Loser indices WL_A for the agriculture sector, WL_I for the industry sector and WL_S for the service sector, which were then added up into the Winner-Loser total index WL_T based on formula (12.8).

Results

Overall GDP Gains and Losses Until 2100

Linear Model

When unidimensionally focusing on estimated GDP growth given a warmer temperature and estimating a linear growth of losses and wins, overall the world will be gaining more than losing until 2100. Based on the WL index of 198 countries of the world and under the assumption of linear gains and losses in light of climate change, less countries ($n = 79$) will win more from global warming until 2100 than more countries (n

= 119) will lose from a warming earth. In particular, 79 countries of the world will gain $WL_L = 78{,}139.08$ in GDP output, whereas 119 countries of the world will lose estimated $WL_L = -52{,}061$.

Prospect Convex Losses and Concave Gains Model

When unidimensionally focusing on estimated GDP growth given a warmer temperature and estimating an exponential growth of losses and concave wins, overall the world will be losing more than gaining until 2100. Based on the *WL* index of 188 countries of the world, more countries ($n = 113$) will lose more from global warming until 2100 than less countries ($n = 75$) will win from a warming earth. In particular, 75 countries of the world will gain $WL_P = 22{,}717.161$ in GDP output, whereas 113 countries of the world will lose estimated $WL_P = -353{,}175.32$.

Hyperbolic Model

When unidimensionally focusing on estimated GDP growth given a warmer temperature and estimating a hyperbolic growth, overall the world will be gaining more than losing until 2100. Based on the *WL* index of 188 countries of the world, less countries ($n = 79$) will gain more from global warming until 2100 than more countries ($n = 109$) will lose from a warming earth. In particular, 79 countries of the world will gain $WL_H = 1{,}037{,}192$ in GDP output, whereas 109 countries of the world will lose estimated $WL_H = -352{,}088$.

Total Estimate

The total Winner-Loser *WL* index was calculated per country based on the mean of the overall Winner-Loser index WL_L for the linear model, the overall Winner-Loser index WL_P for the prospect model and the overall Winner-Loser index WL_H for the hyperbolic model that were then added up into the Winner-Loser total index WL_{TT} and divided by the number of models calculated (n) to gain the mean based on the following formula (12.27):

$$WL_{TT} = \frac{(WL_L + WL_P + WL_H)}{n}, \tag{12.27}$$

whereby WL_L denotes the Winner-Loser index for the linear model, WL_P the index for the prospect model and WL_H the index for the hyperbolic model and $n = 3$. All indices were calculated per country for the year 2100 business-as-usual projection.

When unidimensionally focusing on estimated GDP growth given a warmer temperature, over all calculated models, the world will be gaining more than losing until 2100. Based on the *WL* index of 188 countries of the world, less countries ($n = 77$) will gain more from global warming until 2100 than more countries ($n = 111$) will lose from a warming earth. In particular, 77 countries of the world will gain $WL_H = 354{,}039.6345$ in GDP output, whereas 111 countries of the world will lose estimated $WL_H = -235{,}726.6280$.

References

Acevedo, S., Mrkaic, M, Novta, N., Poplawski-Ribeiro, M., Pugacheva, E., & Topalova, P. (2018, April 12). *How can low income countries cope? The effects of weather shocks on economic activity*. Presentation delivered at The New School Economics of Climate Change Project, The New School.

Altenburg, T., & Assmann, C. (2018). *United Nations Environment Programme Report on Green Industrial Policy: Concept, policies, country experiences (2018)*. United Nations: United Nations Environment Programme & German Development Institute/Deutsches Institut für Entwicklungspolitik.

Altenburg, T., & Rodrik, D. (2018). Green industrial policy: Accelerating structural change towards wealthy green economies. In T. Altenburg & C. Assmann (Eds.), *United Nations Environment Programme Report on Green Industrial Policy: Concept, policies, country experiences* (pp. 2–20). United Nations: United Nations Environment Programme & German Development Institute/Deutsches Institut für Entwicklungspolitik.

Barro, R. (1990). Government spending in a simple model of endogenous growth. *Journal of Political Economy, 98*, 103–125. https://doi.org/10.1086/261726

Bürgenmeier, B. (1994). Environmental policy: Beyond the economic dimension. In B. Bürgenmeier (Ed.), *Economy, environment, and technology: A socio-economic approach* (pp. 166–175). Armonk Sharpe.

Burke, M., Hsiang, S. M., & Miguel, E. (2015). Global non-linear effect of temperature on economic production. *Nature, 527*, 235–239. https://doi.org/10.1038/nature15725

Centeno, M. A., Creager, A. N., Elga, A., Felton, E., Katz, St. N., Massey, W. A., et al. (2013). *Global systemic risk: Proposal for a research community*. Working paper. Princeton Institute for International and Regional Studies, Princeton University, NJ.

Chancel, L., & Piketty, T. (2015). *Carbon and inequality: From Kyoto to Paris*. Paris School of Economics.

Chichilnisky, G. (1996). *Development and global finance: The case for an international bank for environmental settlements.* United Nations Development Programme, Office of Development Studies.
Chichilnisky, G. (2007). *The economics of global environment: Catastrophic risks in theory and policy.* Springer International.
Chichilnisky, G. (2010). *The economics of climate change.* Edward Elgar.
Chichilnisky, G. (2016). Reversing climate change. *Global Policy.* https://www.globalpolicyjournal.com/blog/01/09/2016/reversing-climate-change-interview-graciela-chichilnisky
Chichilnisky, G., & Bal, P. (2019). *Reversing climate change: How carbon removals can resolve climate change and fix the economy.* World Scientific Publishing.
Chichilnisky, G., & Heal, G. (2000). *Environmental markets: Equity and efficiency.* Columbia University Press. https://doi.org/10.7312/chic11588
Chichilnisky, G., Heal, G., & Vercelli, A. (1998). *Sustainability: Dynamics and uncertainty.* Kluwer. https://doi.org/10.1007/978-94-011-4892-4
Chichilnisky, G., & Sheeran, K. (2018). *Handbook on the economics of climate change.* Edward Elgar.
Emmert, H. (2018). *The melting point: The economics and environmental impact of the winter sports industry.* Unpublished working paper. Economics of the Environment, The New School, New York.
Figuerola-Ferretti, I., & Cervera, I. (2018). *Recent credit risk and bubble behavior in the corporate energy sector.* https://repositorio.comillas.edu/jspui/handle/11531/26571
Filip, O., Janda, K., Kristoufek, L., & Zilberman, D. (2018). *Food versus fuel: An updated and expanded evidence.* CAMA Working paper 73. https://papers.ssrn.com/sol3/papers.cfm?abstract_id=3070704
Foley, D. (2007). *The economic fundamentals of global warming.* Working paper 07-12-044. Santa Fe Institute, Santa Fe, NM. https://www.santafe.edu/research/results/working-papers/the-economic-fundamentals-of-global-warming
Gonglewski, M., & Helm, A. (2010). An examination of business case methodology: Pedagogical synergies from two disciplines. *Global Business Languages, 15*(3), 17–31.
Held, D., & McGrew, A. G. (2007). *A new world economic order? Global markets and state power: Beyond globalization/anti-globalization: Beyond the great divide.* Polity.
Helm, A. (2009). *Global marketing classroom: Green marketing.* American Marketing Association Global SIG Newsletter, Global Interests.
Kaabia, O., Dhaoui, A., Abid, I., & Guesmi, K. (2018). *Oil supply-demand shocks and stock prices.* https://www.researchgate.net/publication/324687727_Oil_supply-demand_shocks_and_stock_prices

Kahneman, D., & Tversky, A. (1979). Prospect theory: An analysis of decision under risk. *Econometrica, 47*(2), 263–291. https://doi.org/10.2307/1914185

Kant, I. (1788/2003). *Critique of practical reasons*. Cambridge University Press.

Lips, J. (2018). *Debt and the oil industry: Analysis on the firm and production level*. https://papers.ssrn.com/sol3/papers.cfm?abstract_id=3026063

Nordhaus, W. D. (1994). *Mapping the global commons: The economics of climate change*. MIT Press.

Otto, I. M., Kim, K. M., Dubrovsky, N., & Lucht, W. (2019). Shift the focus from the super-poor to the super-rich. *Nature Climate Change, 9*, 82–84. https://doi.org/10.1038/s41558-019-0402-3

Puaschunder, J. M. (2015). Trust and reciprocity drive common goods allocation norms. In *Proceedings of the Cambridge Business & Economics Conference*. Cambridge University.

Puaschunder, J. M. (2016). Intergenerational climate change burden sharing: An economics of climate stability research agenda proposal. *Global Journal of Management and Business Research: Economics and Commerce, 16*(3), 31–38.

Puaschunder, J. M. (2017a, March 26–27). Climate in the 21st century: A macroeconomic model of fair global warming benefits distribution to grant climate justice around the world and over time. In *Proceedings of the 8th International RAIS Conference on Social Sciences and Humanities organized by Research Association for Interdisciplinary Studies (RAIS) at Georgetown University*, Washington, DC, USA, pp. 205–243.

Puaschunder, J. M. (2017b). Mapping climate in the 21st century. *Development, 59*(3), 211–216.

Puaschunder, J. M. (2017c). Socio-psychological motives of socially responsible investors. *Advances in Financial Economics, 19*(1), 209–247. https://doi.org/10.1108/S1569-373220160000019008

Puaschunder, J. M. (2017d). Sunny side up! From climate change burden sharing to fair global warming benefits distribution: Groundwork on the metaphysics of the gains of global warming and the climatorial imperative. In *Proceedings of the Administrative Sciences Association of Canada Annual Conference*, HEC Montreal, Quebec, Canada.

Puaschunder, J. M. (2017e). The call for global responsible inter-generational leadership: The quest of an integration of inter-generational equity in corporate social responsibility (CSR) models. In *Natural resources management: Concepts, methodologies, tools, and applications: Information resources management association*. IGI Global.

Puaschunder, J. M. (2018a). Climate polices with burden sharing: The economies of climate financing. *Advances in Financial Economics, 19*, 1–13.

Puaschunder, J. M. (2018b). Gifts without borders: Intergenerational glue connecting over distance and time as pure international development in the age of migration. In J. M. Puaschunder (Ed.), *Intergenerational responsibility in the 21st century* (pp. 143–174). Vernon.

Puaschunder, J. M. (2018c). *Nachhaltigkeit und Investment: Psychologische Aspekte von nachhaltigkeitsorientiertem Investitionsverhalten*. In C. T. Schmitt & E. Bamberg (Eds.), *Psychologie und Nachhaltigkeit: Konzeptionelle Grundlagen, Anwendungsbeispiele und Zukunftsperspektiven* (pp. 127–134). Springer.

Puaschunder, J. M. (2019). Value at looking back: Towards an Empirical validation of the role of reflexivity in econo-historic backtesting: Economic market prediction corrections correlate with future market performance. *Oxford Journal of Finance and Risk Perspectives, 8*, 223–239.

Puaschunder, J. M. (2020). *Behavioral economics and finance leadership: Nudging and winking to make better choices*. Springer Nature.

Renz, A. (2018, May 24–25). Sector competition for energy resources in a carbon constrained hoteling model. In *6th International Symposium on Environment Energy & Finance Issues, ISEFI 2018*, Paris, France.

Ricardo, D. (1817/1951). On the principles of political economy and taxation. In P. Sraffa (Ed.), *The works and correspondence of David Ricardo* (pp. 1951–1973). Cambridge University Press.

Roberts, G., Barbier, E., & van't Veld, K. V. (2018). *The beginning of the end of the age of oil*. https://cenrep.ncsu.edu/cenrep/wp-content/uploads/2016/08/Roberts.pdf

Russell, J. A. (1957). *Industrial operations under extremes of weather*. Meteorological Monographs, American Meteorological Society.

Sachs, J. D. (2014). Climate change and intergenerational well-being. In L. Bernard & W. Semmler (Eds.), *The Oxford handbook of the macroeconomics of global warming* (pp. 248–259). Oxford University Press.

Saussay, A. & Sato, M. (2018). *The impacts of energy prices on industrial foreign investment location: Evidence from global firm level data*. https://personal.lse.ac.uk/satom/publication/currentss/

Somanathan, E., Somanathan, R., Sudarshan, A., & Tewari, M. (2014). *The impact of temperature on productivity and labor supply: Evidence from Indian manufacturing*. Discussion Paper in Economics 14-10. Indian Statistical Institute, Economics and Planning Unit.

Täuber, S., van Zomeren, M., & Kutlaca, M. (2015). Should the moral core of climate issues be emphasized or downplayed in public discourse? Three ways to successfully manage the double-edged sword of moral communication. *Climate Change, 130*(3), 453–464. https://doi.org/10.1007/s10584-014-1200-6

The World Bank. (2015). *Green bonds attract private sector climate finance*. World Bank Brief.

The World Economic Forum 2015 Report. (2015). Davos, Switzerland: World Economic Forum.

Trinks, A., Mulder, M., & Scholtens, B. (2017). *Greenhouse gas emissions intensity and the cost of capital*. https://pdfs.semanticscholar.org/c8f5/d6b192df64cde5f578f100d7fb2a6b40515b.pdf

Tversky, A., & Shafir, E. (1992). Choice under conflict: The dynamics of deferred decision. *Psychological Science, 3*(6), 358–361. https://doi.org/10.1111/j.1467-9280.1992.tb00047.x

United Nations Report on the World Social Situation. (2011). *The global social crisis*. United Nations Department of Economic and Social Affairs Report ST/ESA/334.

Wenger, A., Orttung, R. W., & Perovic, J. (2009). *Energy and the transformation of international relations: Toward a new producer-consumer framework*. Oxford University Press.

World Bank 2015 Report. (2015). Washington, DC: World Bank.

World Investment Report. (2015). *Reforming international investment governance*. United Nations Conference on Trade and Development Report. United Nations.

Yarlik, M. V., Filis, G., Lloyd, T., & Degiannakis, St. (2018, May 24–25). Forecasting oil price volatility: The role of mixed-frequency data (MIDAS) model. In *6th International Symposium on Environment Energy & Finance Issues, ISEFI 2018*, Paris, France.

Zhang, D. (2018, May 24–25). Upgrading China's energy structure: The role of financial markets. In *6th International Symposium on Environment Energy & Finance Issues, ISEFI 2018*, Paris, France.

Zhou, S. L., Smulders, S., & Gerlagh, R. (2018, May 24–25). Closing the loop in a circular economy: Saving resources or suffocating innovations? In *6th International Symposium on Environment Energy & Finance Issues, ISEFI 2018*, Paris, France.

CHAPTER 13

Climate Beta Risk in the Climate Change Winners-and-Losers Framework

13.1 Introduction

In previous chapters (see, for example, Chapter 6), a macroeconomic model was presented that estimates a Winner-Loser index per country by calculating deviation from the optimum temperature for agriculture, industry, and service sectors, weighting that deviation by the sector's share of GDP, and aggregating the weighted deviations into an international index. This yields a climate productivity proximity index, which highlights the disparate impact of global warming for almost all countries of the world in the eyes of long-term macroeconomic exposure to climate change.

13.1.1 Overall Model Assumptions

Economic output was measured under projected conditions of a warming earth. For 189 world countries, the Gross Domestic Product (GDP) of agriculture, industry, and service sector composition was retrieved from the Central Intelligence Agency (CIA) World Factbook.[1]

Contemporary climate change projections estimate a mean temperature rise of approximately 4.24 degrees Celsius. The cardinal or optimal temperature per sector was related to the prospected temperature in

[1] https://www.cia.gov/library/publications/the-world-factbook/.

J. Puaschunder, *Governance & Climate Justice*,
https://doi.org/10.1007/978-3-031-87971-5_13

2100 per country. The cardinal age or optimal temperature for agriculture is 28.5 degree Celsius for growing wheat and rice.[2] The cardinal temperature for the industry sector comprising of construction, automotive, airframe, locomotive, petroleum, coal, and chemicals was found to be 21 degree Celsius (Russell, 1957). The cardinal temperature for the service sector lies at 25 degree Celsius (Somanathan et al., 2014).

From the current world temperature mean per country and the estimate of a 4.25 degree Celsius climate change until 2100, the closeness to cardinal temperature per sector was calculated for each country by using the following formula 13.1:

$$T_A = C_A - T_e \tag{13.1}$$

whereby T_A represents the closeness to the optimum cardinal temperature for agriculture calculated by the cardinal temperature for agriculture C_A, which equals to 28.5 degrees Celsius, subtracted by T_e as the temperature estimated for the year 2100 per country.

For calculating each country's time to optimum economic output in the industry sector, Eq. 13.2 was used:

$$T_I = C_I - T_e \tag{13.2}$$

whereby T_I represents the closeness to the optimum cardinal temperature for the industry sector calculated by the cardinal temperature for the industry sector C_I, which equals to 21 degrees Celsius, subtracted by T_e as the temperature estimated for the year 2100 per country.

For calculating each country's time to optimum economic output in the service sector, Eq. 13.3 was used:

$$T_S = C_S - T_e \tag{13.3}$$

whereby T_S represents the closeness to the optimum cardinal temperature for the service sector calculated by the cardinal temperature for service sector C_S, which equals to 25 degrees Celsius, subtracted by T_e as the temperature estimated for the year 2100 per country.

The result for the distance to the optimum temperature for each sector for each country was then multiplied by the GDP contribution percentage of the sector using the following formulas 13.4, 13.5, 13.6

[2] http://agritech.tnau.ac.in/agriculture/agri_agrometeorology_temp.html.

For the agricultural sector, formula 13.4 reads

$$WL_A = T_A * GDP_A \tag{13.4}$$

whereby WL_A stands for the climate change winner and loser index for agriculture per country composed of the distance to the optimum cardinal temperature per GDP sector T_A multiplied by the percentage of agriculture contributing to the overall GDP indicated by GDP_A.

For the industry sector, formula 13.5 reads

$$WL_I = T_I * GDP_I \tag{13.5}$$

whereby WL_I stands for the climate change winner and loser index for industry per country composed of the distance to the optimum cardinal temperature per GDP sector T_I multiplied by the percentage of industry contributing to the overall GDP indicated by GDP_I.

For the service sector, formula 13.6 reads

$$WL_S = T_S * GDP_S \tag{13.6}$$

whereby WL_S stands for the climate change winner and loser index for service per country composed of the distance to the optimum cardinal temperature per GDP sector T_S multiplied by the percentage of service contributing to the overall GDP indicated by GDP_S.

The overall Winner-Loser WL_T index was calculated per GDP sector, leading the sector-specific Winner–Loser indices WL_A for the agriculture sector, WL_I for the industry sector and WL_S for the service sector, that were then added up into the Winner–Loser total index WL_T based on the following formula (13.7):

$$WL_T = WL_A + WL_I + WL_S \tag{13.7}$$

whereby WL_A denotes the Winner–Loser index for the agriculture sector, WL_I the index for the industry sector and WL_S the index for the service sector. All indices were calculated per country for the year 2100 under a business-as-usual projection for climate change.

Overall, a positive WL index result would indicate a long distance to the optimum from now until the year 2100, whereas a negative index would be considered as negative prospect. Basically, the more positive the index, the longer time the country could expect to be having a favorable climate for agriculture, industries, or service production until peak condition as measured by mean temperature under climate change. And the more negative the index, the sooner the country would (have) run out of efficiency time under business-as-usual assumption given global warming. In sum, the Winner–Loser index is an indicator of how much cool time a country still has ahead from now in prospect of an optimum GDP productivity temperature and the assumption that the earth is warming.

13.1.2 Model Variants

In order to derive the overall GDP gains and losses until the year 2100 and paying attention to literature controversies over the mathematical representation of wins and losses, three model variants of the linear model, prospect model, and hyperbolic model are calculated (Burke et al., 2015).

13.1.2.1 Linear Model

Model 1 assumes linear gain and loss functions around the optimum temperature for production to capture the earth's warming-related GDP prospects based on the following formulas 13.8, 13.9 and 13.10. The winners' index W is calculated by multiplying the distance to the optimum cardinal temperature per GDP sector T_A for agriculture and GDP_A for GDP contribution to agriculture, T_I for industry sector and GDP_I for GDP contribution to industry and T_S for service sector and GDP_S for GDP contribution to service using the formulas 13.1, 13.2 and 13.3 multiplied by the respective GDP per sector contributions as exhibited in formulas 13.8, 13.9 and 13.10:

$$W_A = T_A^1 * GDP_A \tag{13.8}$$

$$W_I = T_I^1 * GDP_I \tag{13.9}$$

$$W_S = T_S^1 * GDP_S \tag{13.10}$$

For the losers' index L, a linear growth is calculated by formulas 13.11, 13.12 and 13.13 for each respective sector:

$$L_A = -T_A^1 * GDP_A \tag{13.11}$$

$$L_I = -T_I^1 * GDP_I \tag{13.12}$$

$$L_S = -T_S^1 * GDP_S \tag{13.13}$$

per country and per GDP sector. The overall Winner–Loser WL index was calculated per GDP sector leading to the sector-specific Winner–Loser indices WL_A for the agriculture sector, WL_I for the industry sector and WL_S for the service sector, that were then added up into the Winner–Loser total index WL_T based on formula 13.7.

13.1.2.2 Concave Gains and Convex Losses Prospect Model

Model 2 assumes concave gain and convex loss functions around the optimum temperature for production to capture the earth's warming-related GDP prospects based on the following formulas 13.14, 13.15 and 13.16. The winners' index W is calculated by multiplying the distance to the optimum cardinal temperature per GDP sector T_A for agriculture, T_I for industry sector and T_S for service sector using the formulas 13.1, 13.2 and 13.3 multiplied by the respective GDP per sector contributions as exhibited in formulas 13.14, 13.15 and 13.16:

$$W_A = T_A^{0.5} * GDP_A \tag{13.14}$$

$$W_I = T_I^{0.5} * GDP_I \tag{13.15}$$

$$W_S = T_S^{0.5} * GDP_S \tag{13.16}$$

For the losers' index L, a convex growth is calculated by formulas 13.17, 13.18 and 13.19 for each respective sector:

$$L_A = -T_A^2 * GDP_A \tag{13.17}$$

$$L_I = -T_I^2 * GDP_I \tag{13.18}$$

$$L_S = -T_S^2 * GDP_S \tag{13.19}$$

per country and per GDP sector. The overall Winner–Loser WL index was calculated per GDP sector leading to the sector-specific Winner–Loser indices WL_A for the agriculture sector, WL_I for the industry sector and WL_S for the service sector, that were then added up into the Winner–Loser total index WL_T based on formula 13.7.

13.1.2.3 Hyperbolic Model

Model 3 assumes hyperbolic gain and loss functions with the cardinal temperature equaling the maximum point to capture the earth's warming-related GDP prospects based on the following formulas 13.20, 13.21 and 13.22. The winners' index W is calculated by multiplying the distance to the optimum cardinal temperature per GDP sector T_A for agriculture, T_I for industry sector and T_S for service sector using the formulas 13.1, 13.2 and 13.3 multiplied by the respective GDP per sector contributions as exhibited in formulas 13.20, 13.21 and 13.22:

$$W_A = T_A^2 * GDP_A \tag{13.20}$$

$$W_I = T_I^2 * GDP_I \tag{13.21}$$

$$W_S = T_S^2 * GDP_S \tag{13.22}$$

For the losers' index L, a hyperbolic growth is calculated by formulas 13.23, 13.24 and 13.25 for each respective sector:

$$L_A = -T_A^2 * GDP_A \tag{13.23}$$

$$L_I = -T_I^2 * GDP_I \tag{13.24}$$

$$L_S = -T_S^2 * GDP_S \tag{13.25}$$

per country and per GDP sector. The overall Winner–Loser WL index was calculated per GDP sector leading to the sector-specific Winner–Loser indices WL_A for the agriculture sector, WL_I for the industry sector and WL_S for the service sector, that were then added up into the Winner–Loser total index WL_T based on formula 13.7.

13.2 Results

13.2.1 *Overall GDP Gains and Losses Until 2100*

13.2.1.1 Linear Model

When unidimensionally focusing on estimated GDP growth given a warmer temperature and estimating a linear growth of expected economic losses and wins, overall the world will be gaining more than losing until 2100. Based on the WL index of 198 countries of the world and under the assumption of linear gains and losses in light of climate change, less countries (n = 79) will win more from global warming until 2100 than more countries (n = 119) will lose from a warming earth. In particular, 79 countries of the world will gain $WL_L = 78{,}139.08$ in GDP output, whereas 119 countries of the world will lose estimated $WL_L = -52{,}061$.

13.2.1.2 Prospect Convex Losses and Concave Gains Model

When unidimensionally focusing on estimated GDP growth given a warmer temperature and estimating an exponential growth of losses and concave wins, overall the world will be losing more than gaining until 2100. Based on the WL index of 188 countries of the world, more countries (n = 113) will lose more from global warming until 2100 than less countries (n = 75) will win from a warming earth. In particular, 75 countries of the world will gain $WL_P = 22{,}717.161$ in GDP output, whereas 113 countries of the world will lose estimated $WL_P = -353{,}175.32$.

13.2.1.3 Hyperbolic Model

When unidimensionally focusing on estimated GDP growth given a warmer temperature and estimating a hyperbolic growth, overall the world will be gaining more than losing until 2100. Based on the WL index of 188 countries of the world, less countries (n = 79) will gain more from global warming until 2100 than more countries (n = 109) will lose from a warming earth. In particular, 79 countries of the world will gain $WL_H = 1{,}037{,}192$ in GDP output, whereas 109 countries of the world will lose estimated $WL_H = -352{,}088$.

13.2.1.4 Total Estimate

The total Winner–Loser WL index was calculated per country based on the mean of the overall Winner–Loser index WL_L for the linear model, the overall Winner–Loser index WL_P for the prospect model and the overall Winner–Loser index WL_H for the hyperbolic model that were then added up into the Winner–Loser total index WL_{TT} and divided by the number of models calculated (n) to gain the mean based on the following formula 13.26:

$$WL_{TT} = (WL_L + WL_P + WL_H)/n \tag{13.26}$$

whereby WL_L denotes the Winner–Loser index for the linear model, WL_P the index for the prospect model and WL_H the index for the hyperbolic model and $n = 3$. All indices were calculated per country for the year 2100 business-as-usual projection.

When unidimensionally focusing on estimated GDP growth given a warmer temperature, over all calculated models, the world will be gaining more than losing until 2100. Based on the WL index of 188 countries of the world, less countries (n = 77) are prospected to gain more from global warming until 2100 than more countries (n = 111) will lose from a warming earth. In particular, 77 countries of the world are estimated to gain $WL_H = 354{,}039.6345$ in GDP output, whereas 111 countries of the world are estimated to lose $WL_H = -235{,}726.6280$.

13.3 Modeling Risk Sensitive Climate Change Winners and Losers

In an attempt to phase in risk estimates to refine the model, the concept of climate change risk beta (β) is introduced. The proposed climate change risk beta ($\beta_{climate}$) represents how sensitive a country's economic performance (GDP or returns) is to climate-related shocks, relative to the global average. This parallels financial beta, which reflects a stock's sensitivity to market movements compared to other market risk performance.

For the modeling of risk into the Climate Change Winners-and-Losers Framework, a risk measurement inspired by the Capital Asset Pricing Model (CAPM) is designed into the Climate Change Winners-and-Losers economic growth prospects.

13.3.1 Capital Asset Pricing Model (CAPM)

The CAPM is a standard financial management tool that links an asset's expected financial return to its exposure to market (systematic) risk (Black et al., 1972; Lintner, 1965; Mossin, 1966; Sharpe, 1964). The standard

CAPM risk estimate is based on Eq. 13.27.

$$E[R_i] = R_f + \beta_i\big(E[R_M] - R_f\big), with \beta_i = \frac{Cov(R_i, R_M)}{Var(R_M)}, \tag{13.27}$$

whereby

$E[R_i]$ is the expected return of an investment i, or compensation for holding a risky asset;
R_f is the risk-free rate of return, which represents the return with no default risk and no variance;
β_i is the β of an asset i, which indicates sensitivity to movements in the overall market. Interpretation for different β_i would define:
$\beta_i = 1$, asset moves in line with the market;
$\beta_i > 1$, asset is more volatile than the market (higher risk, higher expected return);
$\beta_i < 1$, asset is less volatile than the market;
$\beta_i < 0$, asset moves inversely to the market (hedge asset);
$E[R_M]$ represents the expected return of the market portfolio based on the weighted average return of all risky assets in the economy;
$E[R_M] - R_f$, being the market risk premium, which measures how much extra return investors demand for holding the risky market portfolio instead of a risk-free asset. It quantifies the reward for bearing systemic, non-diversifiable risk;
R_i actual realized return of asset i,
R_M actual realized return of the market portfolio;
$Cov(R_i, R_M)$ as covariance between the return of asset i and the market return;
$Var(R_M)$ as the variance of the market return or the overall volatility of the market.

The CAPM implies all properly priced assets lie on the Security Market Line and only systemic risk is priced in β (Fama & French, 1992, 2015; Fama & MacBeth, 1973). Overall, CAPM estimates costs of equity and allows for calculating risk-return trade-offs as a benchmarking (Ross, 1976).

13.3.2 Risk-sensitive Climate Change Winners-and-Losers Model

The CAPM has already been used as an intertemporal capital asset value measurement model (Merton, 1973). As an extension of the presented Climate Change Winners-and-Losers model, the concept of beta is now applied to climate change risk phased into countries' expected economic growth or decline perspectives given global warming.

Drawing from the traditional finance beta β that measures how much a stock or portfolio holds risk relative to the overall market, in which

$\beta > 1$ indicates a more volatile situation than the overall market;
$\beta < 1$ indicates a less volatile situation than the overall market;

similarly

Climate Beta ($\beta_{climate}$) could measure how sensitive a country, asset, sector, firm, industry, or economy is to climate-related risks compared to the global average climate change-induced market risk. Climate beta risk $\beta_{climate}$ quantifies how vulnerable an economic entity is to climate change, relative to the market or global benchmark. For the actual calculation, one could integrate physical risks (such as damage from heatwaves, floods, droughts, wildfires, and hurricanes), transition risks (for instance, costs of shifting to renewable energy, carbon pricing, regulatory changes), and liability risks (for example, lawsuits, reputational costs, or penalties linked to climate impact).

The climate risk of an option p (e.g., country, asset, sector, firm, or economy) could be calculated by

$$\beta_{climate,p} = \frac{Cov(R_p, R_{climate})}{Var(R_{climate})}, \tag{13.28}$$

whereby

R_p is the return or growth prospect of a country, asset, sector, firm, industry, economy p;
$R_{climate}$ is the return of a climate risk index as a benchmark of overall climate-related impacts.

If

$\beta_{climate,p} > 1$, there is an exposure to climate shocks and risks (e.g., in climate loser territories);
$\beta_{climate,p} < 1$, there is less exposure to climate shocks and risks (e.g., if hedging against climate risk or transferring some of the expected gains from climate change toward successful mitigation and adaptation);
$\beta_{climate,p} = 1$ denotes average sensitivity to climate risks and impacts.

Potential empirical proxies are the frequency or severity of climate shocks (droughts, floods), insurance premiums or sovereign bond spreads adjusted for climate risk as well as changes in migration flows and financial streams in Foreign Direct Investments (FDIs) that were found to be systemically related to climate risk exposure in previous chapters of the book.

Overall, phasing in climate risks into overall market models helps investors to price climate risks into portfolios. Governments can assess country-level exposure to climate change for strategic decision making. Corporations can evaluate how climate risks affect their cost of capital. Researchers can continue to connect climate change to macroeconomic productivity and financial stability.

13.3.3 *Risk-sensitive Climate Change Winners-and-Losers Model for Countries*

A Climate Change Winner-Loser Index model extension could draw from the initial model as outlined in Chapter 6. Climate risk β would then be phased into the macroeconomic Climate Change Winner-Loser model. While the initial model measures GDP proximity to climate-optimal temperatures, the novel extension would also incorporate systematic climate-related risks quantified in financial drawbacks to economic growth potential.

The initial model as outlined in Chapter 6 calculates deviations from the optimum temperature for agriculture, industry, and service sector production per country c weighted by each productivity sector's share of GDP per country c for each sector agriculture a, service s and industry in. This climate productivity proximity measure is useful for comparing countries long-term on macroeconomic implications of climate change. So far, only pure economic growth prospects have been based on expected temperature changes, cardinal temperature per GDP sector as well as GDP sector contribution per country c.

Now introducing the concept of climate-induced risk, a climate-risk beta, $\beta_{climate,c}$, is phased into the model. $\beta_{climate,c}$ represents how sensitive a country c's economic performance in terms of GDP is to climate-related shocks and risks, relative to the global average. This parallels financial β, which reflects a singular stock's sensitive to market movements by comparing the stock's risk exposure to other market options.

Climate beta β per country c $\beta_{climate,c}$ is defined as

$$\beta_{climate,c} = \frac{Cov(R_c, R_{climate})}{Var(R_{climate})}, \tag{13.29}$$

where

R_c equals the expected economic return (as measured in GDP growth or sectoral performance) of a country c;
$R_{climate}$ is the expected return on a climate index measuring global climate risk exposure or damages;

whereby

$\beta_{climate,c} > 1$ indicates the country c is more exposed to climate risk than the comparative average;
$\beta_{climate,c} < 1$ indicates the country c is less exposed to climate risk than the comparative average.
$\beta_{climate,c} = 1$ means that the country's sensitivity matches the global average.

13.3.3.1 *Risk-sensitive Agriculture Sector Climate Change Winners-and-Losers Model*

Climate beta β per agriculture a production, $\beta_{climate,a,c}$, is defined by the formula

$$\beta_{climate,a,c} = \frac{Cov(R_{a,c}, R_{climate})}{Var(R_{climate})}, \tag{13.30}$$

where

$R_{a,c}$ is the expected return or growth rate of the agriculture sector a in a country c;
$R_{climate}$ is the expected return on a climate index measuring global climate risk exposure or damages;

whereby

$\beta_{climate,a,c} > 1$ indicates the sector agriculture a of a given country c is more sensitive than the average to climate risk due to high exposure;
$\beta_{climate,a,c} < 1$ indicates the sector agriculture a of a given country c is less sensitive than the average to climate risk due to relatively climate-resiliency;
$\beta_{climate,a,c} = 1$ means that the agriculture sector's sensitivity matches the global average.

13.3.3.2 *Risk-sensitive Service Sector Climate Change Winners-and-Losers Model*

Climate beta β per service s production, $\beta_{climate,s,c}$, is defined by the formula

$$\beta_{climate,s,c} = \frac{Cov(R_{s,c}, R_{climate})}{Var(R_{climate})}, \tag{13.31}$$

where

$R_{s,c}$ is the expected return or growth rate of the service sector s in a country c;
$R_{climate}$ is the expected return on a climate index measuring global climate risk exposure or damages;

whereby

$\beta_{climate,s,c} > 1$ indicates the sector service s of a given country c is more sensitive than the average to climate risk due to high exposure;
$\beta_{climate,s,c} < 1$ indicates the sector service s of a given country c is less sensitive than the average to climate risk due to relatively climate-resiliency;

$\beta_{climate,s,c} = 1$ means that the service sector's sensitivity matches the global average.

13.3.3.3 *Risk-Sensitive Industry Sector Climate Change Winners-and-Losers Model*

Climate beta β per industry *in* production, $\beta_{climate,in,c}$, is defined by the formula

$$\beta_{climate,in,c} = \frac{Cov(R_{in,c}, R_{climate})}{Var(R_{climate})}, \tag{13.32}$$

where

$R_{in,c}$ is the expected return or growth rate of the industry sector *in* in a country *c*;
$R_{climate}$ is the expected return on a climate index measuring global climate risk exposure or damages;

whereby

$\beta_{climate,in,c} > 1$ indicates the sector industry *in* of a given country *c* is more sensitive than the average to climate risk due to high exposure;
$\beta_{climate,in,c} < 1$ indicates the sector industry *in* of a given country *c* is less sensitive than the average to climate risk due to relatively climate-resiliency;
$\beta_{climate,in,c} = 1$ means that the industry sector's sensitivity matches the global average.

13.3.4 *Risk-Sensitive Climate Change Winners-and-Losers Model*

In addition, sector-specific climate betas $\beta_{climate,se}$, for agriculture $\beta_{climate,a}$, industry $\beta_{climate,in}$ and service sector $\beta_{climate,s}$ climate-induced risks can be summed up to find the overall climate beta $\beta_{climate,c}$ for a country. Weighting the industry-specific betas with sectoral GDP shares per country forms an overall country beta based on the formular:

$$\beta_{climate,c} = \beta_{climate,a} + \beta_{climate,in} + \beta_{climate,s} \tag{13.33}$$

$\beta_{climate,c}$ is the overall climate beta β estimate per country c, which measures how sensitive a country c is to climate-related shocks compared to the average global response. This value is analogous to financial beta β in the CAPM, where beta β reflects the volatility of an economic shock relative to the market. In the case of climate beta β, it reflects the volatility or vulnerability of GDP returns of a country c in response to global climate change impacts.

$\beta_{climate,a}$ indicates how sensitive the agriculture sector a in a country c is to climate-related shocks. Potential influences are droughts, heatwaves, or shifting rainfall patterns compared to the average global response to global warming in the agricultural sector. It reflects the volatility or vulnerability of agricultural GDP returns in a country c in response to global climate change impacts.

$\beta_{climate,in}$ indicates how sensitive the industry sector in is in a country c to climate-related shocks. Potential influences are cooling, accidents, and wearing down patterns due to heat compared to the average global response to global warming in the industry sector. It reflects the volatility or vulnerability of industry GDP returns in a country c in response to global climate change impacts.

$\beta_{climate,s}$ indicates how sensitive the service sector s in a country c is to climate-related shocks. Potential influences are water preservation, food cooling, and melting glaciers infringing on possibilities to host winter sports events compared to the average global response to global warming in the service sector. It reflects the volatility or vulnerability of service GDP returns in a country c in response to global climate change impacts.

Phasing in the climate beta per country $\beta_{climate,c}$ into the Winner-Loser Index formulas results in estimations for the agriculture a sector productivity gains per country c based on the formula

$$W_{a,c} = D_{a,c} * GDP_{a,c} \tag{13.34}$$

with

> $W_{a,c}$ representing the Winner-Loser Index for agriculture a for country c;
> $D_{a,c}$ representing the distance between the projected 2100 temperature in country c and the optimal cardinal temperature for agriculture (28.5 °C)
> $GDP_{a,c}$ representing the share of agriculture a in country c's GDP.

Similarly, estimations for the service s sector productivity gains per country c based on the formula

$$W_{s,c} = D_{s,c} * GDP_{s,c} \tag{13.35}$$

with

> $W_{s,c}$ representing the Winner-Loser Index for service s for country c;
> $D_{s,c}$ representing the distance between the projected 2100 temperature in country c and the optimal cardinal temperature for service (25 °C);
> $GDP_{s,c}$ representing the share of service s in country c's GDP.

Similarly, estimations for the industry in sector productivity gains per country c based on the formula

$$W_{in,c} = D_{in,c} * GDP_{in,c} \tag{13.36}$$

with

> $W_{in,c}$ representing the Winner-Loser Index for industry in for country c;
> $D_{in,c}$ representing the distance between the projected 2100 temperature in country c and the optimal cardinal temperature for service (21 °C);
> $GDP_{in,c}$ representing the share of industry in in country c's GDP.

Modifying the Winner-Loser framework with attention to climate beta $\beta_{climate,c}$ per country would then lead to the following formulas:

For the agriculture sector a:

$$W'_{a,c} = D_{a,c} * GDP_{a,c} * \beta_{climate,a,c} \tag{13.37}$$

whereby

> $W'_{a,c}$ is the modified Winner-Loser index per agriculture sector a and country c including climate-risk beta β estimations for agriculture sector a performance;

$D_{a,c}$ representing the distance between the projected 2100 temperature in country c and the optimal cardinal temperature for agriculture sector a productivity (28.5 °C);
$GDP_{a,c}$ representing the share of agriculture sector a productivity in country c's GDP;
$\beta_{climate,a,c}$ being the sector-specific climate beta β risk estimate for agriculture a.

The product $D_{a,c} * GDP_{a,c} * \beta_{climate,a,c}$ weighs the agriculture sector and how much a country depends on agriculture production for productivity with climatic disadvantages due to climate change for agriculture. A country heavily reliant on agriculture and far from the optimal temperature for agriculture as well as facing agricultural risks due to global warming will have a higher (more negative) value, signaling vulnerability.

For the service sector s:

$$W'_{s,c} = D_{s,c} * GDP_{s,c} * \beta_{climate,s,c} \tag{13.38}$$

whereby

$W'_{s,c}$ is the modified Winner-Loser index per service sector s and country c including climate-risk beta β estimations for service sector s performance;
$D_{s,c}$ representing the distance between the projected 2100 temperature in country c and the optimal cardinal temperature for service sector s productivity (25 °C);
$GDP_{s,c}$ representing the share of service sector s productivity in country c's GDP;
$\beta_{climate,s,c}$ being the sector-specific climate beta β risk estimate for service s.

The product $D_{s,c} * GDP_{s,c} * \beta_{climate,s,c}$ weighs the service sector and how much a country depends on service sector production for productivity with climatic disadvantages due to climate change for the service sector. A country heavily reliant on service production and far from the optimal temperature for service sector productivity as well as facing service provision risks due to global warming will have a higher (more negative) value, signaling vulnerability.

For the industry sector *in*:

$$W'_{in,c} = D_{in,c} * GDP_{in,c} * \beta_{climate,in,c} \tag{13.39}$$

whereby

$W'_{in,c}$ is the modified Winner-Loser index per industry sector *in* and country c including climate-risk beta β estimations for industry sector *in* performance;
$D_{in,c}$ representing the distance between the projected 2100 temperature in country c and the optimal cardinal temperature for industry productivity (21 °C);
$GDP_{in,c}$ representing the share of industry sector *in* productivity in country c's GDP;
$\beta_{climate,in,c}$ being the sector-specific climate beta β risk estimate for industry *in*.

The product $D_{in,c} * GDP_{in,c} * \beta_{climate,in,c}$ weighs the industry sector and how much a country depends on industry production for productivity with climatic disadvantages due to climate change for the industry sector. A country heavily reliant on industry production and far from the optimal temperature for industry sector productivity as well as facing industry risks due to global warming will have a higher (more negative) value, signaling vulnerability.

Overall, this adjusts the score based not just on temperature deviation and GDP share, but also on sector-specific climate vulnerability, given a more realistic risk-adjusted measure of economic exposure to climate change.

When combining the sector-specific climate vulnerabilities, W'_c results as the Winner-Loser index per country c including climate-risk beta β estimations for economic performance based on the formula

$$W'_c = W'_{a,c} + W'_{s,c} + W'_{in,c} \tag{13.40}$$

whereby

W'_c equals the modified Winner-Loser index per country c including climate-risk beta β estimations for economic performance;

$W'_{a,c}$ is the modified Winner-Loser index per agriculture sector a and country c including climate-risk beta β estimations for agriculture sector a performance;
$W'_{s,c}$ is the modified Winner-Loser index per service sector s and country c including climate-risk beta β estimations for service sector s performance;
$W'_{in,c}$ is the modified Winner-Loser index per industry sector in and country c including climate-risk beta β estimations for industry sector in performance.

This Climate-risk-adjusted Winner-Loser Index weights proximity to optimal productivity not only by economic growth but also country-specific climate vulnerability.

13.4 Discussion

Overall, countries with low climate beta β and longer time ahead until they reach peak conditions for productivity have better conditions than countries with high climate beta β and less time ahead for economic productivity. A country with long time prospects ahead until peak productivity conditions but with high climate beta β will be downgraded by the high exposure to climate risks in comparison to other nations. A country with short time prospect until optimum cardinal temperature for productivity will gain a more favorable positioning compared to other nations if the climate-related beta β is low, hence it faces less climate vulnerability, for instance, due to a land-locked position or successful climate change mitigation and adaptation efforts.

The introduction of climate-related beta estimates in the Winner-Loser Model offers many advantages—for instance, allowing for specified estimates for industries, countries, and regions of the world (e.g., coastal areas vs. land-locked countries). This new method of phasing in risks enables us to compare climate-adjusted risk exposure across countries, not just by temperature and sector but by financial vulnerability too.

Future refinements could comprise of dynamic modeling by using panel data to estimate time-varying climate beta values. A country and industry-specific as well as crop-specific or service-specific timelines would be helpful to measure the exact flow of peak moments in time of various layers of economic growth. For instance, one could derive information what crops will grow well as for being cheap and easy to produce around the world given a specific mean temperature that lets them flourish coupled with what countries offer favorable conditions for what kind of vacation during what generation. These insights could be derivable based on big data estimations and would offer invaluable insights for future economic planning around the world. One could imagine this information being valuable for strategic policy planning, trade relations, and international diplomacy.

The closer integration of risk into the climate change winners-losers model also allows for a more specific stress testing to simulate performance under climate catastrophe scenarios. In the international compound, having information about what countries face more risk than others given global warming will aid in redistribution decisions in order to breed inequality alleviation when it comes to climate change.

Lastly, financial linkages could connect the climate beta to risk premia and sovereign credit ratings in order to factor in climate risks into financial portfolio management. Climate risk estimates could be proxied and sophisticated with damages from temperature and precipitation data linked to natural disasters, insurance losses, global agricultural shocks, world development drawbacks, etc.

References

Black, F., Jensen, M. C., & Scholes, M. (1972). The capital asset pricing model: Some empirical tests. In M. C. Jensen (Ed.), *Studies in the theory of capital markets* (pp. 79–121). Praeger.

Fama, E. F., & French, K. R. (1992). The cross-section of expected stock returns. *The Journal of Finance, 47*(2), 427–465. https://doi.org/10.1111/j.1540-6261.1992.tb04398.x

Fama, E. F., & French, K. R. (2015). A five-factor asset pricing model. *Journal of Financial Economics, 116*(1), 1–22. https://doi.org/10.1016/j.jfineco.2014.10.010

Fama, E. F., & MacBeth, J. D. (1973). Risk, return, and equilibrium: Empirical tests. *Journal of Political Economy, 81*(3), 607–636. https://doi.org/10.1086/260061

Lintner, J. (1965). The valuation of risk assets and the selection of risky investments in stock portfolios and capital budgets. *The Review of Economics and Statistics, 47*(1), 13–37. https://doi.org/10.2307/1924119

Merton, R. C. (1973). An intertemporal capital asset pricing model. *Econometrica, 41*(5), 867–887. https://doi.org/10.2307/1913811

Mossin, J. (1966). Equilibrium in a capital asset market. *Econometrica, 34*(4), 768–783. https://doi.org/10.2307/1910098

Ross, S. A. (1976). The arbitrage theory of capital asset pricing. *Journal of Economic Theory, 13*(3), 341–360. https://doi.org/10.1016/0022-0531(76)90046-6

Sharpe, W. F. (1964). Capital asset prices: A theory of market equilibrium under conditions of risk. *The Journal of Finance, 19*(3), 425–442. https://doi.org/10.2307/2977928

CHAPTER 14

Wealth of Nature: Accounting for the Environment

14.1 Introduction

In the aftermath of financial crises and in the eye of climate change, the call for Corporate Social Responsibility (CSR) and Socially Responsible Investment (SRI) has reached unprecedented momentum. During the 2022 World Economic Forum address of United States Secretary of the Treasury Janet Yellen, the post-COVID-19 economic growth was called for being inclusive and green (United States Department of the Treasury, 2022; Puaschunder, 2023). Rising contemporary East–West tensions being accompanied by war destruction and political freezing of assets have already been discussed to be allocated for greening reconstruction plans with a consideration of environmental protection.

Nowadays corporate and financial sector endeavors are scrutinized to be in line with the United Nations Sustainable Development Goals (SDGs). In the most recent decades, corporate and financial executives growingly consider environmental impacts of economic striving. Stemming out of responsible investment and sustainable finance, responsible investments integrate Environmental, Social, and Governance criteria. Sustainable finance and environmental, social and governance (ESG) aspects in investment decisions have rising prominence. Green finance in particular pays attention to environmental concerns. Climate finance thereby tabs into private investments for the financialization of climate

J. Puaschunder, *Governance & Climate Justice*,
https://doi.org/10.1007/978-3-031-87971-5_14

change mitigation and adaptation and the support of transition of the economy into green and sustainable alternatives.

In particular the post-COVID-19 external shock economic fallout that triggered the biggest-ever internationally concerted wave of economic bailouts and recovery funds has also been intertwined with environmental and social causes as never before in the history of finance (Alpert, 2021; Cassim et al., 2020; Puaschunder, 2023; The White House of the United States of America, 2021a). Paying attention to environmental conditions for financial risk management and the growing role of sustainability for productivity have become subject to leading scientific discourse (Engelbrekt et al., 2022a, 2022b; Ghauri et al., 2023). Beneficiaries of public investments were often pegged to environmental growth and climate stabilization causes as described by the Green New Deal (Barbier, 2009; Earthworks, 2019; Pargendler, 2020; European Commission, 2019).

Starting with the U.S. President Biden administration launching the Green New Deal (GND) but also the European Union Commission sponsoring the European Green Deal and a Sustainable Finance Taxonomy implemented the strategic targeting of environmental causes with economic aid (Barbier, 2009; Earthworks, 2019; Pargendler, 2020; Puaschunder, 2021b, 2021c, 2023; The United States Congress, 2019). The GND is a governmental strategy to strengthen the U.S. economy and foster inclusive growth with respect for the environment (Puaschunder, 2021b, 2021c, 2023). The GND directly aims at sharing economic benefits more equally within society (Puaschunder, 2021b, 2021c, 2023). Financialization of climate stabilization efforts are also part of the GND.

The GND thereby combines former U.S. President Roosevelt's economic approach with more modern ideas of economic stimulus. Governmental funding is thereby given for transitioning to renewable energy and attention to resource conservation via economic efficiency. Healthcare equality and social justice pledges are innovations of the new era GND. The GND thereby aligns economic endeavors with environmental constraints, also concerning the next generations to come (Puaschunder, 2021b, 2021c, 2023). Ethical imperatives and equity mandates lead the economic rationale behind redistribution in the GND (Puaschunder, 2023). Social harmony, access to affordable quality healthcare and favorable environmental conditions are thereby pursued in an understanding of their role as prerequisites for productivity (Puaschunder, 2021b, 2021c, 2023). In all these endeavors, the GND offers hope in making the world and society but also overlapping generations more equitable. Overall, transfer payments and market incentives are used strategically in the GND to bring about positive change.

The European Green Deal (EGD) and the European Sustainable Finance Taxonomy but also the Next Generation EU and the New European Bauhaus are European initiatives to harmonize sustainability with economic endeavors. The European Green Deal is a concerted policy effort to help the European economy transition to a greener, more sustainable productivity. In line with worldwide endeavors to reach climate neutrality, environmental conditions are understood as a key to economic prosperity in a changing world order (Engelbrekt, Bremberg, Michalski & Oxelheim, 2019). Fair economic conditions in a competitive society are also grounded in respect for the environment and foresight for future generations.

The European Sustainable Finance Taxonomy is meant to instigate international environmental economics standards in the future. The Sustainable Finance Taxonomy screens any industry contributions to environmental factors and the wider extent of corporate good governance practices on environmental conditions (Chiu, 2022). Regulatory benchmarking thereby targets at creating positive incentives to promote resilient economic structures with whole-rounded benefits to society.

The COVID-19 crisis also triggered a de-urbanization in the United States—a trend to move to environmentally pleasant surroundings. The Long COVID health implications but also digital remote workspaces have driven demand to live in harmony with the environment in the outskirts of large metropolitan areas. COVID-19 raised awareness for health conditions in congested cities, while the sharing economy also offered workspace closer to nature. Architectural design also currently celebrates biophilic trends to make living spaces resemble nature. The United Nations Framework Convention on Climate Change (UNFCCC) also paid homage to the importance of sustainable clothing in the Conference of the Parties COP26 meeting of 2021. Innovative natural materials include fungus-grown clothing that is a carbon-negative organic alternative to fast fashion.

The 2022 United Nations Conference of the Parties COP27 emphasized the call for climate justice via redistribution including all parts of the world. Climate change inequalities are proposed to be alleviated with a climate taxation-bonds strategy that incentivizes market actors to transform the energy sector and mitigate as well as adapt to climate change.

In the eye of the Ukraine-Russia East–West tensions, already now plans exist to rebuild war zones into a more sustainable infrastructure.

In their entirety, all these novel trends have heralded an extensive interest to consider environmental conditions in economics. This chapter will first address various ways to capture environmental resources and climate conditions in economic forecasting. Environmental factors will be portrayed as a wealth of nations—outlined by concrete examples how the world can benefit from comparative advantages in natural resources as well as climate zones. General theories on the economic costs of environmental degradation will be enhanced with insights into the positive gains from a stable and predictable natural environment. The presented economic efforts to measure and enhance environmental protection concertedly will also be attributed from the angle of climate justice. The chapter rounds up the book with a future-oriented outlook on the endeavors to account for nature in economics, governance, and international law.

14.2 Green Deals

Ever since, natural resources had an impact on politics, for instance if one thinks of borders being drawn after world wars or pressure to form states after oil was discovered in the Gulf nations. The economic investigation of natural resources, however, lacks a clear account of nature in accounting or GDP measurement.

14.2.1 United States

Since the 1960s there has been a growing wave of public policies being evaluated on their whole-rounded socio-economic and ecological impact. From the foundation of the United States Environmental Protection Agency (EPA) in the 1970s, the governmental agency was estimating the impact of economic prosperity on the environment. Strategically measuring environmental outcomes of governance was first practiced by EPA in a Law & Economics-inspired way. EPA directly targeted pollution reduction in order to have positive health improvements, partly as a health cost savings strategy. Problems of externalities and how to account for precaution and damage averted were discussed.

Policies were meant to become more attuned in harmony with the environment as an implicit driver of economic growth. The field of ecological economics grew subsequently, primarily being concerned with the relationships between ecological and economic systems. The implementation of insights on the productive interplay of nature and the economy has foremost been attributed by the Green New Deals in the last decades.

Green Deals are strategies to align economic growth with sustainability. The Green New Deal (GND) initiative started in the United States around 2006 as a governmental concerted effort to share the benefits of economic growth equally within society while paying respect to environmental protection and natural resources conservation.

The beginnings of the GND idea are attributed to Senator Edward Markey and Representative Alexandria Ocasio-Cortez pushing for transitioning the United States to use 100% renewable, zero-emission energy sources including investment into electric cars and high-speed rail systems (Puaschunder, 2020b, 2021a, 2023). In January 2019, a letter signed by 626 organizations in support of a GND was sent to all members of the U.S. Congress (Puaschunder, 2021a, 2023).

The GND encourages to create jobs in green industries, thus boosting the world economy and curbing climate change at the same time (Puaschunder, 2021a, 2023). Economic foundations are grounded on John Maynard Keynes' (1936/2003) multiplier effect, which proves governmental spending to trickle down in the economy and ignite positive transformative change at the same time via innovation and social equity (Puaschunder, 2023). Ethical imperatives and equity mandates are the economic rationale behind GND redistribution. Wealth transfer and harmony with the natural environment are thereby understood as an anchor of social peace, health, and prerequisites for productivity (Puaschunder, 2023). As of today, the U.S. GND funding ranges from 7 to 39 trillion USD.

14.2.2 United States Environmental-Economic Statistics

Implementing the social cost of carbon has already been part of U.S. President Obama administration's plans for addressing climate change. Under the Joseph Biden U.S. Presidency, the U.S. Government widened its attempts to account for nature in economic calculus (The White House, 2023). The National Strategy to Develop Statistics for Environmental-Economic Decisions targets at measuring natural resources in national accounting for strategic policy making and international diplomacy decisions (U.S. Department of Commerce, 2023). Basic natural resources—like land, water, minerals, ecosystems, and habitats as well as livestock—are thereby classified as assets with an economic impact for nations (U.S. Department of Commerce, 2023). Tracking natural capital in environmental stability is credited for being a driver of economic growth and future forecasting comparative advantage (U.S. Department of Commerce, 2023). Linking nature with the economy is meant to broaden the horizon of national accounting and embracing novel forms of inclusive economic growth (U.S. Department of Commerce, 2023).

Led by the White House Office of Science and Technology Policy (OSTP), the Office of Management and Budget (OMB), and the U.S. Department of Commerce, the National Strategy to Develop Statistics for Environmental-Economic Decisions is coordinating the endeavors of 27 Federal Departments and Agencies to capture nature's role in economic prosperity (The White House, 2023). Preserving the environment and stabilizing the climate are goals to accomplish more effectively if national economic statistics account for nature. On the international level, quantifying natural resources for benchmarking and value measurement of natural capital will inform economic growth forecasts and facilitate economic progress. A nomenclature of natural economic values based on standardized scientific procedures to measure wealth of nature will also help allocate the right resources for environmental protection, inform foreign policies as well as help prioritize investments (The White House, 2023).

14.2.3 Europe

Central Europe has a long tradition in environmental protection and conservation due to natural resource constraints. For instance, the German Green wave project in the 1960s and 1970s started with advocating for traffic lights in order to reduce environmental pollution that reduces national wealth and citizens' healthy well-being. The concept of reducing waste and environmental harm for the protection of human lives and livelihood has since then evolved in very many different domains and inspired other continents to follow.

The European pendant of the Green New Deal (GND) is the European Green Deal (EGD) and the related European Sustainable Finance Taxonomy (Puaschunder, 2023). The EGD policy initiatives and programs of the European Commission set out to make Europe the first climate-neutral continent of the world. Concepts like the circular economy, coordinated transition to renewable energy, sustainability, biodiversity, greening of cities, and environmental innovation are at the forefront of the EGD. The European Climate Pact is the European Commission strategy to implement the EGD in the climate change mitigation and adapation domains. As for climate stabilization, carbon tariffs and trading schemes are major vehicles to measure and cope with CO_2 emissions. The EGD is funded with 1.8 trillion Euro and was approved in 2020.

The Horizon Europe structural program sets out to implement a policy agenda on the national and local levels. As the world's largest research program with a total budget of EUR 93.5 billion, the Horizon Europe strategy allocates 35% funding toward green transition and sustainable value creation. Concrete policy areas comprise of clean energy, sustainable finance, sustainable industry and mobility, pollution reduction, building renovation, healthy nutrition, biodiversity, and ecosystem health.

14.2.4 European Union Sustainable Finance Taxonomy

The EU Sustainable Finance Taxonomy holds the most comprehensive classification system to define sustainable or green investment product criteria (Puaschunder, 2023). The European Taxonomy for sustainable activities aims at providing a worldwide standard to classify assets and investments based on their environmental impact (European Union Technical Expert Group on Sustainable Finance, 2020). The measuring of ecological impacts of industries should build a comprehensive classification of all economic activities' contribution to environmental objectives. Quantifying each industry's environmental impact will help in the market transition to sustainable finance and energy.

The European Sustainable Finance Taxonomy provides concrete references to measure and compare climate change mitigation and adaptation activities, including environmental objectives of production industries (European Union Technical Expert Group on Sustainable Finance, 2020; Puaschunder, 2023). The classification of industries' contribution to climate change serves as a basis for setting positive market incentives to change market dynamics. For instance, identifying those industries that cause the problem of global warming helps to create diversified bonds that mandate more harmful industries to contribute more to the climate stabilization solution.

The Sustainable Finance Taxonomy supports climate change mitigation and adaptation efforts by outlining the environmental footprint of industries. The indications of environmental impacts of the various sectors under scrutiny are aimed at helping to transition to a more sustainable economy (Brühl, 2021). In addition, transparency, comparability, and monitoring aim at overcoming greenwashing and directing capital to meaningful greening the economy endeavors (Brühl, 2021). Overall, the common industry classifications and productive harmonization strategies are meant to breed trust in the European Union members (Engelbrekt et al., 2019).

The taxonomy also aids investor decision-making (Brühl, 2021). With first parts of the Taxonomy Regulation having come into effect in 2020, the ongoing progress and implementation of accounting for nature in economics has improved the growing share of financial market options with an environmental mandate.

The EGD and Sustainable Finance Taxonomy were also part of the COVID-19 recovery, which allocated funds toward green and sustainable market solutions. In particular, the Next Generation EU is a comprehensive plan to peg recovery aid to system change that transforms economies into a more inclusive civil project for everyone. The Next Generation EU funding was meant to steer positive market forces and invests over 800 billion Euros to transform Europe into a healthier, greener, and more digital environment.

The New European Bauhaus (NEB) is another EU policy strategy that was launched by the European Commission in 2021. Funding initiatives of the NEB target at finding innovative solutions for greening the environment and transforming lifestyles toward sustainable living. This bottom-up initiative aims to foster inclusive and sustainable agrohoods. Individualized approaches coupled with self-empowerment to create living spaces that are more beautiful and healthier by design and practice are key components of the NEB. Respect for inclusion and aesthetics are rounded up by diversity and arts engagement. All these pillars of sustainable city and environmental space transformation into a more livable environment in the social compound are meant to help boost competitiveness by being more resilient and sustainable.

14.2.5 Global Green New Deal

Governments and multilateral organizations are essential entities to raise funds via taxation on the national level. Country-specific national bonds funded by global governance-institution's enacted transfer payments could feature an interest rate bandwidth around the universal bank lending rate for climate bonds determined by international entities (Puaschunder, 2023). Within the bandwidth, central banks could offer green bonds with specific premium bonds payments for industries based on the industry carbon emission levels (Puaschunder, 2023). Industry-specific interest rates could turn the traditional price-cutting behavior that drives corporations to seek for cheap and often non-renewable energy sources to opt-in for a competitive race-to-the-top for beneficial interest rates that gratify choices for renewable energy solutions (Puaschunder, 2023).

Both GNDs foster global governance in fiscal policy strategies targeted at a carbon tax to fund climate change mitigation and adaptation efforts. Monetary and credit policies—foremost enacted by central banks and monitored by public agencies—indirectly help stabilize economies during market transitions and environmental risks. As for disparate impacts, GND insurance policies back underserved communities' resilience during environmental crises. The GND funding for R&D and governmental infant industry grants target at supporting innovative green market solutions, such as absorbing CO_2 from the atmosphere, ecowellness lifestyle solutions, targeted preventive care, or Green FinTech.

In all these features, the Green New Deal and the European Green Deal, in combination with the U.S. efforts to provide environmental-economic statistics and the European Sustainable Finance Taxonomy, are the most widescale efforts to marry the idea of economic growth in line with the natural resources pool and with respect for environmental limitations (Puaschunder, 2023).

On the intergovernmental level, the Green New Deal program in the United States and the European Green Deal in Europe could help in transferring funds that are pegged to specific causes and concrete national action (European Commission, 2021; The White House of the United States, 2021a, 2021b, 2021c; United Nations, 2020; United Nations General Assembly, 2020, Art. 47; The United States Congress, 2019; Vivid Economics; World Trade Organization, 2020, 2021). Central banks, on behalf of the World Bank or the International Monetary Fund, could take the lead in offering differing bonds regimes on a global scale (Puaschunder, 2023). Central banks could be issuing intermediaries that connect the global governance institutions with the country-specific and industry-specific bonds on the national level.

Global governance institutions play a crucial role in implementing similar ideas to marry economic productivity with sustainability on a global stage. International cooperation on global governance policy making helps build common environmental protection frameworks (UNEP, 2009). In the grand theme of instigating widespread action toward sustainable economic growth, the idea of a 'Global Green New Deal' (GGND) has been proposed. Combining Green New Deal efforts in the Western and Eastern worlds, a Global Green New Deal could enact a binding taxation-and-bonds solution for climate change in the international arena.

The GGND was first presented by the United Nations Environment Program (UNEP) in 2009 as a policy package to revive the world economy (Boyle et al., 2021; O'Callaghan & Murdock, 2021; UNEP, 2009). Economic stimulus should thereby be coupled with decarbonizing carbon-intensive sectors of the economy such as energy, transport, buildings, and agriculture (UNEP, 2009).

The global governance high-level thinking—as pursued in the GGND—could now also serve as international leadership to oversee climate change gains and loss redistribution funding schemes. Global governance institutions, such as the World Bank, IMF, or the United Nations, could act as norm entrepreneurs and action catalysts of a GGND that redistributes the unequally distributed relative economic gains of a warming earth to places that face economically declining economic prospects (Puaschunder, 2023). Global governance institutions can shape the conduct and array of international actors to contribute to a commonly agreed global scheme (Puaschunder, 2023).

Economically driven indices could aid in taking the political nature out of redistribution politics and historically laden international relations (Puaschunder, 2023). Quantification of gains and losses imbued in global warming could serve as a science diplomacy discussion basis for data-driven redistribution mandates. A politically neutral international tax-and-bonds-scheme could be grounded on empirical evidence and comparative economics aside from historical or political agendas (Puaschunder, 2023). The proposed empirically driven redistribution schemes could thereby work with positive market incentives in order to win countries to comply with sustainability pledges in action and attract attention of climate change deniers (Puaschunder, 2023).

Building on Jeffrey Sachs' idea in the aftermath of the COP26 agreement to set up climate tax-funded grants that should be raised by half the nation-states in order to transfer funds for a renewable energy shift, the presented model in this book holds clear data-based guidelines which countries could be funding a common climate stabilization solution and which countries should be considered for transfer payments as recipients of grants in order to offset climate-related losses.

As for a concrete implementation, under the auspice of the United Nations Climate Change Conference meetings, a sophisticated redistribution scheme for climate change mitigation and adaptation could be pursued. The implementation of the GGND on redistributing some of the expected gains from climate change could become a concerted action of multiple entities including global governance institutions, national authorities, and local actors.

On the global governance level, United Nations, the World Bank, and the IMF have an international standing to initiate redistribution of climate gains to territories that have a decreasing GDP prospect in light of global warming via a global bond. The United Nations (UN) could target a binding climate agreement between countries during their Conferences of Parties (COP) meetings for enacting such transfers (Puaschunder, 2023). All UN joining nations would then sign up for implementing a climate gains redistribution via taxes and bonds (Puaschunder, 2023). International intermediary financial institutions could manage and guarantee loans in a common taxation-bonds strategy spanning over the entire world (Puaschunder, 2023).

Prevalent models exist, for instance, of the European Union European Investment Bank acting as an intermediary for credit demand by low-income countries and a credit supply backed by taxes of advanced countries, which could be adopted (European Investment Bank, 2022a, 2022b). The World Bank and/or the International Monetary Fund (IMF) could work out a redistribution scheme via their existing contributions key and loan programs (Puaschunder, 2023). While the United Nations features a democratic 'one-country-one-vote' voting system, the World Bank and IMF have voting schemes that also weigh in the national financial contributions of all participating entities (Puaschunder, 2023). As a universal dispute resolution for non-compliance with the outlined plans, prospectively the World Trade Organization (WTO) or International Law

Commission of the United Nations in New York could serve as panels for oversight, monitoring, and evaluation control (Puaschunder, 2023).

Monitoring and Evaluation (M&E) can help the assessment of the performance of projects and goal accomplishment of institutions and programs by governments, international organizations, non-governmental organizations (NGOs) as well as social media campaigns (Puaschunder, 2023). The local support is key in implementing climate justice via transfer payments as for direct action to the most marginalized communities and those who are hit the hardest by a changing climate. In the continuous assessment of programs, younger generations, whose discounted value of a stable climate is the largest, should be included. Younger generations are also prone to find new cutting-edge solutions in the digital evolution of change. For instance, modern satellite imagery empowering the Geographic Information System (GIS) advancements and geo-engineering are technology-driven scientific advancements to cartograph and change natural conditions for economic outcomes. GIS offers insights about natural resources around the world, enabled through satellite images. Over time tracking of the state of the natural environment allows for deriving inferences of the impact of climate change real time. Investigative journalism but also political decision-making are—by now—influenced by insights derived from GIS. Advancements of geo-engineering are, for example, changing precipitation in desert areas or filtering out CO_2 from the atmosphere to be used for concrete production and carbonated drinks. All these cutting-edge solutions offer innovative ways to cope and combat natural resource depletion and a degradation of climate stability and are avenues of technologically enhanced change for future generations.

14.3 Environmental, Social, Governance (ESG) Finance

Environmental, social, and governance concerns have been addressed in the corporate and financial sector for a long time. Europe and the United States have experienced a transition to Sustainable Finance enhanced by regulatory support and control in the most recent decades (Ahlström & Monciardini, 2022; Lounsbury & Crumley, 2007). Regulatory attempts target at setting international standardization of sustainable finance practice and providing transparent information to improve financial markets in harmony with the environment (Soppe, 2004; Zetzsche & Anker-Sørensen, 2022).

The United Nations Global Compact and the UN Environment Programme (UNEP) Finance Initiative, for instance, launched the Principles for Responsible Investment (PRI) in April 2006 at the New York Stock Exchange (NYSE) in order to raise attention to environmental causes in the finance world (Puaschunder, 2023). The 2008 UN Conference on Trade and Development (UNCTAD) incepted the 'Responsible Investment in Emerging Markets' initiative to also embrace emerging markets with a sustainable finance pledge (Puaschunder, 2023). Global environmental reporting standards were set with the United Nations Environment Programme Finance Initiative (UNEP FI), the Equator Principles, and the Green Bond Principles. Corporate reporting standards trends led to the Global Reporting Initiative (GRI) and Integrating Reporting (IR) as well as the 2015 launched UN Sustainable Development Goals (UN SDGs). For concrete inequality alleviation redistribution strategies, the United Nations Framework Convention on Climate Change (UNFCCC) Conferences of the Parties (COPs) have successfully enacted transition funds for developing nations (Copenhagen Accord).

In addition, central banks and regulators around the world have started to address climate change and environment risks' impact on banking and the financial sector in the most recent decade (Puaschunder, 2023). In response to growing concerns, the Networking for Greening the Financial System (NGFS) was initiated by banking sector executives in 2017 to monitor climate risks from a finance perspective. The Sustainable Finance Disclosure Regulation (SFDR) imposes mandatory ESG disclosure obligations for asset managers and other financial market participants (Puaschunder, 2023). The European Green Bond Standard creates a voluntary European high-quality standard for public and private bond issuers to help finance sustainable investments (Puaschunder, 2023). The EU Climate Transition Benchmarks addresses sustainability-related disclosures for benchmarks (Puaschunder, 2023). The potential to lead the transition to a sustainable economy is addressed in creating instruments, measures, and incentives for a reduction of environmental footprints of economic activities. Climate resilient production standards as well as green FinTech solutions have entered finance circles most recently.

In addition, the United States Stock Exchange Commission supports greening of the economy in financial markets. The U.S. capital market supervision proposed mandatory disclosures regarding climate change and climate risks of corporations to the Securities and Exchange Commission (SEC).

The standardized supervision of environmental impacts of corporations is meant to provide investors with better information on the environmental impact of their choices. The downsides of these market interventions are administrative costs for disclosures—estimated at up to USD 1 million a year for large companies, which could result in corporations withdrawing from the capital market and fewer IPOs (Puaschunder, 2023). Legal scholars are currently also questioning the authority of independent governmental agencies—like the SEC or the EPA (Environmental Protection Agency)—without consulting the Congress. Questions of constitutional freedom of speech infringements to mandate certain topics in reporting were raised. The investor protection mandate is another question of concern in regards to mandatory self-shaming. With the SEC rule making focus, legal enforcement may fall short given limited resources, which implies additional market risks and uncertainty. Other critique addresses the potential to trigger greenwashing and window-dressing on a broad scale with required mandatory environmental impact reporting standards.

In Europe, the European Union (EU) Sustainable Finance Disclosure Regulation (SFDR) imposes mandatory ESG disclosure obligations for financial market participants (Regulation (EU) 2019/2088). The SFDR has a particular emphasis on disclosure of consideration of sustainability risks in the investment process as well as principal adverse impacts (PAIs) on sustainability factors that an investment decision or advice might have (Brühl, 2021; Zetzsche & Anker-Sørensen, 2022).

The EU Climate Transition Benchmark Regulation addresses sustainability-related disclosures, which are added to general regulatory provisions (Regulation (EU) 2019/2089). For instance, sustainability factors are demanded to be considered when assessing financial products by insurance distributors and investment firms (Zetzsche & Anker-Sørensen, 2022). An additional regulation for sustainability in finance in Europe is the proposed revision of Directive 2014/95/EU on non-financial reporting (NFRD) (Zetzsche & Anker-Sørensen, 2022). The Delegated Regulations ((EU) 2020/1816 and (EU) 2020/1817) for ESG disclosure came into effect in December 2020 (Brühl, 2021).

All these benchmarks and comparability of ESG measures are meant to enhance legal oversight, implementation clarity, and transparency while fostering positive market practices via positive incentives. The ESG taxonomy helps classify sustainable investments based on quantifiable accomplishments and integrity of the issuing entities. Transparency and accountability are meant to contribute to financing sustainable growth and providing corporations with crises resilience. Monitoring also allows for detecting emerging new sustainable finance solutions and serves industry forecasts with accurate risk management predictions (Zetzsche & Anker-Sørensen, 2022). Tax breaks for beneficial market participants and early adopters of regulatory guidelines are additional powerful soft governmental tools to foster successful market transitions. Industry standards (e.g., such as IFRS) could also aid in streamlining sustainability reporting standards in the future (Zetzsche & Anker-Sørensen, 2022). As a further development, information brokerages and financial intermediaries should be prompted to integrate sustainability investment decision-making and risk management tools in harmony with the created standards (Zetzsche & Anker-Sørensen, 2022). Software advancements and data-driven supervisory tools can help train qualified and skilled staff for quality control and sanction enforcement (Zetzsche & Anker-Sørensen, 2022).

The European Green Bond Standard (COM, 2021) helps public and private bond issuers to identify and qualify sustainable finance investments. For instance, green bonds qualify if being compliant with the sustainability concept raised by the European Sustainable Finance Taxonomy Regulation. The annual European Green Bond Allocation Report holds broad-based results of the state-of-the-art in the field. Compliance monitoring occurs with external, registered reviewers that are supervised by the European Securities and Markets Authority (Brühl, 2021).

14.4 Climate Wealth of Nations

14.4.1 Environmental Conditions' Impact on Productivity

Economic research focusing on the impact of environmental conditions for productivity can also inform scientific climate justice endeavors (Burke et al., 2015). Burke et al. (2015) estimate how climate change will affect the Gross Domestic Product (GDP) per capita and thereby assume nonlinear impacts of weather changes. As wealthier countries are found to not respond to temperature changes as dramatically as lower income countries, Burke et al. (2015) emphasize to investigate the disparate impact of global warming. Productivity is found to decrease in particular in higher temperature regions (Burke et al., 2015). Elucidating the variegated impact of climate change in different climate zones underlines the rising climate injustice.

In addition, the International Monetary Fund (IMF) conducted a cross-country analysis of the long-term macroeconomic effects of climate change and found country inequalities in global warming effects (Kahn et al., 2019). A 174 countries dataset from 1960 to 2014 reveals that the per capita real output growth is negatively related to temperature changes (Kahn et al., 2019). Kahn et al. (2019) also measure real GDP reductions estimated to be over 7 percent by 2100 due to climate change. Climate change is also projected to have a strong adverse effect on U.S. economic output and productivity (Kahn et al., 2019).

In addition, the International Monetary Fund's Research Department is measuring the effects of weather shocks on economic activity (Acevedo et al., 2020; Topalova, 2018). Starting from the assumption that temperature varies around the globe and therefore has a variegated influence on economic activities, the research highlights regional differences of the impact of global warming (Topalova, 2018). The optimal temperature for productivity ranges around 13–15 °C (55–59°F), which underlines the country and regional disparities starting from general differences in climate around the world (Topalova, 2018). Overall, rising temperature is emphasized to have a long-lasting effect on economic productivity (Topalova, 2018). With a focus on climate change and related weather shocks' mitigation on economic growth, the work group directly addresses contemporary problems and disparities in terms of climate-related labor productivity, investment, and mortality (Topalova, 2018). Detected research obstacles include nonlinear, long-lasting temperature impacts such as tipping points and the Albedo effect heating up the earth faster if reflecting snow coverage is missing (Topalova, 2018). The unpredictability of risk measurement and the lack of motivation to account for the benefits of prevention and precaution upfront are additional challenges reported (Topalova, 2018).

In the measurement of climate-induced productivity differences around the globe, all three GDP sectors of agriculture, service, and industry are prevalent to be affected by climate change (Topalova, 2018). Understanding economic growth prospects' dependency on the environment, also from a capital (production) and labor (human capital) perspective will also enable to derive strategic policy investments and predict health outcomes given global warming (Topalova, 2018). Overall, the research underlines global climate inequality in order for finding multi-modal policies to mitigate and adapt to global warming (Topalova, 2018). Macroeconomic forecasting allows for informed risk mitigation and adaptation policies (Topalova, 2018). The inequalities alleviation efforts targeted include policies, tax, and redistribution bonds, which grants fiscal space, financial transfers, and safety net creation (Topalova, 2018). Resilience, infrastructure development, and migration are additional areas that can be informed with the generated results (Topalova, 2018).

This book presented a model to measure the Gross Domestic Product (GDP) prospect differences under climate change worldwide and found exacerbating climate inequalities. Based on temperature differences around the world, economic growth prospects were estimated for the agriculture, service, and industry sectors. The economic productivity peak temperature per GDP sector was estimated over time and different countries around the world compared. Based on country differences of mean temperature and GDP sector productivity, macroeconomic modeling highlighted countries' productivity forecast for the year 2100 given climate change. As climate gaining countries are also causing the problem with a higher prevalence to Greenhouse Gas emissions and climate winners are also experiencing a migration and financial inflow, redistribution schemes are justified based on scientific grounds. The derived economic calculus was targeted at outlining climate inequalities within society, between nations, and over time in order to find quantitatively justified redistribution schemes. Besides the climate change stabilization taxation-and-bonds strategies proposed in the previous chapters, another state-of-the-art translation of climate change gains and losses in burden-sharing contribution schemes is usually defined in Nordhaus' Regional Integrated model of Climate and the Economy model (RICE model). This regional, dynamic, general-equilibrium model of the economy integrates economic activity with emissions levels as the primary driver of human-made climate change (Orlov et al., 2024).

14.4.2 *Wealth of Nations*

The World Bank Wealth of Nations Reporting measures global wealth since the 1990s.[1] Starting from 1995, the 146 countries database currently exhibits a declining share of natural capital in global total wealth (Onder, 2023). In order to study and highlight differences in natural wealth, the World Bank sets out to account for nature in macroeconomic comparative research (Onder, 2023). Capturing natural resource decline is meant to justify and incentivize environmental preservation (Onder, 2023).

Newest innovation in the recent measuring of changing wealth of nations include adding sustainability indicators, renewable energy, carbon storage, and aquaculture (Onder, 2023).[2] Challenges are that market prices need to be adjusted to reflect the true marginal benefit in terms of shadow prices (Onder, 2023). An accurate measurement also needs to cover substitutions and scarcity modeling for pricing natural resources (Onder, 2023). Economic modeling of reliance, risk, and resilience also face general forecasting obstacles (Onder, 2023).

Future integration of environmental factors in economic models of the ecosystem and landscaping promise as gateways to derive more accurate inferences for economic outcomes and international trade policies (Onder, 2023). Systemic forward-looking long-term scenarios are prospected to include environmental policies (Onder, 2023). A stringent cartography of the wealth in nature around the world helps policy-contingent wealth estimates (Onder, 2023). Benchmarking and generalizations on an international level will foster efficient environmental protection and resilient sustainability on a global scale. The research conducted at the World Bank may also serve geo-engineering frontiers in our world to come (Onder, 2023).

14.5 Future Research

Future research in the rising field of quantifying the wealth in nature could use the European Sustainable Finance Taxonomy for refining the redistribution key between industry contributions to the climate

[1] https://www.worldbank.org/en/publication/changing-wealth-of-nations.

[2] https://www.worldbank.org/en/publication/changing-wealth-of-nations.

change problem (European Union Technical Expert Group on Sustainable Finance, 2020). Different layers of risk and resilience measures could be integrated based on insights derived from regional estimations of the International Monetary Fund. For instance, redistribution keys could give different weights to coastlines and developed and underdeveloped living spaces, that connect economic cost functions of climate change with resilience action.

Contemporary adjustments of coastal lines in light of rising sea levels could become another shifting wealth indicator of nation-states under climate change. Countries having coastlines and urban territories with direct access to oceans are particular areas of shifting sea borders and territorial change. International legal scholars have already started to debate the problem of shifting sea levels and changing borders. Law and economics scholars could now add economic calculus of rising sea levels, especially in coastal areas and countries sharing an ocean. The gained insights could help in refining the outlined redistribution plan to compensate climate losers for their costs and risk incurred due to global warming.

Future Environmental Law research could also target at deriving legal mechanisms to enact environmental justice. The contemporary trend to acknowledge the economic value of nature could be coupled with behavioral insights how nature impacts of human lives (Puaschunder, 2020a). Nature's role for human well-being could also be depicted from an environmental justice perspective. Disparities of access to natural resources could thereby become subject to scrutiny as for differences between living conditions due to environmental impacts. Along these lines, the contemporary programs of the White House to bring nature in economics, the European Green Deal and Sustainable Finance Taxonomy to measure the relation between and impact of economic production on natural resources and climate stability as well as the International Monetary Funds' research and World Bank's Wealth of Nations project could all be scrutinized for attention to disparities in natural resource access. Disparate impact analyses would also be refining the Mapping Climate Justice project highlighting the vast economic impact differences of climate change around the world in connection with stratified impacts and diversified outcomes. The scientific advancement would be the concrete connection of qualitative outcomes of economic disparities based on natural resources and due to environmental conditions. Not only cross-country and within society differences in the impact of global warming are thereby aimed for to be elucidated—but also differences in living conditions over time and inbetween different generations should be thematized. All these novel insights would lead to a Law & Economics-informed discussion of how to alleviate disparities with concrete legal adjustments, economic redistribution as well as policy programs.

Additional legal and economic analyses could draw from ethical and philosophical roots (e.g., featuring Kant, Jonas, Rawls) of climate injustices. Future work could also embark on a Law & Economics analysis of the unequal world regarding natural resources and environmental conditions with particular attention to differing impacts of people on the environment (e.g., Pareto, Kaldor-Hicks). Historical foundations of traditional economic productivity calculus (e.g., Smith, Ricardo) should be applied to the novel context of measuring nature and related empirical economic arguments. Outlining economic differences would motivate to find legal redistribution justifications for inequality alleviation (e.g., Ha-Joon Chang, Sachs). A conventional legal analysis compared with Law & Economics perspectives will not only cover the advantages but also address potential shortcomings of contemporary attempts to bring nature into economics and account for natural resource differences around the world.

The planned future research could also discuss creative cutting-edge redistribution mechanisms and techniques (e.g., ranging from international law, international trading schemes, national taxation and regulatory incentive schemes, bond strategies as well as behavioral economics guidelines) with respect for their legal implementation, political feasibility, and reception in the general populace. Intergenerational equity should be covered from a historical perspective to derive inferences for successful legal and policy frameworks that create incentives to conserve natural resources for future world inhabitants. The discussion could also feature the real-world relevant applicability of inequality alleviation endeavors within society, between countries but also over time inbetween generations. In light of contemporary sustainability concerns, all mentioned avenues can amplify the whole-rounded care for wealth in our common nature.

References

Acevedo, S., Mrkaic, M., Novta, N., Poplawski-Ribeiro, M., Pugacheva, E., & Topalova, P. (2020). The effects of weather shocks on economic activity: What are the channels of impact? *Journal of Macroeconomics, 65*, 1–21.

Ahlström, H., & Monciardini, D. (2022). The regulatory dynamics of sustainable finance: Paradoxical success and limitations of EU reforms. *Journal of Business Ethics, 177*, 193–212. https://doi.org/10.1007/s10551-021-04763-x

Alpert, G. (2021, May 26). International COVID-19 stimulus and relief: International fiscal and monetary stimulus and relief efforts. *Investopedia*. https://www.investopedia.com/government-stimulus-and-relief-efforts-to-fight-the-covid-19-crisis-5113980

Barbier, E. (2009). *A Global Green New Deal: Report prepared for the Green Economy Initiative of UNEP*. United Nations Environmental Programme. https://sustainabledevelopment.un.org/index.php?page=view&type=400&nr=670&menu=1515

Boyle, A. D., Leggat, G., Morikawa, L., Pappas, Y., & Stephens, J. (2021). Green New Deal proposals: Comparing emerging transformational climate policies at multiple scales. *Energy Research & Social Science, 81*, 1–8.

Brühl, V. (2021). Green finance in Europe: Strategy, regulation and instruments. *Intereconomics, 56*, 323–330. https://doi.org/10.1007/s10272-021-1011-8

Burke, M., Hsiang, S. H., & Miguel, E. (2015). Global non-linear effect of temperature on economic production. *Nature, 527*, 235–239.

Cassim, Z., Handjiski, B., Schubert, J., & Zouaoui, Y. (2020, June). The $10 trillion rescue: How governments can deliver impact Governments have announced the provision of trillions of dollars in crisis relief, but translating that into sustained recovery will not be easy, McKinsey Public Sector Practice Report. https://www.mckinsey.com/~/media/McKinsey/Industries/Public%20Sector/Our%20Insights/The%2010%20trillion%20dollar%20rescue%20How%20governments%20can%20deliver%20impact/The-10-trillion-dollar-rescue-How-governments-can-deliver-impact-vF.pdf

Chiu, I. H. Y. (2022). The EU sustainable finance agenda: Developing governance for double materiality in sustainability metrics. *European Business Organization Law Review, 23*, 87–123. https://doi.org/10.1007/s40804-021-00229-9

Earthworks. (2019, January 10). *Group letter to Congress urging Green New Deal passage*. https://www.earthworks.org/publications/group-letter-to-congress-urging-green-new-deal-passage/

Engelbrekt, A. B., Bremberg, N., Michalski, A., & Oxelheim, L. (2019). *The European Union in a changing world order*. Springer Nature, Palgrave Macmillan.

Engelbrekt, A. B., Ekman, P., Michalski, A., & Oxelheim, L. (2022a). The EU and the precarious routes to political, economic and social resilience. In A.

B. Engelbrekt, P. Ekman, A. Michalski & L. Oxelheim (Eds.), *Routes to a resilient European Union* (pp. 1–14). Springer Nature, Palgrave Macmillan. ISSN 9783030931643.

Engelbrekt, A. B., Ekman, P., Michalski, A., & Oxelheim, L. (2022b). The EU and the search for sustainability and resilience. In A. B. Engelbrekt, P. Ekman, A. Michalski & L. Oxelheim (Eds.), *Routes to a resilient European Union* (pp. 245–257). Springer Nature, Palgrave Macmillan. ISSN 9783030931643.

European Commission. (2019). *A European Green Deal*. https://ec.europa.eu/info/strategy/priorities-2019-2024/european-green-deal_en

European Commission. (2021, July 23). *Overview of sustainable finance*. https://ec.europa.eu/info/business-economy-euro/banking-and-finance/sustainable-finance/overview-sustainable-finance-en

European Investment Fund. (2022a, January 18). Poland: EDPR receives EIB Green Loan to support its onshore wind farm projects across Poland, Press Release. https://www.eib.org/en/press/all/2022-011-poland-edpr-receives-eib-green-loan-to-support-its-onshore-wind-farm-projects-across-poland

European Investment Fund. (2022b, January 21). State Investment Management Agency and European Investment Bank launch first support for Lithuanian mid-caps under European Guarantee Fund, Press Release. https://www.eib.org/en/press/all/2022-016-state-investment-management-agency-and-european-investment-bank-launch-first-support-for-lithuanian-mid-caps-under-european-guarantee-fund

European Union Technical Expert Group on Sustainable Finance. (2020). Financing a sustainable European economy: Technical report: Taxonomy: Final report of the Technical Expert Group on Sustainable Finance March 2020. https://ec.europa.eu/info/sites/default/files/business_economy_euro/banking_and_finance/documents/200309-sustainable-finance-teg-final-report-taxonomy_en.pdf

Ghauri, P., Oxelheim, L., & Randoy, T. (2023). The changed geopolitical map: Implications for business policy in a sustainable finance perspective. Working Paper 1465. Stockholm: Research Institute of Industrial Economics.

Kahn, M. E., Mohaddes, K., Ng, R. N. C., Pesaran, M. H., Raissi, M., & Yang, J.-Ch. (2019). Long-term macroeconomic effects of climate change: A cross-country analysis. International Monetary Fund Working paper WP/19/215. https://www.imf.org/-/media/Files/Publications/WP/2019/wpiea2019215-print-pdf.ashx

Keynes, J. M. (1936/2003). *The general theory of employment, interest and money*. Harvard University Press.

Lounsbury, M., & Crumley, E. T. (2007). New practice creation: An institutional perspective on innovation. *Organization Studies, 28*(7), 993–1012.

O'Callaghan, B. J., & Murdock, E. (2021). Are we building back better? Evidence from 2020 and pathways for inclusive green recovery spending.

United Nations Environment Programme. https://wedocs.unep.org/bitstream/handle/20.500.11822/35281/AWBBB.pdf

Onder, S. (2023, May 9). The Changing Wealth of Nations (CWON). World Bank Group. Presentation at the U.S. Committee for IIASA at the U.S. National Academies of Sciences, Engineering, Medicine, Washington D.C.

Orlov, S., Rovenskaya, E., Puaschunder, J. M., & Semmler, W. (2024). Green bonds, transition to a low-carbon economy, and intertemporal welfare allocation: Evidence from an extended DICE model. *AIMS Environmental Science, 11*(4), 628–648. https://doi.org/10.3934/environsci.2024031

Pargendler, M. (2020). The rise of international Corporate Law, European Corporate Governance Institute Legal Working Paper No. 555/2020. https://ssrn.com/abstract=3728650

Puaschunder, J. M. (2020a). *Behavioral economics and finance leadership: Nudging and winking to make better choices*. Springer Nature.

Puaschunder, J. M. (2020b, October 30). The Green New Deal: Historical foundations, economic fundamentals and implementation strategies. *The FinReg Blog: Global Financial Markets Center Duke University School of Law Blog*. https://sites.law.duke.edu/thefinregblog/author/julia-m-puaschunder/

Puaschunder, J.M. (2021a). Alleviating COVID-19 inequality. *ConScienS Conference Proceedings* (pp. 185–190).

Puaschunder, J. M. (2021b). Focusing COVID-19 bailout and recovery. *Ohio State Business Law Journal, 16*(1), 91–148.

Puaschunder, J. M. (2021c, March 1). Monitoring and Evaluation (M&E) of the Green New Deal (GND) and European Green Deal (EGD). *21st Research Association for Interdisciplinary Studies (RAIS) Conference Proceedings* (pp. 202–206).

Puaschunder, J. M. (2023). *The future of resilient finance: Finance politics in the age of Sustainable Development*. Palgrave Macmillan.

Soppe, A. (2004). Sustainable corporate finance. *Journal of Business Ethics, 53*, 213–224. https://doi.org/10.1023/B:BUSI.0000039410.18373.12

The United States Congress, 116th Congress, 1st Session, House Resolution 109. (2019, February 7). Recognizing the duty of the Federal Government to create a Green New Deal. Washington, D.C. https://www.congress.gov/116/bills/hres109/BILLS-116hres109ih.pdf

The White House of the United States of America. (2021a). President Biden and G7 Leaders launch Build Back Better World (B3W) Partnership. https://www.whitehouse.gov/briefing-room/statements-releases/2021/06/12/fact-sheet-president-biden-and-g7-leaders-launch-build-back-better-world-b3w-partnership/

The White House of the United States of America. (2021b). President Biden announces American Rescue Plan. https://www.whitehouse.gov/briefing-room/legislation/2021/01/20/president-biden-announces-american-rescue-plan/

The White House of the United States of America. (2021c, March 31). The American Jobs Plan. https://www.whitehouse.gov/briefing-room/statements-releases/2021/03/31/fact-sheet-the-american-jobs-plan/

The White House of the United States of America. (2023, January 19). Fact Sheet: Biden-Harris Administration releases national strategy to put nature on the nation's balance sheet. https://www.whitehouse.gov/ostp/news-updates/2023/01/19/fact-sheet-biden-harris-administration-releases-national-strategy-to-put-nature-on-the-nations-balance-sheet/

Topalova, P. (2018, April 12). How can low income countries cope? The effects of weather shocks on economic activity. Presentation at the Economics of Climate Change Project. The New School, New York.

United Nations. (2020). *Climate Change and COVID-19: UN urges nations to 'recover better'*. https://www.un.org/en/un-coronavirus-communications-team/un-urges-countries-%E2%80%98build-back-better%E2%80%99

United Nations Environment Programme. (2009). *Global Green New Deal policy brief*. https://www.unep.org/resources/report/global-green-new-deal-policy-brief-march-2009#:~:text=UNEP%20outlines%20a%20GGND%20with,carbon%20dependency%20and%20environmental%20destruction

United Nations General Assembly (2020). *Comprehensive and coordinated response to the Coronavirus Disease (COVID-19) pandemic*. https://www.un.org/pga/74/wp-content/uploads/sites/99/2020/09/Omnibus_Final-clean.pdf

United States Department of Commerce. (2023, January). National strategy to develop statistics for environmental-economic decision: A U.S. system of natural capital accounting and associated environmental-economic statistics. Office of Science and Technology Policy, Office of Management and Budget, Department of Commerce. https://www.whitehouse.gov/wp-content/uploads/2023/01/Natural-Capital-Accounting-Strategy-final.pdf

United States Department of the Treasury. (2022). Remarks by Secretary of the Treasury Janet L. Yellen at the 2022 'Virtual Davos Agenda' Hosted by the World Economic Forum. https://home.treasury.gov/news/press-releases/jy0565

Vivid Economics. (2021). *Greenness of Stimulus Index*. https://www.vivideconomics.com/casestudy/greenness-for-stimulus-index/

World Trade Organisation. (2020). *Communication on trade and environmental sustainability*. https://docs.wto.org/dol2fe/Pages/SS/directdoc.aspx?filename=q:/WT/CTE/W249.pdf&Open=True

World Trade Organisation. (2021). *Members review draft MC12 declaration on trade and environmental sustainability*. https://www.wto.org/english/news_e/news21_e/tessd_21jul21_e.htm

Zetzsche, D. A., & Anker-Sørensen, L. (2022). Regulating sustainable finance in the dark. *European Business Organization Law Review, 23*, 47–85. https://doi.org/10.1007/s40804-021-00237-9

Index

J. Puaschunder, *Governance & Climate Justice*,
https://doi.org/10.1007/978-3-031-87971-5

C

D

H

I

Y

Z

FSC
www.fsc.org
MIX
Papier | Fördert
gute Waldnutzung
FSC® C083411

Kolibri 360 GmbH
Ferdinand-Jühlke-Straße 7
99095 Erfurt, Deutschland
produktsicherheit@kolibri360.de